Praise for EATING THE SUN

"*Eating the Sun* could not be timelier and firmly establishes Oliver Morton as one of the world's finest science writers."

—Steven Shapin, author of
Leviathan and the Air-Pump: Hobbes, Boyle, and the Experimental Life

"A fascinating and important book."

—Ian McEwan, author of *Atonement,*
Saturday, and *On Chesil Beach*

"When you are done with this book, you will see the world differently and understand it better. Going directly to the most important question of our time—the origin of the carbon/climate crisis—and delving deeply into it, *Eating the Sun* transcends science writing as we usually think of it, offering something more like the mental reorientation that we get from writers like George Orwell."

—Kim Stanley Robinson, author of *Red Mars,*
Green Mars, and *Blue Mars*

"Oliver Morton tells a magnificent story of the early epochs of our earth and the gradual greening of its surface. From primeval sludge to the diminishing rainforest, he brings us bang-up-to-date with our current crisis over greenhouse gases and global warming. A stylish, engaging, and very important read."

—Janet Browne, Professor of the History of Science at
Harvard University and author of *Charles Darwin: The Power of Place*

"In beautifully written, lucid prose, Morton clearly explains the way in which plants, algae, and some bacteria are able to capture a small portion of energy from the sun and thus make possible all life on earth, including ours. Insightful and easy to follow, this book should be read by everyone interested in understanding the origin and maintenance of life."

—Peter H. Raven, Director of the Missouri Botanical Garden

"A rare delight. . . . Oliver Morton writes so engagingly that [*Eating the Sun*] reads as a well-crafted biography of the earth on behalf of the plant kingdom."

—*Prospect* magazine (London)

Also by Oliver Morton

Mapping Mars:
Science, Imagination and the Birth of a World

EATING THE SUN
How Plants Power the Planet

OLIVER MORTON

HARPER

An Imprint of HarperCollins*Publishers*
www.harpercollins.com

HarperCollins books may be purchased for educational, business, or sales promotional use. For information please write: Special Markets Department, HarperCollins Publishers, 10 East 53rd Street, New York, NY 10022.

First published in Great Britain in 2007 by Fourth Estate, an imprint of HarperCollins Publishers.

FIRST U.S. EDITION

Library of Congress Cataloging-in-Publication Data is available upon request.

ISBN 978-0-00-716364-9

08 09 10 11 12 OFF/RRD 10 09 08 07 06 05 04 03 02 01

CONTENTS

List of illustrations   ix
Author's note   x
Introduction   xi

PART ONE – *In the span of a man's life*
Chapter One: Carbon   3
Chapter Two: Energy   49
Chapter Three: Light   93

PART TWO – *In the span of a planet's life*
Chapter Four: Beginnings   145
Chapter Five: Fossils   193
Chapter Six: Forests and Feedback   229
Chapter Seven: Grass   269

PART THREE – *In the span of a tree's life*
Chapter Eight: Humanity   317
Chapter Nine: Energy   373

Glossary   413
Bibliography   420
Further Reading   436
Acknowledgements   443
Index   446

LIST OF ILLUSTRATIONS

Diagram 1: The Keeling curve   7
Diagram 2: The green and the red   61
Diagram 3: The Z-scheme   81
Diagram 4: The chloroplast   86
Diagram 5: Overview of photosystems I and II   131
Diagram 6: The Vostok record   297
Diagram 7: Wedges   379

AUTHOR'S NOTE

The most technical parts of this book are concentrated towards the beginning. Readers whose interest is not held by these passages, who are eager to read about the earth and its carbon/climate crisis, or who just enjoy jumping about should feel free to skip forward. To help these intrepid and non-linear readers, the more technical ideas introduced early on are also defined in a glossary at the end.

INTRODUCTION

Here's what happened today. What really happened.

Dawn broke first in the Pacific: because our international dateline is in the middle of our largest ocean that's where the day's dawn always breaks first, its tangential light reflected from a million waves and a few container ships into an empty sky. What wasn't reflected lit up the upper layers of the ocean, a soft new light for the fish and that which they feed on.

When it made landfall in the north, the sun swept over the tundra like water up a beach; a couple of hours later, at the other end of the world, it broke like a wave against the mountains and pastures of New Zealand. Soon it was filling the rice paddies of the Philippines and the shallows of the South China Sea. And every time the sunlight hit something green – something truly green, not something painted green or dyed green: something with a greenness that grew – the most important process on the planet began again.

When the light shone on the greenness, the greenness welcomed it, and comprehended it, and put it to use. The greenness was chlorophyll, a pigment. It was arranged in pools and the sunlight's energy bounced from one molecule to the next like a frog across lily pads before reaching the subtle trap at the pool's centre, the three-billion-year-old trap where the light of the sun becomes the stuff of the earth. As the trap's jaws snapped shut on the sunlight, the spring that powered those jaws pulled electrons from a nearby water molecule, breaking it up into hydrogen and oxygen. The hydrogen was used, along with the stream of electrons that flowed up through the trap, to turn carbon dioxide into organic matter. The oxygen was discarded.

In every plant reached by the dawn this extraordinary mechanism came to life millions of times over. There are hundreds of thousands of pigment pools and sunlight traps in every green cell, hundreds of thousands of cells in full-grown leaves. And once awakened by the light, the flow of electrons through the leaves did not stop until darkness fell. The carbon dioxide to which those electrons were channelled was turned first into a sugar and then into all sorts of other molecules. Some of them were used to thicken the plants' stems, to lengthen their leaves, to enrich the soil beneath them and to colour the flowers still held tight in their buds. The rest were used to fuel the processes that make such growth possible. Light made life: that is what photosynthesis means.

If the light-driven flow of electrons stopped, on this day or any day, so would everything else that you care about. So would almost everything that evolution has wrought. If this process stopped, so would the world. The planet wouldn't stop turning: dawns would still arrive with impressive regularity. But they wouldn't matter. No more datelines. No more dates.

The manufacture of life from light is not the only thing going on as our dawn sweeps over Asia. For every leaf that greets the dawn there's an ant that eyes the leaf; for the growing grass there are hungry calves, and for the fattening calves there are hungry men; for the sugar-swollen root there's a sugar-sucking fungus. For growth there's decay. All over the world, by day and night, animals and bacteria and fungi and the plants themselves are using the oxygen which photosynthesis spits out into the atmosphere to turn organic material back into carbon dioxide and water. In so doing, they liberate the energy the plants stored away. The two processes come close to cancelling each other out.

Today, though, things are out of balance. Today is a spring day – at least it is spring in the northern hemisphere, which is home to most of the world's plants, as well as most of its people. And on spring days, photosynthesis wins. Carbon is pulled out of the air and into the green faster than it can be returned. Today, the world is growing.

By the time it reaches me in England, the dawn has come halfway around the planet. Between the sloped roof of my neighbours' house and the high, unbroken side elevation of the ugly hall that sometimes plays host to my wife's dance troupe, the sun lights upon a line of four young sycamores, the tallest of them maybe ten metres high. I don't think anyone told the sycamores to be where they are, on a sliver of soil by the ramp down to the hall's car park. I think they just grew there. Twenty years ago, at a guess, they fluttered there as seeds. Since then they've grown into full-blown trees. And they've done it just by doing what every other plant that's been lit up over the past twelve hours has been doing: eating the sun.

They're not eating as greedily today as they will in a month from now. It is still spring, and they have only a few leaves. A month ago the branches were bare; a couple of weeks ago, they were in bud, their black winter lines touched with a slight fuzziness, a dream of the green to come. Now, though the shortest of the four is still only in bud, the tallest has small leaves held close to its upraised branches. And on the other two trees the leaves have opened out quite fully, especially at the far ends of the branches, where they flop into the air like spindly five-fingered flags. Sycamores, especially small sycamores, come into leaf early, to get a jump on the competition, eager to pull themselves a little higher into the sky and out of others' shade.

I mention the differences between these four sibling trees to point out that there is individuality here. For the most part, this book will be about the universal – about the physics and physiology of trapping sunlight and putting it to use, about the structure of the molecular machinery shared by everything that's green, about the historical role that the photosynthesizing plants play en masse in the planet's life. But it's worth remembering that the organisms that embody these universalities all have their little quirks – a particular pattern of branches, a slight suddenness in coming into bud, a shape that reflects the shadow of an ugly building, an accommodation to the drainage offered by a parking ramp. Within the universal, the unique.

On the leaves of each of these variously developing trees, and on every other leaf that greets the sun, there are tiny openings through which the air that will be strip-mined of its carbon is let in. With the sugars built from this carbon, and with the energy stored in those sugars, the little leaves are made bigger; with sugars stored in the roots over winter and brought upwards in the sap new leaves are begun. Within a couple of months these sycamores, which a few weeks ago were just sets of sticks pointing upwards, will have increased their surface area a hundredfold by covering themselves with leaves. The trees will absorb many times more sunlight than the little patch of land on which they stand. Next year they will do the same – but they will spread a little wider and start off a little taller, having turned another year of air and sunshine into new rings of wood in their trunks and branches.

In the Atlantic, to the southwest of me, plankton are blooming off the coast of Africa, floating in cloudy eddies large and distinct enough to be seen from space, like Florentine marbling on the ocean's blue parchment. The plankton eat the sun in just the same way that the tundra and the forests and the rice paddies and the sycamores do, using inorganic carbon dissolved in the water that surrounds them. Unlike the trees, though, the plankton will not survive for year after year; they will not build up an architecture of trunk and branch and twig and leaf. The only growth they know is multiplication – which they carry out with great alacrity. For all that the bloom is a passing fancy of wind and current, rather than a solid landscape like a forest or savanna, it will end up weighing tens of thousands of tonnes. Transient as their lives may be, over the course of the year the plankton in the sea will suck up almost as much carbon dioxide as all the trees, grasses and other green things on the land.

By the time dawn has reached America, most of the people on the earth have woken up and gone to work; some in the east have retired for the night. During their working day, almost all of those who can see will have seen something green. In built-up London I can count well over fifty trees on the hundred-metre walk from my flat to the Light Rail station. We might not give a conscious thought to the ever-present green, but at some level we will enjoy it. The greenness of life is so important and all-pervading that evolution has tuned our eyes to discriminate among its various hues more precisely than among those of any other colour, and so shaped our brains that we take solace in it. The green, we know without thinking, is good.

We don't just enjoy seeing the green. It shapes the possibilities of our lives. More than two billion of us will have tended to the eaters of the sun in some way today. We will have hoed the ground for them, planted them, fed them fertilizers. We will have picked their fruits, dug up their nutritious roots, fed them to our livestock and ourselves. We will have made their carcasses into fabrics and furniture and firewood. We will have tended to some of them simply for their beauty – and to others because we know no finer surface over which to run while kicking a ball.

And even if we ignore today's plants completely, if we cut ourselves off in concrete and steel, we will still rely on yesterday's. On this day we will burn over thirty million tonnes of fossil fuel to generate our electricity and drive our cars and fire our factories and warm our homes. And all that power and warmth comes from sunlight eaten long ago. Energy trapped 300 million years ago by trees simpler but grander than my sycamores ended up stored in coal; plankton like those now blooming off the Azores were transformed into oil and gas. The carbon in the carbon dioxide we give off by burning them is carbon taken from the ancient atmosphere they breathed.

There's a catch, though. The rate at which we reclaim energy from the distant past has produced an accounting error; the profits and loss in our carbon accounts no longer balance. Which brings us to the last place, more or less, to see our dawn; Hawaii. Set apart from all the continents, Hawaii is one of the best places on the planet from which to measure the average composition of the atmosphere. Such measurements have been made there almost every day for nearly fifty years. While the measurements may jump around a little from one day to the next, over time there are trends. Today's measurement will probably be a little lower than the one made a week ago, because today – and every day for the past few weeks – the great greenness of the northern hemisphere has been taking in carbon dioxide through photosynthesis much faster than it is released by respiration, or the burning of fossil fuel. The world is breathing in.

Over the course of the spring, the carbon-dioxide level will drop and drop as billions of tonnes of carbon are taken from the air and put into plants. Only in the autumn, when the leaves fall and the grasses lose their song, will the carbon-dioxide level start to grow again, the great exhalation of a world eating its stores of food while waiting for the sun's return. From Hawaii, you can see the carbon rise and fall in this annual cycle. You can watch the breathing of the world, year in and year out. And you can also see that, each summer, after the plants have had their fill, there is a bit more carbon dioxide left in the atmosphere than there was the year before. That's the residue of the fossil fuels: ash from energy stored away long ago. Ash building up faster than it can be swept away.

As the dawn moves past Hawaii, the day is almost done. On this day, and the next day, and every day, a scarcely conceivable 4000 trillion kilowatt hours of energy reached the top of the earth's atmosphere as sunshine. Some was reflected back into space and some was absorbed by the atmosphere. Some warmed the land and the sea, its warmth driving the winds and the ocean currents. Only a small fraction of one percent of that sunlight was captured by the pools of chlorophyll. But this tiny fraction of a vast number is still vast: the scrap of sunlight eaten by the plants today represented a similar amount of energy to that stored in all the world's nuclear weapons put together. And over the course of the day, that energy served to turn hundreds of millions of tonnes of carbon dioxide into food and living tissue.

And as a result the world stayed alive.

That's what really happened today.

We all learn at school that photosynthesis is the defining property of plants. It is what makes plant life strange and alien and wonderful, silent and strong and unfeeling. Left to themselves the plants need almost nothing more than light and air and water; they need no prey, no care, no others.

But photosynthesis is not just a thing that plants do. It is a thing that planets do, too. More specifically, it is a thing that this planet does, and the thing that marks it out from all the others in the solar system.

This planet is covered not in dust, or lava, or dirty, cratered ice, like the others. It is covered in tiny photosynthetic machines with which it has been ceaselessly resupplying itself for three or possibly almost four billion years, craftily fashioned pools of pigment with traps at their centres. Their colour is the single most noticeable feature of the land's surface as seen from space. These machines, more numerous than the sands of the sea or the stars of the sky, are vital parts of the organisms they sit in. But at the same time they belong to the planet as a whole. They have built themselves into the physical and chemical cycles of the earth; they are quite as fundamental to them as the falling of the rain or the rhythm of the waves. They are interactions of light and matter far more complex and purposeful than anything a non-living planet would have to offer; but they are as basic to the nature of the earth as the colour of a rock or the dustiness of a sky. And they are much longer-lasting.

The individual machines are ephemeral; the hard work of taking light from the pools of pigment and funnelling it into chemical reactions takes its toll on their mechanisms. Plants spend a significant amount of the energy they take from the sun keeping their photosynthetic machinery in good repair, and when light levels drop in winter, the factories in which these machines do their work are discarded in their unnumbered trillions as no longer worth the upkeep. But the idea behind the machines, the idea written down in DNA and broken out afresh each spring, is not so fragile. It is always there, and always available. Instantiations perish; the design does not.

All the machines are variations on the same preserved design, a plan that pre-dates all the species now living on the earth. Machines built to this plan were there when the grasses first began to ripple in the winds, and when the first flowers bloomed for the delight of the insects around them. They were there when forests and swamps covered continents to which our animal ancestors were still strangers. They were present in the seas when all the land was barren, and when all the world was ice. And they were there before.

This book is about how those machines work, and how they have shaped the world. And it is about how we can use that knowledge to understand and deal with the imbalance in our carbon account – the carbon dioxide that we are importing from the planet's past by burning fossil fuels, and with which we are changing the climate. It is about how, and whether, we might use our knowledge to take further control of the machines that made the earth what it is.

A book, like science, like language, like life itself, is a thing that cuts the world up and then puts it back together. Life breaks down molecules and reconstructs them. Science cuts the world into inter-locking concepts that can be understood through physical laws. The elucidation of photosynthesis is as inspiring a tale of such achievement as twentieth-century science has to offer.

This book takes the dismantlings of science that have made the green world understandable and tractable and reassembles them into a picture of the whole process that transcends the boundaries of any single scientific discipline. It is a picture larger and stranger than we are normally used to, in which planets are seen through molecules and molecules through planets. Once glimpsed, it can be seen anywhere. Walking over open meadows, or cutting basil in a windowbox, or putting a log on a fire, or lying beneath a tree in a park, you can see the whole wonderful chain from pigment to planet.

To try and describe all this, I've cut this book into three parts from which to make a whole. Each part is defined by a subject matter and a span of time. Part One, which fits into the scale of a human life, is the story of how scientists used the astonishing tools of the twentieth century to replace a view of photosynthesis that had come down from the Enlightenment with something far richer and deeper. Part Two is on the scale of the planet's life. It tells the story of how the molecules discovered in Part One came to dominate the earth's chemistry, to reshape its atmosphere, and to drive vast changes in its climate and habitability. Part Three is on the scale of a tree's life – a matter of a few centuries, past and future. It is the story of what our use of fossil fuels is doing to the carbon cycle, and through it to the climate.

It's also about how our understanding of photosynthesis might help us choose a wiser future. The science that enriches our wonder at the world also offers us ways of making things better. Our understanding of photosynthesis, at the level of the pigments and at the level of the planet, offers ways to use up the carbon-dioxide ash that's clogging our air and to find new methods of drawing energy from the sun.

A tree can live for hundreds of years, and this last part of the story, which began two centuries ago, is far from over. But it is the most pressing aspect of the process at the heart of this book, and its telling is tied firmly to the preceding stories of human ingenuity and planetary evolution. The only place where all these scales can really come together is in our minds; and it is on the present and near future that our minds should be focused.

PART ONE

In the span of a man's life

The mind, that ocean where each kind
Does straight its own resemblance find;
Yet it creates, transcending these,
Far other worlds, and other seas . . .
ANDREW MARVELL, 'The Garden'

Tinkering as a child seems to me, in retrospect, to have been as enjoyable, intense and meaningful as doing research in later years. The feelings have not changed, only the budget and the kind of questions that are being asked. The amazing thing is that one is paid for what one likes to do and occasionally is even appreciated by the outside world.

GEORGE FEHER

CHAPTER ONE

Carbon

Scripps, planets and proteins
Andrew Benson's chemical education
Martin Kamen and the Rad Lab
Carbon-14
The reversals of war
Melvin Calvin and the path of carbon
Rubisco

Yet still the unresting castles thresh
In fullgrown thickness every May.
Last year is dead, they seem to say,
Begin afresh, afresh, afresh.
PHILIP LARKIN, 'The Trees'

Scripps, planets and proteins

The Scripps Institute of Oceanography must be, simply by grace of its setting, one of the nicest places in the world for a scientist to work. Just north of La Jolla Cove, just south of the cliffs of Torrey Pines, its buildings are worked into the slopes that lead down to the Pacific like foliage tucked into a tumbling rockery. In the soft climate of southern California the distinction between inside and outside is blurred. The buildings' corridors take every opportunity to open out into balconies; staircases wander from one building to a seemingly separate one below with no need to worry about exposure to the elements, stepping down the slope to the steady rhythm of the surf below.

Scripps owes much of its modern pre-eminence to Roger Revelle, once described by a colleague as 'a combination of charismatic visionary and con man'. A spectacularly entrepreneurial scientist, he used the funding boom which followed the Second World War to expand the laboratory's remit as far as he could, and the further post-Sputnik boom to establish the University of California's San Diego campus, which stands on the higher ground behind it. Revelle used to joke that the field of oceanography covered everything that anyone at Scripps wanted to study, and today Scripps boasts as eclectic a research agenda as you could hope to find, with scientists studying everything from the giant kelp in the cove offshore to air bubbles trapped in the Greenland icecap.

Among the things Revelle himself studied was the radioactive carbon-14 produced in nuclear tests – studies dating from his time

5

leading the science team studying the first tests on Bikini Atoll. In 1957, in the course of this work, he gave voice to one of the defining truths of our age. In the late nineteenth century the great chemist Svante Arrhenius had pointed out that carbon dioxide warmed the earth by trapping outgoing heat in the atmosphere – the greenhouse effect – and that humans were putting a lot of carbon dioxide into the atmosphere. That addition, Arrhenius suggested, should warm the planet, something he saw as being basically a good idea. In the first half of the twentieth century, though, oceanographers had argued that the oceans would quickly soak up the carbon dioxide humanity added to the atmosphere, forestalling any such warming.

While thinking through his carbon-14 work, Revelle realized there was a basic chemical flaw in this argument which meant that the oceans could not absorb carbon dioxide anything like as quickly as humans could produce it. Carbon dioxide must have been accumulating in the air since the beginning of the industrial revolution. As a result, Revelle wrote, 'Human beings are now carrying out a large-scale geophysical experiment of a kind that could not have happened in the past nor be reproduced in the future.' That experiment is the climatic background against which the history of the twenty-first century will unfold.

Today, when details of new carbon-dioxide measurements and trends can make front-page news, it is hard to realize that when Revelle wrote his now-famous words there were no reliable measurements of the world's carbon-dioxide level, and few scientists had any interest in making any. Revelle made it his business to hire one of the only people attracted by the problem, a young geochemist from the California Institute of Technology (hereafter Caltech) in Pasadena. With Revelle's encouragement and support, Dave Keeling was able to put carbon-dioxide monitoring equipment in inaccessible places where the value might reflect the global average, rather than anything going on nearby; places like Antarctica and Mauna Loa, one of the Hawaiian volcanoes (which is not, at the moment, emitting gases itself). Within a few years, Keeling had shown that all round the world levels of carbon dioxide were steadily rising year on year. He also showed that levels in the atmosphere tracked changes in the biosphere, the earth's active, living component. Carbon dioxide rose and fell as plants grew in the spring and leaves then rotted in the fall. Keeling's findings illustrated both the global impact of industrial carbon emission, and the global effects of photosynthesis.

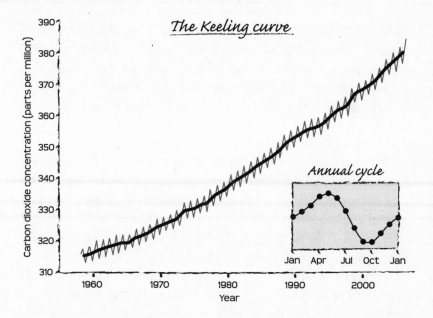

The Keeling curve

Annual cycle

Almost fifty years on, Keeling was still at Scripps when I visited a few years ago, and still measuring carbon-dioxide levels; he had devoted the bulk of his professional life to monitoring atmospheric carbon dioxide and trying to understand the ways in which it changes over time. A charming, courtly man, slightly hard of hearing, he had a large, airy office at the south end of the Scripps campus, lined with drawers full of data; younger colleagues working on similar issues had desks in the larger room outside his door.

Dave Keeling's story is the story of a planet; Andrew Benson's centres on a protein. Benson's office, a few hundred yards north of Keeling's, has a very different feel: a small room on the top floor of one of the Institute's bigger, blockier concrete buildings, cluttered with reminders of a scientific life devoted to a great many different subjects. Keeling chose a hedgehog career, centred on knowing one big thing. Benson has been more eclectic. He has worked on the biochemistry of ageing in salmon, and on the energy-rich waxes in the Lantern Fish and the Orange Roughy (don't eat them whole, he warns, or that wax will go through you like a dose of salts). He's studied specialized topics – the extraordinary concentration of arsenic in the kidneys of giant clams on the Great Barrier Reef – and universal ones, such as the composition of the membranes that define the surfaces of cells, a subject on which almost everyone in the field thinks he is wrong. The bare walls of his room – they have the grain of the wood that was used to form the concrete still preserved in their texture – are hung with art from the Pacific northwest and from Japan. Some of Benson's closest friends are Japanese, and their national aesthetic is one that pleases him. There is no contemplative calm, though, in the stacks of papers and other research materials that brim over on the shelves and fill the cabinets.

Benson is a short, spry man in his eighties, born a decade earlier than Keeling. A Californian native, he has something of the midwestern farm boy about him, perhaps because his father, a Minnesotan doctor of Scandinavian extraction, chose a peculiarly midwestern-looking part of California to settle in, soft stream-rich country near Modesto. He remembers his boyhood, well supplied with woods to play in and uncles to learn from, as an idyllic one. The lifetime that came after that has left him with the face of a man who knows his own mind, who keeps a clear head, and who has spent a fair number of his days in the open air. He's fun to talk to, a bit testy now and then, sometimes a little sly. Like most elderly men, he has his bugbears and eccentricities; like most elderly chemists, one of them is the paralysing spread of environmental regulations into the laboratory, which, he avers, must surely mean that young people today can't begin to learn or teach their craft.

Benson has spent much of his life as a biochemist, someone who uses the tools of chemistry to study the processes of life. When he began his career, though, he was something which, though it sounds similar, is really rather different. He was an organic chemist – an *Organiker*, as this ritualistic order was known when German set the tone for science and chemistry was king.

Organic chemistry is the chemistry of compounds which contain carbon. The carbon atom's gift is that it is peculiarly good at making various sorts of chemical bond. It can make straight chains, kinky chains, branched chains and a variety of rings in ways that no other element can manage. These complex molecules were first seen in living things, which is how they came to be called organic. But for well over a century chemists have delighted in making carbon-bearing molecules that life has never bothered with, and those synthetic molecules are called organic, too.

Organic chemistry is thus no longer unique to living things. But it is still indispensable to them, which means that life, or at least the sort of life we find on earth, needs a source of carbon. For animals that source is other creatures. We eat them, break down some of their molecules to use as components with which to build up our own proteins, fats, nucleic acids and suchlike, and burn the rest up as a source of energy. Once the useful components and energy have been taken out, what's left is water and carbon dioxide. And as far as animals are concerned, once the carbon is in the form of this colourless, odourless, unreactive gas there's nothing more that can be done with it.*

---

* As I write this, it strikes me how unfortunate it is that English gives only a technical name to something as basic to life as carbon dioxide, something as fundamental as blood and breath. Because it is indiscernible without instruments and has thus never furnished our perceptual world, I'm forced to write about it under a name that, even if it doesn't alienate, certainly can't carry any freight of emotion. 'Water' is as rich in imagery as a word can be; 'oxygen', coined just over two hundred years ago, has far fewer associations, but still has some general aura of necessity and energy and freshness. 'Carbon dioxide' is just a chemical. There's no way out of this – I'm certainly not going to try your patience by replacing the term with some daft nonce word like 'themypoid' or 'pickliff'. But it's salutary to remember that language itself can contrive to mask the richness and relevance of the world that science reveals.

The oceans can take this inert carbon dioxide and use it as the basis for further 'inorganic' chemistry, creating electrically charged ions – bicarbonates and carbonates in which carbon is combined with oxygen. It was by considering the details of these inorganic transformations that Revelle first realized that carbon dioxide must be building up in the atmosphere. But neither the ocean, the air nor the rocks of the earth have the power to turn carbon dioxide back into organic carbon compounds. Only living things can do that – and some 999 times out of a thousand, photosynthesis is the way in which they do it. All the organic carbon molecules you are made of rely on some plant (or possibly a photosynthetic bacterium) having used sunlight to take some carbon out of the inorganic world and refashion it into a form suitable to the world of the living.

That plants revive the world in this way has been known since before the words 'carbon dioxide' were ever uttered. In the late eighteenth century, a number of Enlightenment scientists showed that plants took inert carbon from the air and, with the help of sunlight, 'fixed' it into their living tissues. Their work is discussed in chapter 8. In the nineteenth century, this assimilation of carbon from the atmosphere was seen as the basis of plant physiology, and by the end of the century it had been given the name 'photosynthesis', a word intended to distinguish the unique way that plants use inorganic carbon from the ways in which animals assimilate organic carbon. The requirements for photosynthesis in plants were established as water, carbon dioxide, sunlight and chlorophyll, a green pigment; its primary products were carbohydrates – sugars, often stored as starch – and oxygen. This list of attributes didn't capture everything about photosynthesis – by the late nineteenth century it was known that there are photosynthetic bacteria which use sunlight to fix carbon without giving off oxygen – but it was good enough for practical plant physiology.

Photosynthesis was not just an abstract idea; by the second half of the nineteenth century it could be visualized beautifully with the help of stencils and iodine. A leaf with a stencil superimposed on it would photosynthesize only in the areas that sunlight could reach, and thus only in those regions would starch accumulate. Exposed to iodine, the starch would blacken, producing an image of the stencil. Photosynthesis was as real as photography. What distinguished the two was that no one knew how the natural process worked. No one knew how the carbon was fixed, how energy from the sunlight drove the fixing process, or how the oxygen was produced. The tools with which to answer those three questions – questions around which the first three chapters of this book are arranged – had not been invented. Nor had the disciplines in which such questions could be articulated. Biochemistry, central to the question of how the sun's energy drives the chemistry of photosynthesis, was a twentieth-century invention; so too was biophysics, which explained how the light was captured in the first place; molecular biology, a later innovation still, finally laid bare the mechanisms underlying the whole process, including the production of oxygen.

Now those questions have been answered; and all the work that went to answering them has taken place within Andrew Benson's lifetime. As proof of the point, Benson points at a shelf in his office that carries a book printed in 1917, the year that he was born. Benson says *Untersuchungen über die Assimilation der Kohlensäure* ('Investigations into the assimilation of carbon dioxide') by Richard Willstätter and Arthur Stoll, is still 'elegantly informative'. But their later book on how chlorophyll might actually work, 436 pages of ideas and experiments aimed at finding out how the carbon gets fixed into plant tissue, now seems utterly irrelevant – 'forlorn and futile', as Benson puts it.

Carbon dioxide's journey from air to starch was eventually traced with the same quintessentially twentieth-century tool that Revelle used to monitor carbon-dioxide uptake in the ocean: carbon-14. There was a time, before the first nuclear explosions, when Benson had in his safekeeping all the man-made carbon-14 in the world. And later, when Dave Keeling started measuring the rate at which carbon dioxide builds up in the planet's atmosphere, Benson discovered which of the plants' proteins carbon dioxide must pass through if it is to be used by living things, the molecular gateway

between the organic and inorganic worlds. Working with two pioneers who would never complete the task, and then with one of
the century's most renowned chemists, Melvin Calvin, who almost
denied him the honour that was his due, this sharp old man discovered one of the basic facts of life on earth.

Andrew Benson's chemical education

As a boy, Benson had been interested in all sorts of science: in his
father's portable X-ray screen; in the insects living in the woods; in
the test tubes of his chemistry set and the more exciting reactions
of not-entirely-burned-out fireworks after the Fourth of July; in the
stars. He'd ground his own telescope mirror – a common rite of
passage among scientifically enthused teenagers living under clear
skies at the time. His father took him to the University of California,
Berkeley for an interview with Wendell Latimer, the dean of the
chemistry department, which at the time was the most prestigious
department on the campus, and one of the best chemistry departments in the world. Latimer approved of him, and Benson enrolled
at the university in 1937.

While the boy Benson played with his fireworks, chemistry, particularly what was known as physical chemistry, was suffering thun-
derclaps of its own. Until the early years of the twentieth century,
chemists had treated atoms as simple building-blocks. Every chemical element was identified with a specific type of atom; all the atoms
of that element would be identical in every way. As far as being a
practical chemist was concerned – and chemists had a penchant for
being practical, not least because of the huge industrial importance
of their discipline in the production of dyes, drugs, explosives and
eventually artificial fertilizers – there was little point in speculating
about the insides of these building-blocks. Indeed there was a strong
supposition that they were indivisible and had no insides.

Around the turn of the century, physics found its first ways of looking into the structure of the atom. The new understanding of matter, energy and their effects on each other that spilled out of those investigations changed the ways that first physics, and then its neighbouring disciplines, saw the world. The change spread out, discipline by discipline, altering the ground on which the scientists stood and the goals of which they could dream.

In the new physics atoms were made up of, in order of discovery: electrons, light particles that carry a negative electric charge; protons, much heavier and positively charged; and neutrons, of similar mass to the proton but chargeless. By the time Benson came to Berkeley, the physicists had developed a 'quantum mechanical' picture of the atom in which a heavy nucleus of protons and neutrons was surrounded by shells of electrons. The nucleus gave the atom its mass; the surrounding electrons gave it its chemical identity – its willingness to partake in various different sorts of reaction. In the 1910s and 1920s, Gilbert Lewis, the chairman of the Berkeley department and probably the most famous American chemist of his day, provided a thoroughgoing picture of how the disposition of electrons in atoms explained the affinities of the different elements.

The most dramatic features in the new landscape of chemistry were the sea cliffs at its edge, the frontiers where parts of the subject were being eroded away by the waves – and wave equations – of physics. As physics solved basic questions about the atom, and about the ways in which electrons mediated the relationships between atoms by binding them together, it removed them from the realm of future chemical enquiry, or so it seemed. 'Physical chemistry', Lewis wrote in the 1920s, 'no longer exists.' But the young Benson was not particularly drawn to physical chemistry; he was more interested in the sort of chemistry that builds intriguing new molecules than in the sort which explains the fine details of atomic interactions. His enthusiasm was for organic chemistry, with its entrancing synthetic lattices of carbon, hydrogen, nitrogen and oxygen.

Graduating from Berkeley in 1939, Benson went on to pursue a doctorate in organic chemistry at Caltech, down in Pasadena, where

13

Linus Pauling was building a spectacular career on chemistry's physical shore, extending Lewis's insights into the quantum mechanical view of the world that physicists formulated in the 1920s. It was said that if everyone else had decided to go home and just leave Pauling to get on with things, Caltech's chemistry department would still have been one of the best in the country.

Benson's graduate years in Pasadena were happy ones. Caltech, though small, was an incredibly ambitious institution, engaged on a whole range of scientific frontiers. Every week Benson would go over to the optical shop and marvel at the apotheosis of one of his childhood hobbies: the endlessly precise polishing of the five-metre (200-inch) mirror destined for the observatory at Mount Palomar.* His research centred on the synthesis of a chemical called difluoro-diiodothyronine, a process which started off with a gallon of burning anise oil in a vat of fuming nitric acid and ended up, twenty reactions later, with a few drops of liquid in a vial. By and large it went well. There was lots of time for climbing expeditions to the Sierra Nevada and snorkelling outings to South Laguna. There was time to fall in love and get married to a girl whose family lived nearby.

Three years after arriving, Benson didn't just have a doctorate and a wife; he also had a new job. At the end of his final oral exam, Pauling asked him, out of the blue, 'Andy, can you write on the board the differential equation for the decay of a radioactive isotope?' The question had nothing to do with the work Benson had just described, and Benson wasn't particularly hot on nuclear theory. But Pauling was always asking odd things, and Benson knew enough to answer the question, so he chalked the equation up on the blackboard and chalked the experience down to some sort of caprice on Pauling's part.

In fact, the question had been a quite deliberate one. Within a week, Benson received an offer to go back to Berkeley as an instructor, an offer that he later realized had been set up by Pauling. He

* Almost seventy years on, Benson still treasures his membership card to Caltech's faculty club, signed by Edwin Hubble.

was to work with a pair of scientists who needed some organic chemistry back-up on a fascinating research programme. Their plan was to use radioactive carbon – hence Pauling's desire to be sure Benson wasn't completely unversed in physics – to try to understand a great mystery that had, until then, passed young Benson by: the transfer of carbon from the atmosphere to living tissue. With a doctorate and a young family Benson headed back up to Berkeley to work with Sam Ruben and Martin Kamen.

Martin Kamen and the Rad Lab

Martin Kamen was well-educated, prodigiously gifted, and keen to enjoy those gifts to the full. Just four years older than Benson, he was born in Toronto in 1913, the son of immigrants from Russia who soon moved on to Chicago. His father had a photography business, in which his mother also worked, and invested some of the money they made in real estate; they were soon fairly comfortably off. Young Martin, much doted on, grew into a scholar with a broad range of interests, mainly in philosophy and the humanities. He was also a talented musician, a child prodigy on the violin before changing to the viola in his late teens. He entered the University of Chicago, a few blocks from the family home, in the spring of 1930, intending to major in English.

Unfortunately for the Kamens, at more or less the same time their fortunes took a severe blow in the aftermath of the Wall Street Crash. The property portfolio evaporated, and young Martin found himself under pressure to follow a course of studies with more practical potential. His father thought chemicals sounded like the road to riches, and as Martin was not averse to the idea, chemistry became his new major. The same change of fortunes saw Kamen adding jazz pieces to his burgeoning chamber-music repertoire, the better to pick up a dollar or two a night in Chicago's speakeasies.

Chemistry could indeed make you rich; at the World's Fair in 1933, Kamen was amused to discover that a man he had taken for a tramp shuffling up to the registration desk at a convention of chemists was in fact Leo Baekeland, the Belgian chemist whose invention of the photographic paper used by Eastman Kodak and then of Bakelite had made him a multi-millionaire. But Kamen himself showed no inclination to head off into industry. He wasn't much interested in the synthesizing-new-compounds parts of chemistry that entranced Benson. He was attracted by the how-it-all-works parts, the parts with a lot of mathematics, the parts on the shores of physics. He wanted to pursue deep problems about the building-blocks of the physical world. He was a *Physiker*, not an *Organiker*.

Staying on at the university as a postgraduate student, Kamen started doing research into the chemistry of radioactive elements. Radioactivity was one of the key phenomena of the science of atomic structure, and thus thoroughly in the purview of physics. But radioactive decays and transformations could change atoms of one element into atoms of another, and sorting out such transformations took chemical expertise. In Chicago, a large amount of the university's work on radioactivity was going on in the chemistry department, which was in fact better equipped for such investigation than the physicists were (a fact that naturally didn't stop the physicists looking down on the chemists).

In traditional chemistry, the atoms of each element were identical, and thus all atoms of the same element weighed the same amount. In atomic physics, atoms of the same element could weigh different amounts. The chemical identity of an element depended on the number of electrons in the shells surrounding each of its atoms, and the number of electrons depended, in turn, on the number of protons in the nucleus. But though all atoms of the same element had to have the same number of protons, they did not have to have the same number of neutrons. And so, contrary to the chemical orthodoxy of the nineteenth century, two atoms of the same element could have different masses.

Carbon atoms, for example, all have six protons in their nuclei, but can have five, six, seven or eight neutrons. Atoms that differ in their number of neutrons are distinguished as an element's different 'isotopes', a term deriving from the Greek for 'same place'. The isotopes are chemically all but identical, occupying the 'same place' in the chemists' periodic table of the elements, but physically distinguishable. Isotopes thus allowed the tools of physics – such as sensitive ways of weighing molecules – to reveal things about how atoms behaved that chemistry alone could never fathom. That ability was to be the basis of first Kamen's and then Benson's careers.

Weight is not the only distinction between isotopes. Some isotopes – including carbon-12 (six protons, six neutrons) and carbon-13 (six protons, seven neutrons) – are stable. There is just as much carbon-13 on the earth today as there was when it was formed 4.5 billion years ago. But the six-proton, five-neutron carbon-11 nucleus is incapable of keeping its act together, and will fall apart in a matter of minutes, giving off radiation as it does so. That quick decay makes it easy to tell carbon-11 from other isotopes. Of course, on a 4.5-billion-year-old pile of atoms such as the earth you wouldn't actually expect to find any short-lived carbon-11 lying around. But that doesn't mean it couldn't be made in laboratories. By the 1930s, it could.

The new technologies of radioactivity, which provided sources of neutrons, or protons, or neutron-proton pairings called deuterons, made it possible to change the numbers of protons and neutrons in atomic nuclei. It was possible to turn one isotope into another (meaningless in the old chemistry), or one element into another (unthinkable in the old chemistry) or even to make elements that didn't exist outside the lab (don't even start). 'Radiochemists' like Kamen busied themselves studying the processes that produced new isotopes of old elements and finding the most practical compounds to irradiate for different purposes. It was an amazingly up-to-date field in which to begin a career. When Kamen switched to chemistry as an undergraduate in 1930, the neutron had not even been discovered; by 1936, the changes neutrons could bring about in nitrogen nuclei were the central concern of his doctoral dissertation.

For all the freshness of the science, though, Kamen did not much enjoy his graduate years at the University of Chicago. He chafed at the hierarchy within the chemistry department, and resented the degree to which the young researchers were exploited by their elders. In 1935 his mother, whom he had seen as his one unfailing source of strength, died in a car crash. By the time he received his doctorate he was eager to leave the city he had grown up in, though he knew he'd miss the White Sox. He headed for Berkeley, California, where Ernest Lawrence was pioneering what would come to be known as 'Big Science' at the University of California's Radiation Laboratory: the Rad Lab.

Science undertaken on a quasi-industrial scale, built around very expensive pieces of equipment and the new ways of organizing the workforce that come with them, was a new phenomenon in the 1930s, and like many new phenomena of the twentieth century something of a Californian speciality. It was growing up in well-funded and ambitious new institutions such as Stanford, Caltech – think of the great Palomar mirror – and the University of California, Berkeley. In the mid-1930s Lawrence's Rad Lab was the greatest flowering of Big Science yet seen. The instruments the Rad Lab was built around – quite literally, in the case of its later buildings – were cyclotrons, a type of particle accelerator invented by Lawrence in 1929. If you want to bang protons into atomic nuclei you first need to accelerate them; the greater the acceleration, the more you can do with the particles. Lawrence's cyclotrons were the best accelerators of the age – built, at least to begin with, from components being made in the cluster of electrical engineering firms near Stanford University, across the Bay, among the orchards that would become Silicon Valley forty years later. The name 'cyclotron' was derived, partly in jest, from the trade-name of the Radiotron vacuum tubes the early machines made use of. From the electron on, -tron was for decades the designated suffix of techno-chic.

By the standards of today's particle accelerators, which are measured in kilometres, the cyclotrons started off tiny. Lawrence's first machine was just ten centimetres across. However, the magnets needed to keep the particles confined to their circular paths quickly grew impressive. The second-generation cyclotron, just thirty-five centimetres across, required a magnet that weighed over two tonnes. By the time Kamen arrived in 1936 the state of the art was a machine just under a metre in diameter (thirty-seven inches) which required an eighty-six-tonne magnet, something unwieldy and complex enough to eat up time and expertise at a phenomenal rate.

The machines' performance was always being fine-tuned to provide greater acceleration or greater current. As a result they were always breaking down and needing maintenance. And they were used for a number of purposes other than physics. Lawrence's brother John was pioneering the use of radioactive isotopes made with the cyclotron beams in radiotherapies.

These competing demands meant that the simple pursuit of science was often squeezed out, and the Rad Lab was often scooped in questions of fundamental physics. The early cyclotrons could easily have been the first machines to split the atom, but they weren't – it was done with much simpler equipment at Cambridge University's Cavendish laboratory. When the first creation of a new radioactive isotope was announced by Frédéric Joliot in Paris, Lawrence and his team realized they had been making such isotopes in their own lab for ages. Emilio Segrè, a radiochemist who would later join the Rad Lab full-time, discovered one new element, technetium, in a piece of discarded shielding from Berkeley that he took back to his lab in Palermo. For all this, though, the Rad Lab was at the forefront of research thanks simply to the power of its machines, and Lawrence's unrivalled ability to sell the merits of that power to people who might give him funds.

It was a hotbed of research – 'hot' in more than one respect: stray cyclotron radiation induced radioactivity in everything from gold teeth to the zippers on trouser flies – and Kamen got into the bed with gusto. He loved the power of the cyclotrons. He loved the sense of a future coming into being, the intellectual excitement, the lack of stuffiness, the socialist politics, the workplace camaraderie. He loved Esther Hudson, a woman he met shortly after his arrival, and soon married her. He loved the access to a pool of musical talent as impressive as the intellectual resources of the university. He played with Yehudi Menuhin and started a lifelong friendship with Isaac Stern at the same time as working with Gilbert Lewis and Ernest Lawrence and Robert Oppenheimer. And on top of all this he had a salary. Most workers at the Rad Lab had to find funds from elsewhere; some worked for free. Kamen's duties as the man in charge of producing radioactive isotopes, mostly for medical purposes but also for pure research, meant he got a paycheck.

While the main cyclotron product was radioactive sulphur for radiotherapies, there was an increasing interest in a range of other isotopes from biochemists trying to study the chemical processes going on inside living creatures. The isotopes' radioactivity meant that they could be used to label chemical compounds. Imagine you want to track what a cell does with the phosphorus that it ingests in order to find out which of the many compounds containing phosphorus get made first. You can dose the cell with radioactive phosphorus, wait a fixed amount of time, and then analyse the cell's contents. Some of them will be radioactive; those will be the ones into which the cell incorporated the phosphorus you fed it. Normal chemistry had no way to distinguish between phosphorus atoms that the cell had just encountered and those it had been hanging on to for ages. Radiochemistry did. A radioactive tracer would in principle move along a biochemical pathway as noticeably as a goat moves down the digestive tract of a python.

Science is always looking for new ways to see things; the clear descriptions and diagrams that end up in textbooks hide the pervasive uncertainty about what's going on that characterizes so much of research. The mind's eye of a scientist spends most of its time peering around darkened rooms, straining to see objects the shape of which, even the existence of which, is unclear. A tool that lets you see what previously was only imagined is worth just as much as a tool for making something that could never be made before. Maybe more. The cloud chambers which allowed Kamen and his colleagues to see the particle beams they produced were in many ways as vital as the cyclotrons that made the beams in the first place. Radioisotope tracers gave biology a similarly powerful new way of seeing what was going on. As Archibald Hill, a Nobel-prize-winning physiologist, said to Lawrence when he visited the Rad Lab around this time, 'Some day people may look back on the isotope as being as important to medicine as the microscope.'

One person keen to use this new microscope was an ambitious and highly gifted Berkeley chemist called Sam Ruben. He and Kamen came to form a remarkable team, exploiting the potential of the new radioactive tracers with, as Andy Benson later recalled, 'the frantic energy of maniacs'. They were a contrasting couple to look at. Ruben was tall, a high-school basketball star, with a receding hairline and a fine, handsome face; Kamen had a broad brow and a strong jaw frequently dark with stubble, and his shock of dark hair just about came up to Ruben's chin. But they were well matched in the dirtiness of their lab coats, and in their passion for their work. In their lab in the 'Rat House' – an annexe to the chemistry department that had at one time been used to breed experimental animals, and in which their free-living but less academically inclined descendants maintained a proprietorial interest – they would argue fiercely over every detail of their experimental plans. Then they would stay up all night helping each other make sure the plans worked. Perhaps most importantly, they complemented each other near-perfectly in their expertise. Kamen knew how to make the radioisotopes and deliver them in chemically useful forms; Ruben understood the organic chemistry needed to try to distinguish the various types of molecule the radiation ended up in. This was the team to which Pauling would send Andrew Benson in 1942.

Kamen and Ruben were interested in how living things assimilate carbon. The tracer they started off using was radioactive carbon-11, a spirited but short-lived little isotope. It had a half-life of just twenty-one minutes – which means that however much you started off with, within twenty-one minutes you'd have only half that amount left. In practice, this meant that experiments with carbon-11 could not take much more than two hours, because after that only one percent of the original radioactive carbon atoms would remain, and the sample would thus be a hundred times less radioactive. Their research would start at about eight in the evening, when the physicists would normally have shut up shop. Kamen would take over the controls of the cyclotron – a U-shaped console of welded steel festooned with dials and switches and control wheels. It looked pretty much exactly as a reader of the 1930s science-fiction would have imagined the cockpit of a spaceship. If the machine refused to cooperate, he would call Ruben with the scattily laconic catchphrase 'Cyc's sick, Sam'. If it worked, then after irradiating his target and separating off the radioactive carbon dioxide, Kamen would run down the hill from the cyclotron building to the Rat House.

In the Rat House, Ruben would have all the experimental materials he needed set up and ready to go. Hot plates were hot, Bunsen burners were lit; the atomic clocks, after all, were ticking. A red light would tell him when the cyclotron up the hill had been turned off, alerting him to Kamen's imminent arrival.* Ruben would meet Kamen at the door, grab the radioactive carbon, and turn its bearer away – fresh from the cyclotron Kamen would be radioactive enough to trigger the various Geiger counters set up for the experiments and ruin the results. On at least one occasion when he had been splashed with a radioactive contaminant, Ruben himself stripped off and worked in the nude, which in a chemistry lab is imprudent as well as immodest.

Many scientists study photosynthesis because they have been fascinated by it since an early age. Ruben and Kamen were not of their number. In the early days of their collaboration, their focus was on animal metabolism. But pretty soon they came to focus almost exclusively on plants, for a number of reasons. One was that carbon-11 could be got into plants quicker than it could be into

---

* The cyclotrons were used to being monitored from a distance; in the early days Lawrence would keep a radio tuned to the frequency at which the cyclotron operated by his bedside, waking up if its soothing hum were interrupted. His wife hated it.

animals (to get it into animals you had to grow a plant and then have the animal eat it). Another was that there was a biologist involved in the animal work for whom they didn't much care – Ruben felt the man had stolen some of his ideas – and if they concentrated on plants he could be frozen out. Their contact in plant biochemistry, Zev Hassid, was a much more welcome collaborator.

And for chemists used to the inanimate, plants were just much easier to work with. Animals were difficult. On one occasion a pigeon that was going to be fed radiocarbon-doped leaves got loose, smashing all sorts of glassware in the lab before an enraged Ruben finally caught it and pulled its head off with a pair of pliers, an action that left him too devastated to continue with the experiments. Plants were more peaceable.* But practicalities and personalities were not the whole story; the more the two ambitious young scientists thought about it, the more they realized that, if they could show what plants actually did with the carbon dioxide they fixed from the atmosphere, they would have made a world-class scientific breakthrough.

Carbon-14

The radioactive tracers being used by Kamen and Ruben were not the only way of using isotopes to make distinctions normal chemistry could not. Harold Urey, a Nobel laureate running an isotope chemistry lab at Columbia University in New York, was taking another approach, using stable isotopes. On earth, for every ninety atoms of common-or-garden carbon-12 there's an atom of carbon-13,

* A little later Ruben suggested reintroducing pigeons to their work in a less troubling role. When the two of them were considering using some particularly promising equipment in a lab in Stanford, Ruben woke his partner up in the middle of the night to enthuse about the possibility of using carrier pigeons to get the carbon-11 samples across the bay before they decayed. The idea was never put into action.

which is no more radioactive than carbon-12 but has one extra neutron in its nucleus; for every 500 atoms of oxygen-16 there's an atom of oxygen-18; and as Urey himself had shown, in the work that won him the Nobel, for every 6400 hydrogen atoms with just a single proton as a nucleus, there was one with a proton bound to a neutron, an isotope marked out by a name, not just a number: deuterium.

Urey's approach to biological chemistry was to produce samples of carbon with more than the usual amount of carbon-13. Molecules that incorporated this 'enriched' carbon would be slightly heavier than their unenriched counterparts: carbon-13-dioxide, for example, weighs two percent more than the usual stuff. By careful weighing of the various compounds produced when isotopically enriched carbon compounds were fed into a biochemical pathway, you might be able to track the path of the heavy carbon from reaction to reaction. The laboratory procedures required fastidiousness – but you had all the time in the world to get them right. Unlike here-today, gone-to-two-to-the-72nd-power-tomorrow carbon-11, carbon-13 would stick around.

In the fall of 1939 Lawrence heard of Urey's claims and began to worry. Radioisotopes were a key part of his Rad Lab's claim to fame. If they were devalued as research tools, the lab's prestige, and thus its ability to raise funds, would suffer. And it was certainly true that for the elements most commonly found in life – carbon, hydrogen, nitrogen and oxygen – the Rad Lab had little to offer. Carbon-11 was the best of the bunch, and the work by Kamen and Ruben was getting some attention, but its short half-life was proving hard to work with: nitrogen-13 was yet worse, with a half-life of just ten minutes: oxygen-15 managed just two. Tritium (hydrogen-3), which had recently been made for the first time by Lawrence's protégé Luis Alvarez, would eventually turn out to be much longer-lived, but at that point its half-life had yet to be measured. So Lawrence called Kamen into his office and told him to create some new, longer-lived radioactive tracers. It was to be his, and the lab's, top priority; Kamen had more or less unlimited access to the latest one-and-a-half-metre (sixty-inch) cyclotron, as well as the thirty-seven-inch machine, and could call on the help of anyone else he needed.

Kamen set about exploring as wide a range of possibilities as he could think of using all the powers Lawrence had put at his disposal. He flanked the sixty-inch cyclotron with a couple of five-gallon flasks of ammonium nitrate; the cyclotrons gave off neutrons whatever they were doing, and if those neutrons proved adept at displacing protons from the seven-proton seven-neutron nitrogen nuclei in the ammonium and the nitrate they might make six-proton eight-neutron carbon-14.* Another approach was to bombard boron atoms (five protons, six neutrons) with alpha particles (two-neutron, two-proton quartets). When the sixty-inch machine then had to be 'put into dry dock' for repairs, Kamen used a beam of deuterons (one neutron, one proton) from the thirty-seven-inch machine to irradiate the hell out of a graphite target that was more or less pure carbon. Urey, champion of the rival stable-isotope approach, sent his radioactive-isotope competitors at the Rad Lab almost a gram of carbon painstakingly enriched in six-proton, seven-neutron carbon-13 to see if a first extra neutron in the nucleus made it easier to add a second. He thereby proved that some scientists live up to everything that's best about what they are meant to be.

From his speakeasy days on, Kamen had been no slouch when it came to all-nighters, and the carbon-14 problem frequently had him working round the clock. On 19 February, the night of a torrential storm and a multiple homicide elsewhere in the east Bay, he was picked up by the police while stumbling home after three days straight on the cyclotron, his clothes unkempt, his chin stubbled, his eyes wild. When, released by the police, he finally awoke the next evening, Kamen called Ruben at the Rat House, where he'd dropped off the irradiated graphite he'd been working on. Ruben

---

* If you think putting ten gallons of the stuff you make fertilizer from into an environment rich in low-level radioactive material sounds like a recipe for making a 'dirty bomb', you aren't the first. Lawrence was later to become rather worried about this.

told him that carbon dioxide made from the graphite seemed to show some interesting activity. Kamen rushed to campus; they put the carbon dioxide through a long succession of reactions, turning it into solid carbonate and back into gas again and again; the activity persisted. They performed further reactions to rule out contaminants. Long after any carbon-11 would have vanished the activity remained. It was not far above the background level, but it was definitely there.

They had found what they were looking for, but couldn't quite believe it. They talked the finding over with Gilbert Lewis himself, whose offhand vote of confidence – 'If you boys think it's carbon-14, then that's what it is' – flattered but didn't quite reassure. They went to see Lawrence, who was at home trying to stave off a cold in anticipation of the ceremony at which the Swedish consul would present him with his Nobel prize. Lawrence jumped from his bed and danced round the room, immediately talking of alerting the newspapers. The young researchers blanched; what had seemed sure in the lab seemed rather less certain when faced with the prospect of the press. They went back and looked at the sample again. The radioactivity was still there.

The discovery was formally announced in the most dramatic setting possible a few days later: at Lawrence's Nobel prize ceremony. The chairman of Berkeley's physics department described carbon-14 as, 'on the basis of its usefulness . . . certainly much the most important radioactive substance yet created'. All eyes turned to Kamen – Ruben, fearless in many ways, had been too nervous to attend. Again Kamen was overcome by the worry that they might be wrong.

They weren't. A month or so later, there was a rebellion at the sixty-inch cyclotron. Kamen's flasks of ammonium nitrate had been getting in the crew's way for ages and had now started leaking; they had to go. So Kamen and Ruben hauled them off and analysed the contents. They precipitated solid carbonates out of the liquid and checked them with a gently clicking Geiger counter. The machine went completely silent; the carbonates were so radioactive the Geiger counter couldn't keep up. Kamen had vastly underestimated the capacity of nitrogen to soak up neutrons and turn itself into carbon-14, and now a practical path to the large-scale production of a long-lived and much-needed radioisotope lay open. There was talk of cyclotrons devoted to nothing but making neutrons with which to transmute nitrogen into this wonderful new substance.

Carbon-14 is an illustration of one of the great scientific themes of the first half of the twentieth century; the way in which the consequences of the discovery of structure within the atom rippled out into surrounding disciplines. Chemistry, the nearest neighbour, took the first brunt of the shock. In disciplines further afield the effects were felt a little later, but not necessarily any less strongly. Biology, for example, was faced not just with a new physics but also with a new chemistry; it was as though the ripples, far from damping down over time, actually became stronger.

It was not just the ideas that were powerful – it was also the techniques. New ideas about how atoms worked mattered a lot; but the techniques for understanding and changing their behaviour mattered as much or more. The new physics made possible new ways of doing things. Andrew Benson caught some of that sense of agency when he argued, years later, that his friend and colleague Kamen invented, rather than discovered, carbon-14. Kamen was not just looking for something in nature; armed with new ideas and new techniques, he was setting out to make something he could use as a tool, a new type of atom tailored to the needs of biological research. The primary reason Kamen's creation of carbon-14 is now seen as a discovery is that nature later turned out to have been producing carbon-14 all along. Cosmic rays are ceaselessly hitting the earth's upper atmosphere like particles fired from a cyclotron. The cosmic rays produce neutrons, and those neutrons hit nitrogen atoms, transforming a tiny fraction into carbon-14. (Photosynthesis then assimilates some of that carbon-14 into plants – and millennia on, the amount of carbon-14 left in that plant material will reveal how old it, or anything made of it, is. And so the ripples spread further. In the conceptual atlases most of us steer by, archaeology is a long way from nuclear chemistry. But carbon-14 dating techniques would revolutionize the field in the 1960s and 1970s.)

Whether invention or discovery, there's no doubt that carbon-14 was a great boost for the Rad Lab in general and Kamen and Ruben in particular. But from Kamen's point of view, there was a hitch. When he and Ruben had begun their partnership, they had agreed that papers dealing mainly with radioisotopes would give Kamen's name pride of place as first author, while papers that were principally biochemical would list Ruben's name first. But Ruben was desperate for tenure in the extremely competitive chemistry department: he felt that Lewis might blame him for some experiments they had worked on together that had not gone according to plan; he worried that the dean was not on his side; and he thought there was anti-Semitism in the department as a whole that would make him easier to pass over. In the light of these perceived obstacles Ruben really wanted the kudos of first authorship, and Kamen – who wanted a tenured partner with ever more access to the chemistry department's impressive resources – gave it to him.

The decision brought Kamen a sharp rebuke from Lawrence, who saw Kamen's second place as a slight to the Rad Lab. How could ever-bigger cyclotrons be sold if credit for the discoveries of the existing ones wasn't claimed? It was resolved that things would be put right with a second, longer paper with Kamen's name in front. But Ruben had that second manuscript typed up by the dean of chemistry's secretary – a formidable power in her own right – and she reversed their names. In years to come, Kamen came to suspect that, to many, these details of authorship suggested that their great discovery was really Sam's alone, and that that perception cost him dear. In his memoir, *Radiant Science, Dark Politics*, Kamen shows himself keenly aware of which of his colleagues and contemporaries won Nobel prizes and which did not. Kamen lived out his life in the latter camp. But the radioisotope he invented went on to be the key to tracking carbon on its photosynthetic journey out of the atmosphere and into the biosphere.

The reversals of war

One morning in 1939, shortly before Andy Benson graduated from Berkeley, the instructor for his optical physics class turned up 'almost pale with fevered excitement'. The instructor – Luis Alvarez, the Rad Lab researcher who created the first tritium – had just heard that German scientists had proved that uranium atoms could tear themselves in two, letting off significant amounts of energy in the process. It had long been known that there was a lot of energy locked up in atomic nuclei. This discovery showed a way that it could be released. In a world getting ready for war, it was news well worth some fevered pallor. The uranium atom was going to change history. It was also going to change the Rad Lab.

When Benson returned to Berkeley to work with Ruben and Kamen in the spring of 1942, both changes were well under way. Six months before, the photosynthesis work had been one of the Rad Lab's proudest boasts. It was on the track of a basic discovery that would prove the value of radioactive tracers, and thus of the machines that made them. It was featured in the October 1941 issue of *Life* magazine. Ruben and Kamen thought they had evidence that the first product the plants made out of carbon dioxide was an organic compound with two oxygen atoms attached to a carbon atom at one end – what's called a carboxylic acid. This wasn't the view taken by others in the field, but it made a kind of sense; after all, a carbon-dioxide molecule was carbon with two oxygens attached, so if you stuck one on the end of a receptor molecule of some kind, a carboxylic acid with two oxygens at the end sort of made sense.

They had also had the important insight that the reactions involved must be arranged in some sort of cycle. After carbon dioxide was added to the unknown receptor molecule in order to make an unknown carboxylic acid, some of this carboxylic acid would be turned into carbohydrate. But some of it would have to be turned back into the receptor molecule, so that the process could

29

start again. This insight would later prove crucial. Ruben also started speculating that phosphate groups – stable sets of phosphorus and oxygen atoms – might play a role in powering the process. This too turned out to have some truth in it.

But history was against them. Even before America entered the war in December 1941, the Rad Lab's expertise and resources were a central part of what was to become the Manhattan project. After verifying the fissile nature of uranium with their own neutron beams – uranium fission was another discovery that the Rad Lab could have made, but didn't – Lawrence's scientists went on to isolate a new, artificial element that behaved in a similarly alarming way: plutonium. Lawrence had also turned his attention to the problem of isolating the isotope of uranium that would split apart under neutron bombardment – uranium-235 – from the far more common isotope, uranium-238. He'd dismantled the old thirty-seven-inch cyclotron and used its parts to build what he called a 'calutron' ('calu-' for *cali*fornia *u*niversity); the different uranium isotopes followed slightly divergent paths through the calutron's magnetic field, and so could be separated from each other. This is the principle behind 'mass spectrometry' – a magnet can sort out particles of different masses in a beam just as a prism can sort out the different wavelengths in a ray of light. Physicists see more or less any device that spreads things out according to their energy as a spectrometer, whether it is a prism fanning white light out into a rainbow or a detector measuring the kick of cosmic rays.

The first successful isotope separations at the Rad Lab took place on the day the Japanese navy attacked Pearl Harbor. By the time that Benson arrived the next year, much of Kamen's time was taken up with war work, and while Ruben still managed to keep their photosynthesis work moving on, he was hampered by a lack of carbon-14. Lawrence had become worried about the chemical instability of the ammonium nitrate from which it was most easily made. All the chemists assured him that when kept in solution the nitrate was entirely safe, but Lawrence never quite believed them, and large-scale production had not got under way. But progress was being made.

At the beginning of the following year, though, Ruben's focus switched from photosynthesis to phosgene, a chemical weapon that the army wanted to know a lot more about. Deploying phosgene, or defending troops against its use, required better understanding of its chemistry and its behaviour, and Ruben was in charge of the project. Kamen – by then concentrating largely on ways to improve uranium isotope separation, since the calutrons did not really do a good enough job – produced phosgene from carbon-11 and radioactive chlorine – as nasty a substance as you could imagine. Benson suggested ways that the phosgene might inflict its harm, and experiments to test the hypothesis started. Lab rats returned to the Rat House.

Benson still did some photosynthesis work on his own; Ruben had given him a vial containing all the man-made carbon-14 in the world – it's quite possibly still on one of the shelves of his office at Scripps. He confirmed in his own mind the carboxylic acid hypothesis Kamen and Ruben had been working on, and wrote up a paper on it. But then the war interrupted him, too. Benson's upbringing had been strongly influenced by Quakers, and he was registered as a conscientious objector. Curiously, his beliefs didn't stop him from joining in Ruben's work on phosgene, but his official status made it hard for the university to hold on to him as an instructor after his initial contract was over. In the middle of 1943 he was forced to leave Berkeley and head for the hills as part of what was called the Civilian Public Service programme. His photosynthesis paper was left unpublished. Being away from the lab was a disappointment, but logging, fighting forest fires and building dams was the sort of outdoors activity that he relished. It could have been worse.

For his colleagues it was. In the summer of 1943, Ruben ran a series of experiments at Mount Shasta and on the beaches of Marin County to study the way poison gases spread. The gases were simulated by mercaptans, the substances that give skunks their stink; sniffing graduate students monitored their movement. Driving back from Shasta in September, even more sleep-deprived than usual,

Ruben crashed his car and broke his wrist. On Friday 24 September he had lunch with Kamen at the Faculty Club – the two of them were still hoping to get back to their photosynthesis work. Ruben, his arm in a sling, looked wan and tired. Not for the first time, Kamen told him to get some rest.

The next Monday, Ruben was back in the Rat House with two students. He needed to get a new sample of phosgene from one of the glass ampoules in which it was provided by the government. Andy Benson had developed a meticulous method for this hazardous manoeuvre; Peter Yankwich, one of Ruben's graduate students, insisted on doing it out of doors, rather than in an enclosed space. This time Sam did it himself, at the bench. He was tired, and with his arm in a sling perhaps a little clumsy, but that's probably not what killed him. The ampoule seems to have had a pre-existing flaw; when Ruben carefully immersed it in liquid air to freeze the phosgene, the glass simply shattered. Boiling liquid air sprayed droplets of phosgene out into the lab; Ruben, the closest, breathed in more than the students. According to some accounts, he had the presence of mind to write notes telling people not to enter the lab and post them to the doors before getting outside and lying down among the eucalyptus trees. Immobility, Ruben knew, was the best countermeasure.

The students survived. Sam died in hospital the next morning, drowned by the fluid that filled his lungs. It was his eighth wedding anniversary. He and his wife, also a chemist, had three children. In his rush to do ever more science, he had never filled out the forms that would make the family eligible for a federal pension.

Kamen was numbed, overwhelmed. His wife had asked him for a divorce a few months earlier. Now he had lost his other other half. The next year, things got worse. Kamen had been a supporter of the Spanish Republicans, interested in the possibility of unionizing labour in the labs; he was a friend of Oppenheimer. He was spending more and more of his time with 'a new and exciting group of leftist intellectuals and bon vivants', as he put it in his memoirs, to ward off depression and loneliness. And he was also spending more and more of his time under surveillance by the Manhattan Project's security staff.

At a party to welcome Isaac Stern back from a tour entertaining the troops in 1944 Stern introduced Kamen to the Russian vice-consul, who said he was interested in contacting John Lawrence, Ernest Lawrence's brother; a Russian in the consulate in Seattle had leukaemia and it was thought that he might benefit from radio-therapy using cyclotron-produced radioactive phosphorus. Kamen mentioned the request to Lawrence. By way of thanks, the vice-consul and another Russian took Kamen out to a fine dinner at which his family's Russian roots were discussed.

Kamen didn't learn the full extent of the suspicions that built up around him over these incidents until years later, when some of them were leaked to the *Chicago Tribune* along with allegations that he was an 'Atom bomb spy'; the scandal led him to attempt suicide. All he knew at the time was that one day in July 1944 he was summoned to the office of Donald Cooksey, Lawrence's closest college friend and most trusted lieutenant. Cooksey awkwardly told Kamen that 'he talked too much in the club' and that he had to leave the project immediately.

That afternoon he walked out of the Rad Lab for the last time. He was thirty years old, his first marriage over, his closest colleague dead, his job gone, his patriotism questioned, his career apparently finished.

He got through it. He found a job inspecting the welding on Liberty ships being built in the Oakland shipyards. A friend allowed him to work for free in a chemistry lab at the university, in a building and a field safely distant from the Rad Lab and radiochemistry. His friends in music never left him. After the war he got a job overseeing the cyclotron at Washington University in St Louis, a machine devoted to biological and medical work with no security implica-tions. From there he rebuilt a first-rate career in science, working on a variety of biochemical problems to which radioactive tracers could be applied, then on other sorts of biology. He remarried, became a widower, and remarried again. He was persecuted by the

House Unamerican Activities Committee and withstood it; he had his passport revoked and got it back.

Eleven years after being thrown off the Manhattan Project, Kamen finally saw a copy of the report that had precipitated his fall. It was presented as part of the defence case when Kamen made the bold decision to sue the *Chicago Tribune* for libel in 1955. The judge ruled it inadmissible, but thought Kamen had the right to see it. After all the heartache, it was a flimsy thing – some innuendo, some lies, and the fact that he was a left-wing Jew from Chicago with a wide circle of friends.

Kamen won the case.

Melvin Calvin and the path of carbon

In 1946 Andrew Benson returned to Berkeley. While Ernest Lawrence was no help to Kamen in his struggles to clear his name, either during or after the war, he was still keen to see carbon-14 used to crack the problem of photosynthesis, and after the war he was able to restart the research on a grander scale. The care and feeding of physicists and their machines had, in a grim flash, become a national priority, and Big Science was getting ever bigger. At Berkeley, new types of accelerator were being developed, and new buildings at the top of the hillside campus were under construction to house them.

Lawrence recruited Melvin Calvin to head up the use of radioisotopes as biological tracers, with a particular emphasis on photosynthesis work. Calvin's qualifications were obvious. He was one of the chemistry department's stars, and had worked with Gilbert Lewis on the ways that some chemical compounds react with light, research that seemed as if it might be relevant. He was keen to apply chemistry to the biggest questions – in the 1950s he would soon become one of the first people to take an experimental approach to the origin of life, bombarding chemicals he thought might have been present on the early earth with radiation from Lawrence's synchrotrons. He was endlessly inquisitive, capping the Thursday seminars Gilbert Lewis presided over with question after question, some of which would be most incisive. To his younger colleagues, some of Calvin's endless questions seemed embarrassingly dumb; but Calvin was not embarrassed, and knew he wasn't dumb. He was impervious in self-belief, and extremely ambitious.

Lawrence provided Calvin with money from the newly established Atomic Energy Commission, with chemists who'd been working on uranium compounds as part of the Manhattan Project, with carbon-14 – which could now be produced in copious amounts using the neutrons that flooded out of nuclear reactors – and with lab space in various buildings that the Rad Lab was relinquishing. Calvin sought out Benson, the survivor of the original Rat House, who was at the time working on the synthesis of anti-malarial drugs. Benson took on responsibility for the photosynthesis section of Calvin's growing research team. He was put in charge of the building which had previously housed the thirty-seven-inch cyclotron – its former occupant, quaint compared to the machines then planned and under construction, had been donated to Berkeley's University of California sister campus down in Los Angeles.

Benson threw linoleum over the orange-yellow uranium salts ground into the maple floorboards and started laying out a lab to Caltech's high standards. He was surely thinking of Ruben as he did so. The workspaces were well lit, with black glass worktops and the best, deepest porcelain sinks he could lay his hands on; there were ventilation hoods to suck up fumes. It was a far cry from the bare board walls, exposed wiring and unshaded light bulbs of the Rat House. Green algae were grown in big flattened flasks that Benson had designed in such a way that the cultures could be conveniently stirred, easily flushed out and evenly illuminated from both sides; they quickly came to be called the lollipops, which was an apt description but rather ticked Benson off. In the basement, a room was carefully shielded from external radiation sources so that the rather weak radioactivity of the carbon-14 could be measured, click by click, as it registered on the team's Geiger counters.

It was a young lab – Calvin, the 'old man', was just thirty-five when it was founded, and most of his crew started off there in their early or mid twenties. Calvin and his wife, Gen, hosted lots of parties, and there was a constant stream of weekend expeditions to go skiing, or rock climbing, or hiking, heading off to Death Valley, driving to Pacific Grove to see the monarch butterflies, or whatever. Alice Lauber, a secretary whose remit ranged from counting the clicks of Geiger counters to all-purpose den-mothering, remembers being worried at her interview that she knew almost no chemistry; when she revealed she liked to ski the job was hers. Calvin wouldn't come along on the trips – a heavy smoker who was severely over-weight, he suffered a heart attack in 1949. But on the Monday after some skiing or climbing expedition Calvin would get into the lab early, just to make sure everyone had got back safe and sound.

The problem that Calvin, Benson and the team in the Old Radiation Laboratory (ORL) faced was how to identify precisely the molecules into which the algae in the lollipops were turning their carbon-14-laced carbon dioxide. The technique they hit on was called paper chromatography. You take a mixture of organic compounds that need separating out and identifying – an extract of the innards of some photosynthetic algae, for example – and drop a sample on to a big sheet of paper. You then bathe the paper in carefully controlled mixtures of organic solvent and water. The different compounds in your sample will creep through the paper in different directions, depending on which of the solvents suits them best. And they will move different distances in their chosen direction depending on how well that solvent suits them. The different compounds all find the places they're most comfortable in, producing a self-selected pattern called a chromatogram. The procedure was time-consuming and offensively smelly, giving off odours that sickened the small number of physicists left in the building. But it was revealing. Different sugars would always move to their pre-ordained places; glucose here, pentose there, triose off to this side, ribulose over there. If you didn't recognize something, you could always label the mystery patch Godnose, and more than once they did.

The ORL team added its own wrinkle to this technique. Once they had the chemicals spread all over the paper, they would slap a piece of X-ray-sensitive film down on top of it. Chemicals that contained carbon-14 would fog the film through their radioactivity. It wasn't perfect – you still couldn't say what, exactly, the chemicals containing the carbon-14 were – but at least you knew where they went on a chromatogram, which told you something about their chemical properties. A little light was let into the darkened room of the scientific imagination. And you also had a purified sample, albeit a small one – you just had to cut out the bit of paper responsible for the fogging. Benson had inherited from his doctor father a firm belief in the powers of the pocket knife as an all-purpose tool.

Benson built a big white worktop in the centre of the lab, right where the cyclotron had been, so that the big, unwieldy chromatograms could be spread out and studied by a clutch of collaborators. It was here, like explorers poring over maps, that Benson, Al Bassham, a graduate student who became, in time, one of the team's leaders, and the rest of the ORL members plotted what Calvin called 'The path of carbon in photosynthesis' – the shared title of a long series of papers in the *Journal of the American Chemical Society* that recorded their progress.

Or, as often as not, their lack of progress. The Calvin group had a reasonably relaxed attitude to false starts and blind alleys. An observer from another, more careful laboratory recalls being amused at the way a new compound would be introduced with caveats and question marks in one Path of Carbon paper, graduate to being an accepted, no-question-mark part of the process in the next Path of Carbon paper, and then vanish from sight in all subsequent offerings. But that was the way Calvin operated. He didn't mind being wrong frequently, as long as he was right in the end, when it counted, and he was ceaselessly in search of the novelty that would get him to that final goal. Of all his questions, the most common was the all-encompassing 'What's new?' After his teaching and admin and lecturing was over for the day he'd wander through the ORL and the other labs in his empire and ask the question of almost everyone he came across. And he'd really want to know the answer. Then he'd do much the same thing at eight the next morning, before the day was under way. Benson and his colleagues soon learned never to tell Calvin quite everything, to make sure there'd be something in the cupboard for the next time he asked – or that he wouldn't take some early finding and blow it up into a paper or a scientific talk.*

Within a couple of years, the splotch that represented the first product formed when carbon dioxide was taken up had been pretty conclusively identified; it was phosphoglycerate, a form of carboxylic acid – Kamen and Ruben had been right – with three carbon atoms. The sooner the team stopped their algae from photosynthesizing, the more of the carbon-14 would be in the form of what looked like phosphoglycerate, because there hadn't been time for the cells to turn this primary product into the array of secondary and tertiary products from which they made complex carbohydrates, proteins and everything else.

The first step down the path of carbon had been taken – but the identification of phosphoglycerate presented a new problem. There was no way of making phosphoglycerate with just carbon dioxide and water; there had to be another organic molecule involved, an acceptor on to which the carbon dioxide could be stuck. And as Ruben and Kamen had realized, that meant that as well as turning some of their phosphoglycerate into glucose, the algae also had to

---

* When the biochemist Hans Kornberg was visiting the ORL in the 1950s Calvin presented data Kornberg wasn't fully confident of at a meeting; Kornberg objected, quite strenuously. A few days later Kornberg was woken up in the middle of the night by Atomic Energy Commission security men who told him he'd breached lab security by leaving secret documents open on his lab bench. Kornberg, confused and worried, protested that he didn't have any secret documents – none of the ORL work was classified. It turned out that, impressed by the way he'd stood up to the boss, someone else at the ORL had found an official stamp with which to mark the first page of Kornberg's laboratory notebook 'Secret'. The security people weren't happy with this explanation and wanted to report the 'breach' to Washington; when Bert Tolbert, Calvin's unflappable deputy, was presented with the evidence he resolved the situation by tearing the relevant page out of the notebook and throwing it away.

be turning some of it back into this mysterious acceptor molecule, since otherwise they would run out of places to put new carbon dioxide.

The obvious place to start looking for the acceptor was among small molecules that contained just two carbon atoms. The logic behind this was that the acceptor had to be one carbon-dioxide molecule short of being phosphoglycerate, and since phospho-glycerate contained three carbon atoms, the acceptor had to have two. So a lot of effort went into looking for a two-carbon acceptor, to little avail. At the same time, though, more of the molecules that were being created out of the original phosphoglycerate were being discovered, molecules that were presumably on the path from phos-phoglycerate to glucose. Oddly, one of them was actually larger than the glucose it would end up helping to make. This seven-carbon sugar was so unexpected that when Al Bassham told Benson about it Benson suspected it was a mistake. But Bassham didn't make mistakes on that sort of thing.

Benson found one of these larger sugars particularly interesting: ribulose diphosphate, a five-carbon sugar with phosphate groups stuck to each end. When algae were grown without access to carbon dioxide, the levels of ribulose diphosphate tended to climb. Eventu-ally, it dawned on the team that this was because ribulose diphos-phate was not just a downstream product of carbon assimilation. It built up in the absence of carbon dioxide because it was the raw material for a reaction that needed carbon dioxide in order to proceed. The hunt for a two-carbon acceptor had been a blind alley; the acceptor was this five-carbon sugar. The algae were forcing it to react with carbon dioxide in a way that produced two molecules of three-carbon phosphoglycerate at once. Most of the reactions cells carry out have their counterparts in the practice of organic chemists. This one didn't. It was something no one had ever imagined.

Once the surprise was taken on board – the paper in which it was announced, 'The path of carbon in photosynthesis, part XXI', was published in 1954 – the cycle by which the receptor was regener-ated, and the roles of the bigger sugars within it, started to make sense. It was like a dance; if you understood which foot to start off on, and knew how to count, you were halfway there.

The best place to start this particular dance is with three molecules of carbon dioxide being jemmied into three molecules of ribulose diphosphate. That produces six molecules of phosphoglycerate. Take one of those molecules and put it to one side: that's the product, the stuff that will be used by other metabolic pathways as they make sugar and starch and fats and amino acids and all the other small molecules from which big things are built. As such, it sits the dance out.

You're left with five molecules that have three carbons each; you need to rejig them into three molecules with five carbons each, so that the cycle can start again. You should feel free to take it on trust that the photosynthesizing cells manage this five-threes-to-three-fives metamorphosis with elegant dispatch – but if you want a sense of the fancy footwork, here it is. Two of the three-carbon molecules join together to make a six-carbon sugar; this is then split into a four-carbon sugar and a two-carbon sugar. Each of these products is added to another of the original three-carbon molecules. The two-carbon sugar and its three-carbon sugar, when joined, make a five-carbon sugar, which is what we wanted. The four-carbon sugar and its three-carbon sugar come together to form one of Bassham's weird seven-carbon sugars. Take a two-carbon sugar away from this seven-carbon sugar and you're left with a second five-carbon sugar; add that two-carbon sugar to the last of the five original three-carbon sugars and you have your third and final five-carbon sugar. And there you are, back where you started, with three five-carbon sugars primed and ready for three more carbon-dioxide molecules.

And so the Calvin-Benson cycle turns, an odd-and-even square dance that was already old when the planet was half its present age.

Rubisco

The Calvin team's formal name was 'The Bio Organic Chemistry Group'. The main players – Calvin, Benson, Bassham and Bert Tolbert – were chemists, used to making and breaking down smallish molecules, and studying their activity or lack of it. They weren't biologists, though as the group grew biologists were recruited.* They weren't even biochemists – but they were headed that way.

Biochemistry is chemistry with which to find out about life. It's not about products. It's about processes. It's interested in how living creatures bring about the chemical reactions between organic molecules that are necessary to their lives. The answer to this question is normally 'with an enzyme'. Enzymes are protein molecules that bring other molecules together and persuade them to react with each other – biological catalysts. They are, in the chemical sense, organic molecules. But they are beyond the synthetic scope of organic chemistry; they are not things you can make from scratch. Organic chemists in the 1940s and 1950s spent their time synthesizing and degrading things that are either relatively small (that seven-carbon sugar of Bassham's, for example, is made of thirty-four atoms – chlorophyll of 137) or relatively simple, such as the long chains of identical units in polymers like Bakelite and nylon. Proteins, on the other hand, consist of thousands of atoms, carefully folded in on each other, with every atom of every element in its own appointed place. There was and is no way of synthesizing such a thing in any laboratory without using protein-making machinery borrowed from living cells. So in the 1950s biochemists tended to study enzymes either in the context of the creatures they came in – in algae, yeasts, bacteria, tissue cultures and so on – or in preparations made by breaking down such creatures and taking out

---

* One of them, Clint Fuller, greatly improved the lab's workings when he discovered that their cultures of algae were thoroughly contaminated with yeast.

the chemicals of interest. Their aim was to identify the enzymes responsible for all the various reactions that made up the metabolic pathways and to work out how – and how well – they did what they did.

Their radioactive tracers had let Calvin's band of chemists trespass into the realm of the biochemists without becoming biochemists themselves. They had put together a picture of the carbon cycle central to photosynthesis without really paying any attention at all to the enzymes that made it possible. Once the cycle was reasonably well understood, though, Benson started taking an interest in them, and in particular that which catalysed the addition of carbon dioxide to ribulose diphosphate, the only enzyme in the whole process that did something an organic chemist would never have dreamed of doing.

With a visiting Belgian scientist, Jacques Mayaudon, Benson started building up a picture of the enzyme. And the more he saw of it, the more it looked familiar. Down at Caltech – which he still visited frequently, not least because his wife's family lived in Pasadena – Benson had come across a scientist called Sam Wildman who was studying the proteins to be found in leaves. Caltech, at the time, was the best place in the world to be studying proteins. Linus Pauling, having applied a physical approach to chemistry and come up with the definitive theory of the chemical bond, was marching on into the realm of biology. At Caltech proteins were being zapped by X-rays, teased apart with centrifuges, purified with electric currents. The ripples spreading out from atomic physics were creating a new 'molecular biology'.

As part of this programme, Wildman had been taking cells from leaves, wringing out everything that wasn't a protein, and studying what was left. There were thousands of different types of protein, but one protein greatly outnumbered the others. Wildman called it 'fraction one protein' because it was the first protein you found when you put the leaf extracts into a centrifuge – which meant it was a particularly large, heavy molecule, as well as a very common one. In some leaves it made up fifty percent of the total protein content. In some algae, it made up a quarter of the protein in the whole organism. The more Wildman learned about the protein that came from the leaves and the more Benson learned about the protein that incorporated carbon into the world's carbohydrates, the more Benson became convinced they were one and the same. The protein responsible for knitting the dead carbon dioxide of the earth's atmosphere back into the living tissues of its vegetation had been found. And, rather fittingly, this most crucial of proteins was also the most common of proteins. There are more copies of this one molecule on the earth and in the oceans than there are of any other catalytic protein.

The peculiarities of this protein – in particular an odd affinity for oxygen, and a preference for the lighter carbon isotope, carbon-12, over the heavier carbon-13 – are crucial for our understanding of the earth's past, and also for driving various changes in the environment, both in the deep past and the present. We'll have more to say about that in due course. For the time being, let us content ourselves with simply giving it its name. For years it struggled, as do many scientific entities, under a series of obscure tags, from the reasonably plausible 'carboxydismutase' to the ludicrously officious '3-phospho-D-glycerate carboxylase oxygenase (dimerizing) EC 4.1.1.39'. It was not until Sam Wildman came to retire, in the late 1970s, that it got its name. Striving for what one must imagine was a slightly contrived joke about biscuit-making, a colleague of Wildman's contracted one of the protein's formal names – ribulose bisphosphate carboxylase – into 'rubisco'. The term caught on immediately. The world's most popular protein finally had a name.

The effect of rubisco on Benson's career was far quicker – and far less amusing. Benson himself professes not to be quite clear on what happened, but from his accounts and others, the gist is apparent. Though Calvin himself was involved in the work that showed ribulose diphosphate to be the carbon-dioxide acceptor, he was at the same time much more involved in an elegant theory as to how the energy that drove the carbon cycle might be liberated by a photochemical effect not that dissimilar to those he had studied in his early days with Lewis. It was a gorgeous theory, and in some ways a much fuller one than the simple cycling of phosphoglycerate and ribulose diphosphate; it explained things that work could not, such as the role of chlorophyll. When Calvin presented his ideas to the American Association for the Advancement of Science, Cornelis van Niel, perhaps then the country's greatest microbiologist and a deep thinker on matters photosynthetic, greeted it with tears of joy. The only problem was that it was wrong. What looked to be a crucial confirmation turned out not to be; on closer examination, the other supports crumbled, too. Calvin had over-reached, again. At the same time, Benson was well established as the leader of the parts of Calvin's programme that were actually producing results.

It proved an intolerable situation, and Benson was effectively banished from the ORL; according to one account, Calvin's wife, who had helped Calvin to stop smoking and lose thirty kilos after his 1949 heart attack, told Benson that if he stayed, Calvin would have another one. The paper Benson had written with Mayaudon that identified the fraction one protein as the key protein for carbon uptake by plants, which like all papers from the Bio Organic Group had to be OK'd by the laboratory authorities, never made it out of the lab. By the time it was published, three years later, others had made the same leap and it was old news.

Clint Fuller, the first microbiologist to join the Calvin lab, later pointed to a peculiarly poignant token of the rift. In a memoir published in the journal *Photosynthesis Research* in 1999 (after Calvin's death), Fuller wrote:

> I would like at this point to express a personal note that represents my own feeling and the recollections of many of the scientists who with me experienced the research years at the ORL in Berkeley on photosynthesis. Calvin's autobiography, *Following the Trail of Light*, represents an extremely singular view of the research carried on in the laboratory particularly in the area of the path of carbon ... In all the 175 pages of his autobiography there is not one sign of Andy Benson or a mention of him. There is not one picture of Andy in a

book that contains fifty-one photographs ranging from graduate students to the King of Sweden. There is not the citation of a single paper with Benson as author or co-author in an extensive bibliography of over 150 references. Benson's name appears nowhere in the text and consequently is absent in the 12-page index. This appears to be an undeserved slight to a great scientist both personally and professionally who had contributed in a major way to all of Calvin's research and technology in the field of photosynthesis ... I know that all of us who were colleagues at Berkeley agree that it was Andy and Al who contributed greatly to our own success in future endeavors. I have no idea what may have caused this unfortunate event, but I think that history should record that the contribution of Andy Benson is not properly recognized in Calvin's autobiography.

Benson headed off to new projects on the east coast, but though the science was good, the life there didn't suit him. In the early 1960s he returned to California, this time to the south – to Scripps. He was there when, in 1961, Calvin received the Nobel prize, alone, for the discovery of what was by then called the Calvin cycle. When I talked to Benson in his office forty years later, I asked him if he thought Calvin had deserved the prize. He was thoughtful, and sly. 'It's a hard question. How can anyone say whether someone deserves a Nobel prize? Did he deserve it as much as me? No.'

In a way, this doesn't matter. One of the curious things about the history of science is that those involved in it tend to think it would turn out more or less the same way regardless of who did what. Nature, they say, is what it is; rubisco, which would be as bulky and odourless by any other name, would also do exactly what it does whoever discovered it. Had Sam Ruben lived, he might have followed the path of carbon a few years quicker than Melvin Calvin and his crew did; he might have won a Nobel prize that bit sooner, and even, possibly, with Kamen and Benson at his side. And the way in which the path of carbon was imagined might have been slightly different. But the square dance in the cells that turns three-carbon sugars to five-carbon sugars and back again would not have been any different.*

Discoveries feel determined. They are there to be made, and if one person doesn't, another will. This doesn't lessen the achievement; indeed it can give it spice. The thought that 'this is the way the world is – and I am the first to see it as such' is an intoxicating one. It is not unique to science – a poet may have the same feeling, or a painter – but the scientist who feels this way has the feeling in full measure, because she knows that it is in the nature of science that what she first sees as a truth will, if she is right, eventually be received as such universally. It will change the way the world is seen by everyone. No artistic insight can make that claim so universally. But the other side of this power is that a truth we accept as truly universal loses the need for an author. It becomes part of the way the world is, regardless of who saw it first, and in time the identity of whoever it may have been who first looked out from that particular peak in Darien is lost.

For all that it will fade, there is still a cost to being excluded from your place in a science's history, as Benson was. To be thrown out of a team that you were leading in great things, banished from a place that had become your mind's home, and which you had largely built with your own hands, is a terrible thing. Not winning a Nobel prize you deserved is also a sorry fate – it certainly hurt Martin Kamen deeply. But there have been worse things in Benson's life. He has suffered other losses – those of children to mental illness and suicide – that surely cut far deeper. And yet he carried on, and carries on still. He kept friends and made others, discovered useful new techniques and intriguing new molecules. He developed a few eccentric hobby horses, such as a proselytizing faith in methanol as a stimulant of plant growth (part of a general pro-methanol attitude

---

* As a result the What-If questions that fascinate many historians (even if some of them won't admit it) are curiously uncaptivating when it comes to the history of science; when they entice people at all it is in their application to theories, rather than discoveries. It is much easier to see theories as products of minds and cultures that could be otherwise and so we can, for example, amuse ourselves by asking how different the phenomena we understand through natural selection might look if the theory had been conceived in a context other than that of Victorian Britain.

– 'as long as you have enough folic acid it's perfectly safe. You could make your martinis with it, but my wife won't have it'). Trips to the northwest to study the ageing process in salmon – they do it far less gracefully than he has – led to a rekindling of a youthful interest in anthropology when he met the native families that the anthropologist Franz Boas first studied a century ago. Benson now counts them as friends, and on one of his trips he introduced them to Boas's great-grandson, then a high-school science whiz in La Jolla whom Benson had come upon by chance.

When we talked in his office, the sun sinking towards the Pacific, Benson was obliging in answering questions, and warmed to his recollections with some vigour. He's a man with points of view and he'll take some pleasure in putting them across. He relished being able to take out old X-ray films from his drawers and point out the dark blotches created by carbon-14-labelled ribulose diphosphate, back in the early days. He's proud of the past, and having long kept fairly quiet about it, he's not unhappy to let his role in the path of carbon become more widely known. The fact that, increasingly, people within the realm of photosynthesis studies talk about the Calvin-Benson cycle, rather than the Calvin cycle, is something that gladdens him (some people go so far as to call it the Calvin-Benson-Bassham cycle, and that suits him fine too). But Benson doesn't live in the past. He does his duty by it, but he's more interested in what's now, and what's next. In talking to me he was taking time from the final preparations for a research trip up to the northwest, and that was what mattered most to him.

After we talked, I wandered about Scripps a little longer, just enjoying the place, taking a bit of time to watch the surf rise and the shadows lengthen. Retracing my tracks, I saw Benson from a distance in one of the car parks, wrestling a piece of equipment into the back of a beautiful white Mercedes that he bought in Denmark in 1961 and which has served him faithfully ever since. Getting something done, it seemed to me, suited him more than reminiscing.

He saw me watching and smiled, happily. He waved, and he headed off.

CHAPTER TWO

Energy

Autumn and the conservation of energy
Robin Hill and Cambridge biochemistry
Where the oxygen comes from
Daniel Arnon and the isolated chloroplast
The Z-scheme
Peter Mitchell and the role of the membrane

Modern painters in their fervour
Seek to startle the observer
    By reliance on their peacock hued appeals
But I find more consolation
In the dusty rubrication
    Which the background of the galaxies reveals
And though rare sweets and ices
Compounded at high prices
    A transitory rapture may impart
The glycosides of madder
Make me infinitely gladder
    And rejoice the inmost cockles of my heart
        C.L.G. 'The Sweets of Science'*

* Published in *Punch*, 15 July 1936 and inspired by papers in *Nature*, including one by Robin
Hill on the biochemistry of the dye-plant madder.

Autumn and the conservation of energy

In the garden of Vatches Farm, a short bicycle ride down the Barton road from Cambridge, the stored sunlight of a so-so English summer is starting to sink into the autumn soil. The apples fermenting beneath the trees give the air a sweet, slightly grubby smell, a distinctive store-cupboard mustiness that always seems a little odd out of doors. It tells us the season's over. The October afternoon is beautifully bright, and there are sweet grapes on the vine growing against a sheltered wall, but winter is on its way. Much of the garden's annual crop has been harvested: most of the walnuts were taken in a week or so ago, just before the squirrels got to the rest of them; the gooseberries are long gone. But some of the fruit has fallen ungathered to the ground, lending the autumn afternoon its beguiling cider-press scent.

That scent is a sign of life. Microbes are turning the fruits' sugars into alcohol, and this fermentation is providing them with the energy that they need to live. Everything the molecular machinery inside them does – the synthesis of new molecules, the intake of needed nutrients, the repair of DNA, the eventual creation of new daughter cells – requires energy, and the fermentation which gives off that sweet scent provides it.

The sugars driving this process and providing the energy were created by photosynthesis. The reason the sugar-making powers of the Calvin-Benson cycle are important to the rest of us is that, ultimately, we get our energy from breaking down those sugars and their derivatives, just as the yeast in the apple does. And the energy we get out has to have been put in. When the Calvin-Benson cycle turns carbon dioxide, from which living things can derive no energy, into carbohydrates, from which they can, it requires an energy source to do so. The Calvin-Benson cycle is not free-wheeling; it needs to be driven. Photosynthesis is about more than just the cycle in which carbon is fixed. It's about the energy that drives that fixation.

One of the founding principles of thermodynamics – the science of energy, work and heat – is that energy is always conserved. It can be transformed from one form to another, but it is never created or destroyed. This principle of the conservation of energy, the First Law of thermodynamics, was given something like its modern form by Robert Mayer in the 1840s – shortly before the concept of energy, as we understand it, was introduced into science.

Mayer, a German doctor, became intrigued by the observation that the blood he was letting from the veins of feverous sailors in the Dutch East Indies was considerably redder than he would have expected. Since redness in blood is a sign that it is carrying oxygen, the observation implied an unusually small appetite for oxygen in the sailors' tissues; much of the oxygen the heart pumped out was coming back unused.

Mayer deduced that the sailors were using less oxygen than they would in cooler climes because, thanks to the tropical temperature, their bodies didn't need to generate so much heat. This led him to consider the more general idea that all the body's heat came from using oxygen to break down its foods. Lavoisier, the greatest chemist of the eighteenth century, had argued that respiration in animals was a specialized form of the combustion seen in the inorganic world, a reaction between food and oxygen, and this insight, while still not universally accepted, was a common belief among chemists of Mayer's day. Mayer's approach to it, though, was both fresh and quantitative. The total amount of heat that the forces in the body generate, Mayer decided, must be exactly balanced by the capacity for generating force that was latent in its food, a capacity released when the food reacted with oxygen. The capacity to produce force

– the capacity about to be named energy – was passed from one to the next without addition or subtraction.

Mayer was ready to follow his ideas wherever they led him, and this idea led him to the sun. It had long been known that plants needed sunlight. But in the green mansions of Java's jungles, Mayer saw that the sunlight was not just the source of sustenance to the plants, but also the source of the energy which animals later derive from eating those plants. The plants were converting the energy from a physical form – light – into a chemical form, one that both they and their herbivore predators could make use of.

Mayer was something of an outsider in the scientific world (if he hadn't been, he wouldn't have been tending to the health of sailors in the far-flung tropics). He had been kicked out of university and had no institutional home, little interest in experiment and a shaky grasp of mathematics that did not serve his qualitative ambitions well. His ideas about the conservation of a capacity for force were largely ignored when he published them in 1842. Within years other people had made similar observations and calculations, developing them with a mathematical and physical insight that Mayer lacked. Mayer, for his part, attempted to follow the chain of energy further back still by asking how the sun acquired its energy; his theory relied on an endless stream of falling meteors keeping the solar surface incandescent, and met with little success. In 1850 he attempted suicide, and was sequestered in a succession of asylums. He regained his sanity, in part, through self-denial: he no longer allowed himself to follow the scientific trains of thought which had led to his madness, and concentrated on the mundane.

But as the science of thermodynamics advanced, Mayer developed a retrospective fame. He was seen to have had one of the big ideas first; by the 1860s he was becoming an honoured pioneer, and when he died in 1878 he was an undeniably grand old man, one of the framers of the First Law. He had shown the vision to apply his ideas about conservation of energy not just to the work of steam engines and electric currents – the focus of a great deal of thermodynamic concern, for obvious reasons – but to the world as a whole. As a physiologist, he thought in terms of the processes of life, and he offered a way of seeing those processes as being continuous with physical flows of energy. Eschewing the vitalist belief that there is a special, unique life-force animating us, he saw the workings of life as part of a greater flow of energy through the cosmos.

Inspiring as it is, the energy-based view of the world Mayer helped to formulate has what can be seen as a dark side, encapsulated in the Second Law of thermodynamics; while energy is conserved in a closed system, the usefulness of that energy inevitably declines. Think, as nineteenth-century thermodynamicists would at the drop of a hat, of a steam engine. For work to get done there have to be hot bits and cold bits; every time the engine does some work, a hot part gets cooler and a cold part warmer. The amount of energy doesn't change, but its distribution does. Unequal distributions of energy – orderly arrangements of hot things and cold things – allow work. Disordered ones don't. And other things being equal, disorder will always win out. The universe has a tendency to uselessness; it is drawn towards a drab lack of order.

Thermodynamics named the uselessness that arose from the tendency to equilibrium 'entropy'. Entropy is a physical property of systems, just as energy is. Energy is what lets you do work; entropy is the price you pay for doing that work. If there were no entropy, the same energy could be used to do work again and again and again: as the First Law says, energy never goes away. But the Second Law says that that can't happen, because when you use energy you wear its usefulness down and build its uselessness up. The Second Law says that, in closed systems, entropy is permanently on the rise. The great nineteenth-century physicist Ludwig Boltzmann – who, unlike Mayer, did in the end commit suicide – put this notion of unavoidable disorder into probabilistic terms. Ordered systems are unlikely: disordered ones more probable. Boltzmann's version of the Second Law said that a closed system would evolve to its most probable state, meaning to its least ordered, least useful state. Its entropic uselessness can only increase, an unlikely order will not re-emerge. Autumn will not re-set itself to spring.

This idea of a world in which uselessness builds up underfoot like leaf-litter beneath trees is one of the major cultural burdens of nineteenth-century physics, a melancholy counterpoint to its engines of heat and light. It underwrote a cosmic pessimism, dominated by the spectre of an inevitable 'heat death' – a universe in which energy is distributed utterly evenly and entropy is at a maximum, in which the sun is as cool as the earth and the sky, in which change can take place no more and time has lost its meaning.

Life seems, at first glance, to be an exception to this law of ever-increasing entropy, its order springing up fresh and new after every winter. In fact, though, life isn't breaking the Second Law. It's using it. It's wilfully making use of entropy in order to survive. Living things have to do a lot of work to keep themselves alive, ceaselessly making and repairing the mechanisms within them. The structures of life, from DNA helices to Austrian physicists, are wonderfully, gloriously improbable – and life has to produce a great deal of entropy in order to keep them that way.

If the earth were a closed system this ceaseless production of entropy could not persist. But the earth is open – it can absorb energy from the sun, and radiate waste heat out into the universe. This provides the contrast necessary to drive the great engine of life. As Boltzmann put it: 'There exists between the sun and the earth a colossal difference in temperature ... The energy of the sun may, before reaching the temperature of the earth, assume improbable transition forms. It thus becomes possible to utilize the temperature drop between the sun and the earth for performing work, as is the case with the temperature drop between steam and water.'

And that is where photosynthesis comes in. Without photosynthesis, the sun's light would serve simply to warm the earth; its energy would be transformed straight into heat. The sun would still power onshore breezes by heating the land more than the sea, and ocean currents by heating the tropics more than the poles, but it would not be able to drive life.

The world of wind and rain is driven, like a steam engine, by temperature differences. And those involved are pretty small: a few degrees between cooler land and warmer ocean at night, a few tens of degrees between tropics and poles. In the world of photosynthesis, the energy in the sunlight is not used to heat things up, but to drive a series of chemical reactions. It creates the 'improbable transition forms' – improbable meaning, among other things, low in entropy – that Boltzmann imagined but could not describe or enumerate. Because the plant's work is done chemically, not thermally like the wind's or the ocean's, the amount of work that energy allows is not limited by the difference in temperature between sun-warmed leaf and its cooler environment. It is limited by the difference between the temperature of the leaf and the temperature of the sun, which means that the amount of work that can be done with a sunbeam channelled through photosynthesis is ten times greater than the amount of work that sunlight could get done if it just warmed up water.

This energy is used to bind new carbohydrate molecules together through the Calvin-Benson cycle – bonds which, when broken down, release that energy. This is what drives almost all the metabolisms on earth, and thus all the wonders of life. With photosynthesis storing up free energy the world can make entropy with impunity, safe in the knowledge that it can go on bringing improbable order to the universe for as long as there is a sun in the sky and plants to make use of its light. This view of a world opened up to the unlimited energies of the cosmos inspired Mayer, Boltzmann and many of those who came after them. In Russia, the pioneering twentieth-century biogeochemist Vladimir Vernadsky made this thermodynamic view central to his idea of the 'biosphere'. In the nineteenth century the term biosphere meant either the sum total of all the living matter on the planet or the shell around the planet's surface which contained that living matter. In Vernadsky's writings, steeped in the concepts of thermodynamics, it took on a much more dramatic meaning: it was 'a region of transformation of cosmic energy'. Vernadsky's biosphere was a planet-sized engine of change powered by the sun.

Modern variations on Vernadsky's views will crop up in some of this book's later chapters; while I think they are important, I'll admit they are also often fairly speculative. The source of life's power to make changes, though, is not a matter of speculation. The 'cosmic energy' transformed is sunlight, just as Mayer imagined when he first saw the surprising redness of his sailors' blood in the greenery of the tropics. But how does it work? How does the sunlight's energy get packed into the chemicals of life? Mayer didn't know; Boltzmann said it was a complete mystery; Vernadsky knew no better.

The question raised by Mayer took more than a century to be answered. When it finally was, it was in large part thanks to the man who planted the apple trees at Vatches Farm: a shy, brilliant and eccentric Englishman named Robert Hill.

Robin Hill and Cambridge biochemistry

Hill's family, having christened him Robert, actually took to calling him Robin, and so will we. He is remembered as a scientist, and he showed leanings that way from an early age. But he was also a painter, and in love with the flowers of the garden and the countryside. He would remain a countryman, an artist and a craftsman all his life; they weren't callings in which he became eminent, as he did in science, but they informed his personality just as surely.

As a teenager at boarding school in the 1910s, Hill became keen on meteorology, and in his twenties he designed and built a peculiar fish-eye camera that, when laid on the ground, was capable of taking pictures of the whole sky at once. The images are not just useful but, in some cases, beautiful – his colleague Derek Bendall, who though retired still has a desk in Cambridge's biochemistry department, has on his wall a beautiful picture of the ornate spreading vaults of Ely Cathedral as captured by a Hill camera. Hill also developed a more down-to-earth interest in the production and use of dyes made from woad, madder and other traditional plants, a skill that took on greater significance when, from 1914 on, the dyes produced by the German chemical industry were no longer available.

His interest in plants that produce dyes and in pigments for painting lasted throughout his life. One of the reasons his lab was so famously messy, with chemical scraps from decades gone by strewn around the place, may have been simply that he wanted to see how the colours aged. The flowerbeds at Vatches Farm were always well stocked with woad and other dye-plants, and in later years, when Hill played up some of his eccentricities, he would delight in casting a disconcerting look at visitors to his lab and muttering 'I'm growing madder, you know'. The effect was un-doubtedly heightened by his wonderful asymmetric eyebrows, the right-hand one turned down, the left flicked up, growing ever more vigorous and ever more twisted as he aged.

Leaving school in 1917, Hill had to face the prospect of the Great War. He hoped for a place in the meteorological field service, but proved to be too fit; however his chemical skills convinced the War Office to place him in a special anti-gas unit, and so he spent what was left of the war in London, some of it in laboratories, some of it in trials that were held in the south of the city, on Clapham Common – 'an unpleasant, coughing, sucking job', as he wrote to his parents, but one that he came through in better shape than poor Sam Ruben.

After the war he was able to take up the place he had been offered at Cambridge University; it was to be his home for the rest of his life. He joined Frederick Gowland Hopkins' Department of Bio-chemistry, soon to be recognized as one of the great biochemistry departments of the world. Biochemistry was still a new discipline; the department had been endowed only in 1920, and the term itself was only first used in 1902. The discipline had grown out of microbiology and the study of yeasts, out of chemistry and its use in the service of medicine, and out of physiology and its concern for the processes of the body. The circumstances of its instauration were not always easy. Hopkins, who owed his prominence to pion-eering work on vitamins, had to endure more than a decade of bureaucratic stratagems to get the Cambridge department funded and running, working in a cramped basement equipped with an elderly centrifuge that would wander around the room in a rather disconcerting manner.

Hopkins animated his department with a need to understand the most basic processes of life, rather than to solve problems that might be applicable to health care. The physiologist Walter Fletcher, who had worked with Hopkins on the chemistry of muscle fibres and who, as the first secretary of Britain's Medical Research Council, was his most powerful ally in the struggle to get the Cambridge department started, decried its lack of practical bent. In 1927 he wrote to a colleague:

> I told Hopkins that, having somehow bagged the credit for inventing vitamins, he spends all his time collecting gold medals on the strength of it, and yet in the past ten years he has neither done, nor got others to do, a hand's turn of work on the subject. His place bristles with clever young Jews and talkative women, who are frightfully learned about protein molecules and oxidation-reduction potentials and all that. But they all seem to run away from biology. The vitamin story is clamouring for analysis . . . yet not a soul in Cambridge will look at it.

But the department was not running from biology; it was running to a new sort of biology, one less concerned with the ingredients that fed life's machinery – and thus matters dietary – and more concerned with the machinery itself. So, for example, Marjory Stephenson, quite possibly one of those talkative women Fletcher was vexed by, came to study vitamins but ended up a pioneering biochemical microbiologist, devoting herself to the ways that microbes respond to their environment. And Hill became one of those 'frightfully learned . . . about oxidation-reduction potentials', which turned out to be the key to understanding how plants deal with the sun's energy.

Hill arrived wanting to work on plant pigments, such as those from which he had made dyes. On this, though, Hopkins had other ideas. Hill later said that Hopkins disapproved of plants on the basis that, since they lacked excretory systems, they must be rather dirty creatures, stewing in their own waste. That said, he had an eminent specialist in plants in his department already, a Mrs Onslow; it may be that he just felt a more established field would be a more propitious start. If so he was right.

Hill was set to work on exactly what Fletcher had complained about: 'protein molecules and oxidation-reduction potentials and all that', in this case haemoglobin, the oxygen-carrying protein in blood. This seems a long way from plants, but it would in fact equip Hill with the practical skills and theoretical insight he needed to start cracking the problem of energy flow in photosynthesis.

Haemoglobin's affinity for oxygen, and also its colour, is explained by the iron atoms tucked away within it, atoms held in a type of molecular cage called a porphyrin. A porphyrin is a flat, squarish molecule with space in the middle into which an extra atom can be set like a loosely mounted jewel in an open wire-work brooch. Haem, the porphyrin that holds the iron in haemoglobin, is only one example of the ways in which cells trap metals in such cages in order to make use of their properties. Chlorophyll, the pigment that absorbs sunlight in plants, is based on a porphyrin cage for magnesium atoms and it is that magnesium which makes the pigment green.* But that is of only incidental interest here; chlorophyll was not particularly important to Hill's scientific work, and nor is it a good pigment for painting with.

In his first year in Hopkins' department, Hill showed how, using the right solvents, haem groups could be teased out of the slots in the haemoglobin molecule where they normally sat and then put back in. Or, he discovered, instead of putting the haem back, he could replace it with porphyrin cages containing copper or zinc, compounds that could be told apart by their different colours. J. B. S. Haldane, the

---

* Melvin Calvin's experience with porphyrins and their photochemistry, gained in Manchester with Michael Polanyi, was one of the things that made him an obvious choice to head the Berkeley photosynthesis group.

The green and the red

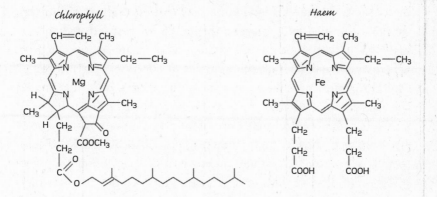

Chlorophyll

Haem

great biologist who was one of Hopkins' first appointments to the department, memorialized the work's potential in comic verse:

> Thanks to some work of Mr Hill's
> Iron is banished from our pills.
> Pale people wishing to be pink
> Can make fresh pigment out of zinc.
> Or, if their doctors think it proper,
> Iron may be replaced by copper.

At the opening of the Department of Biochemistry's building in 1924, the members of the research staff were expected to display some of their recent work. Hill thought his brightly coloured porphyrins would make a pretty good show. They did, and they also attracted the attention of David Keilin, a distinguished Polish scientist who was in the process of making a fundamental discovery about how energy is handled within living organisms. Keilin was peculiarly well liked in Cambridge, his generally poor health never dimming a kindly and gently infectious enthusiasm. He was to become a strong professional and quasi-paternal influence on Hill. And his work, though at that point he did not yet know it himself, was to have direct bearing on the storing of sunlight.

Keilin's main concern was research into parasites, but in looking for haemoglobin in a weird little creature called the horse bot-fly he had come across an intriguing biochemical oddity.* Keilin was studying the flies' wing muscles through a microspectroscope – an instrument that spread out light into a spectrum, thus revealing which wavelengths of light the samples in question were absorbing. The absorption of light by haemoglobin gives a quite distinctive spectrum. But it was nowhere to be seen. Instead Keilin found a previously unseen spectral pattern of four distinctively grouped lines – lines which showed that something was absorbing light of four specific wavelengths. Looking at other insects he found the same thing – and then went on to see the lines in microbes, too. But they weren't always quite the same. In the muscles of a moth frantically beating its wings the lines would be strong; in a resting moth, much less so. Similarly, in yeast that was sitting quietly suspended in liquid, and had thus used up the oxygen around it, the bands appeared strong; if the yeast was supplied with oxygen through some vigorous shaking, they disappeared.

Keilin named the substances that produced these bands cyto-chromes – 'cell-colours' – and showed that their appearance and disappearance was due to their 'reduction' and 'oxidation'. When electrons are added to a chemical, it is said to be reduced; when they are stripped away, it is said to be oxidized. The two processes, by definition, always happen together, because if the electrons are to leave one chemical they needs must end up at another: electrons will not make their way in the world alone. These reactions in which

---

* 'Weird', I suppose, depends on your point of view. If you're used to the baroque life-cycles that parasitologists specialize in, the bot-fly may be a very straightforward customer. The female lays her eggs on the hair of a horse towards the end of summer. The release of larvae from these eggs is triggered by the horse licking its hair, at which point the larvae leap out and burrow into the tongue, in the tissues of which they then wander for weeks, occasionally drilling little holes to the surface through which to breathe. The larvae then moult and migrate to the stomach, where, having changed from a body shaped for tunnelling to one designed for clinging, they attach themselves to the stomach lining for nine months. At the beginning of the next summer they allow themselves to be excreted, burrow into the soil, pupate and emerge as flies, ready to mate and start the cycle over again. Readers delighted by this sort of ingenuity would do well to pick up a copy of *Parasite Rex* by Carl Zimmer.

one thing is reduced and another oxidized are known as redox reactions.

Redox reactions, like all chemical reactions, can be understood through the laws of thermodynamics: a reaction will come about when what's called the 'free energy' of the products is less than the free energy of the things needed to make the products. The laws of thermodynamics are in general an inducement for energy to be spread out more evenly and entropy thus increased; they want the unlikely and orderly to be reduced to the predictable and dull. To impute such intentions to energy flows feels peculiarly unscientific, but it is also very hard to avoid: 'thermodynamics smacks of anthropomorphism' as Max Planck, the physicist who started the quantum revolution, is said to have complained. Everything gets forced into metaphors of preference. Objects want to roll downhill, chemicals want to give up free energy, electrons want to flow across voltage differences, uranium atoms want to tear themselves apart. The need to lower free energy and raise entropy makes thermodynamics the science of desire.

The desire that drives a redox reaction is the desire for electrons. The more two chemicals differ in this desire, the more eager the electrons will be to flow from one to the other. Take iron and oxygen. Iron has a fairly loose and easy-going attitude towards the electrons furthest from its atomic nuclei, while oxygen atoms take a strongly acquisitive stance. So electrons will flow from the iron to the oxygen: result – rust.

This difference between oxygen and iron which encourages electron flow is a difference in 'redox potential'. Just like the potential in a circuit of wires and resistors, this redox potential can be measured in volts. When electrons flow across a redox potential, just as when they flow as current between the terminals of a battery, they provide energy that can be used to do work. Just as work can be harvested from a current in a wire, or from water flowing downhill, so work can be harvested from the redox reactions in a cell.

The cytochromes Keilin had found are responsible for channelling a flow of electrons that releases the energy stored in food. Hill and others were later to show that a similar system channelled the flow of electrons in photosynthesis. In the case Keilin studied, the energy started out stored in the form of chemical bonds: pairs of shared electrons. Break one of these bonds in the right way and you have an electron you can put into what is now called an 'electron transfer chain' of cytochromes. The first cytochrome picks up an electron, thus becoming reduced; the second cytochrome, with a greater affinity for electrons, then oxidizes the first cytochrome, relieving it of the electron, thus becoming reduced in turn. The ease with which the cytochromes pick up electrons is down to the iron that they contain; their differences depend largely on the details of the ways in which the iron is caged. At the end of the chain is the molecule that wants the electrons most: oxygen.

When you take away something with a negative charge from a molecule, such as an electron, it's a good idea for that molecule to lose something positively charged, too, in the interests of balance. So as electrons are stripped out of food, electron-free hydrogen atoms – hydrogen ions – are normally produced too; in general, to say something loses hydrogen is to say it has been oxidized, because if a hydrogen ion is pulled off an electron will be lost too.* These hydrogen ions are used by the cell at the end of the electron transfer chain, where the electrons that have been passed from cytochrome to cytochrome are united with the hydrogen ions and with oxygen to make water. As Lavoisier argued and Mayer believed, the source of energy in animals and plants is an oxidation – an energy-releasing reaction of organic matter with oxygen analogous to a flame. But in living things this reaction, respiration, makes use of clever cyto-chrome intermediates to channel the energy to useful ends.

The flow of electrons down the respiratory cytochrome chain will continue for as long as there's food at the top and oxygen at the

---

* This means that the greater the proportion of hydrogen in an organic compound, the more energy it can give up when oxidized, which is why the electricity industry makes increasing use of methane (CH_4), also known as natural gas. In redox terms, methane is the most reduced carbon compound there is, stuffed with electrons that can be removed. It thus provides more energy per carbon atom than any other carbon-based fuel.

bottom. If there's not enough food to provide electrons, or enough oxygen to soak them up, the flow stops. That was why Keilin's cytochrome bands appeared most strongly in yeast if there wasn't any oxygen around: the whole chain filled up with electrons, because there was nowhere for them to go, and so all the cytochromes were reduced, in which state they absorb light. The ability of the cytochromes to play fast and loose with electrons and their absorption of light were closely linked – both depend on the iron caged within them, and the details of the caging. Keilin's microspectroscope let him look into the darkened room of the cell and see what the electrons were doing.

Though it was not his main research topic (and indeed Keilin was not himself a member of the biochemistry department, but rather of the Molteno Institute for Parasitology) Hill worked with Keilin on cytochromes for much of the late 1920s, studying a strain of brewer's yeast in which the spectral bands were particularly clear. It had the added attractions of being a bright pretty pink colour and having a nice smell, and they used to snack on it from time to time – another of those idiosyncratic aspects of laboratory life that would nowadays be frowned on.*

By this stage Hill's work looked like the basis of a promising career. Too shy to be much good as a lecturer, he didn't get a university position – indeed, like many eminent Cambridge scientists, including some who have won Nobel prizes, he didn't even produce a PhD – but he received a succession of grants that let him keep working in the biochemistry department. His shyness was an impediment to his personal life, though, and in the early 1930s he became quite depressed. A friend recommended that he get away, and he headed off to Singapore. It was apparently just what he needed; he found a new enthusiasm for the world, taking delight in studying the magnificent tropical vegetation, using his specially

* Even if health and safety standards allowed it, Cambridge scientists could no longer spread that particularly toothsome yeast on their hot toast as a rosy substitute for Marmite; the strain was lost in the Second World War.

designed camera to take photographs of towering thunderclouds. When he came back to Cambridge he became engaged to Priscilla Worthington, the younger sister of the friend who had suggested the trip to Singapore, and when they were married they settled at Vatches Farm in the village of Barton, a few miles outside Cambridge.

Hill had admired, perhaps loved, Vatches for some time. As an undergraduate he had more than once cycled out to Barton to sketch it. Following in his bike-tracks on one of those perfect afternoons with which Cambridge so often and somewhat misleadingly sees in the new academic year, it was easy to understand the attraction. The main building is a long whitewashed farmhouse just across the road from the duck pond on the village green. The south-facing garden – now divided between a few different properties, as the farm was broken up after Robin and Priscilla died in the early 1990s – is a lovely mixture of lawns, flowers and orchards, lit by warm, low sun. Under the old fruit trees he used to tend, the fallen apples perfume the air. Between two small lawns in the eastern part of the garden (for which its new owners have justly won awards) stands a striking, proud beech tree that Hill must have planted not that long after he and Priscilla bought the place. Nearby is what seems to be a relic from the trip to Singapore: an amelanchier of some sort, the current owner tells me, which in spring fills the air with a scent of almonds and coconut. At the end of the main lawn a farm gate set in a line of trees opens on to farmland. A grassy track leads through a field already ploughed for winter wheat; a mile or so away, the dish of the university's radio telescope at Lord's Bridge Road slowly sweeps the skies.

Where the oxygen comes from

Somewhere in his recovery from depression, in his subsequent marriage – which was to be a long and happy one – and in the acquisition of Vatches Farm, Robin Hill seems to have found a new security and sense of purpose. He looked into the possibility of having a formal position in plant biochemistry created at the university, and though this came to nothing he was now sufficiently his own man to make plants increasingly central to his work, whatever Hopkins thought. The newly acquired gardens at Vatches became a continuation of his laboratory, though thanks to his gardener were far more orderly. Hill grew some of the plants on which he would experiment in one of the farm's greenhouses and plants for the kitchen in its beds as well as his beloved woad and madder. His farm notebooks of the late 1930s are full of ideas about what to grow, what to fix, what to spray, where to spread manure, whether to buy a new Aga, what French texts on horticulture to acquire, and so on, interrupted from time to time by thoughts about plant biochemistry. The only dark note is a couple of pages given over to a talk to be given to air-raid wardens about the effects of various poison gases, and what to do in the case of attack.

His understanding of cytochromes, their redox potentials and the electron-transfer pathways they provide would prove crucial to Hill in his understanding of the energy flows in photosynthesis. But as the notebooks show, Hill's first work on photosynthesis used his previous experience not so much as a theoretical template, but as a practical tool. The set-up Hill had used to make sensitive measurements of how well haemoglobin picked up oxygen under different conditions could also be put to use as an exquisite system for detecting the presence of oxygen. It was just a matter of turning your mind around – of seeing the haemoglobin as part of the apparatus, rather than the subject of the investigation, and oxygen as something to look for, not something given. It was an inversion typical of the way Hill thought. One of the attractions of the camera he designed for taking flat fish-eye images of the whole sky was that, if you looked at the image you'd taken through the camera with which you'd captured it you could see it undistorted; the system was reversible. He taught his children that a good way of seeing the essence of something afresh was to bend over and look at it upside-down between your legs.

One of the aims of biochemistry was to find ways to free reactions from their cells, the better to study them; the shining example was the nineteenth-century discovery that you could squeeze gunk out of yeast cells that would ferment sugars into alcohol just as the cells themselves would. (It won't perfume the autumn air as sweetly, but that's to some extent beside the point.) In the same spirit previous researchers had produced oxygen using dehydrated leaf extracts, rather than whole cells; the fact that it could be abstracted from the cell in this way made oxygen production, Hill wrote, 'the only property specific for green plant tissue which is at the moment open to biochemical investigation'. His ability to measure oxygen production with haemoglobin gave him the tools for such investigation.

Armed with a sensitive way of measuring oxygen, Hill started looking for ways to isolate the relevant parts of plant cells. In the nineteenth century, studies using microscopes had shown that the green pigments in plants were all to be found on discrete structures within the plant's cells; these bodies, home to the plant's chlorophyll, were called chloroplasts. It was clear that the chloroplast was the seat of photosynthesis; by finely focusing beams of light it was possible to show that the process only took place if the chloroplasts themselves were illuminated.*

In the lab, Hill worked on new techniques for isolating chloroplasts from a range of plants found in his garden and elsewhere, using sugar solutions and a mortar and pestle. He found that if these chloroplasts were mixed with a powdered leaf extract they could be made to produce oxygen when light was shone on them. Then, to his surprise, he found that the same happened if he added a yeast extract. It eventually turned out that all that was needed was that the added extract contained salts of a particular form of iron.

* Such focusing is not something that only takes place in the laboratory. There are mosses adapted to life in caves and rabbit holes that carry out exactly the same trick, having evolved translucent cells to act as lenses and focus the light from the mouth of the hole on to the chloroplasts behind them. Hill loved these mosses, delving into rabbit holes for them and attempting to grow them in suitably arranged sections of drainpipe at Vatches Farm.

If you had chloroplasts, light, water and what was known as ferric iron you could produce oxygen; nothing more was needed.

Hill knew what the role of the iron had to be; it was acting as an electron acceptor in a redox reaction. Ferric iron is a thoroughly oxidized form of the metal from which the outer electrons have already been stripped. It is thus eager to soak up electrons from elsewhere – to be reduced. When something is being reduced, something else is being oxidized. In a system that consisted basically of chloroplasts, water, ferric iron and light, the something else had to be the water. Some mechanism in the chloroplasts was pulling water molecules into their constituent hydrogen and oxygen, and releasing electrons in the process.

This was not the conventional photosynthetic wisdom. For more than a century it had been assumed that the oxygen released by plants came from the carbon dioxide that they absorbed. The commonsense logic behind the assumption was that if respiration and combustion produced carbon dioxide, photosynthesis must use it up. The alternative was hard to imagine. After all, water is remarkably stable stuff. We don't see it exploding or burning, even when we boil it. It is part of the unchanging, everyday furniture of our lives in a way an invisible, insensible compound such as carbon dioxide just isn't. Anyone who takes biology at high school will be taught where the oxygen produced in photosynthesis comes from, but it's my experience that, in later life, they will forget this; when asked they'll say that the oxygen comes from the carbon dioxide. It just seems to be more natural.*

Hill's conclusion was dramatic, but not unprecedented. A French scientist named René Wurmser, who had a strong interest in the thermodynamics of redox reactions, had published a book arguing that the oxygen given off in photosynthesis must come from water, not carbon dioxide, in 1930. (He also argued that the internal structure of the chloroplast must play some role in segregating the

---

* Maybe part of the reason is that in carbon dioxide the clear hint of oxygen is there in the name itself, just waiting to be let free – at least it is if you're an English speaker.

various chemicals involved in making energy from sunlight, an insight that would be borne out thirty years later.) Wurmser's opinions were not widely known outside France, but a strong experimental argument in their favour was soon provided by Cornelis van Niel, then working at Stanford University's Hopkins Marine Station.

Van Niel did not study plants, but microbes; his early research was done at the microbiology laboratory in Delft, Holland (which was probably also the source for Keilin's pretty and palatable pink yeast). There, under the supervision of a visionary microbiologist named A. J. Kluyver, he had imbibed the belief that microbiology was not just a service science, there for the convenience of pathologists and brewers, but something more noble – the path along which biochemists must pass to understand the basic processes underlying all life. An appreciation of redox reactions was a key part of this view: as Kluyver wrote in *Microbial metabolism: evidence for life's unity*, 'The most fundamental character of the living state is the occurrence in the parts of the cell of a continuous and directed movement of electrons.' For decades, van Niel spread this attitude through a highly influential summer course in 'general microbiology' which attracted all manner of people, from the just-graduated to the well established, to his seaside laboratory on Monterey Bay. An unassuming man, he was an inspiring teacher.

In his own research van Niel was applying his mentor's ideas to photosynthesis, a bacterial process that he thought might serve to link microbiology directly to physics – light, the input – and chemistry. Plants are not the only things that photosynthesize. There are many sorts of algae, some multicellular, some single-celled. These 'green algae', 'red algae', 'brown algae' and various others not known by their colours are what are now called 'eukaryotic' creatures, a term that van Niel introduced into general use towards the end of his life. Eukaryotic cells, of which all plants and animals are composed, contain regions set off from the rest of the cell by membranes, such as the nucleus, the mitochondria – in which respiration takes place – and, in the case of plants and algae, the chloroplasts which are the site of photosynthesis.

Van Niel, though, was interested in bacteria. These creatures (along with a superficially similar but biochemically very different group called the archaea) lack a clearly defined nucleus, and none of them has mitochondria or chloroplasts; but despite this some bacteria are capable of photosynthesizing. Indeed a significant part of the earth's photosynthesis is carried out by what were once called 'blue-green algae' and are now called cyanobacteria, a change of name designed to set them apart from the eukaryotic green, brown and red algae. These cyanobacteria photosynthesize in the same biochemical way that plants and algae do, using water and carbon dioxide and producing oxygen and carbohydrates.

Other photosynthetic bacteria do things differently. In some of them – the green sulphur bacteria and purple sulphur bacteria – the production of carbohydrates from carbon dioxide is accompanied by the transformation of hydrogen sulphide into sulphur. In other photosynthetic bacteria – for example the 'green non-sulphur bacteria' – the fixing of carbon is accompanied by reactions that turn one pre-existing organic molecule into another, for example acetate into acetone.

These reactions, van Niel realized, are all variations on a redox theme. The difference between the various bacteria lay in evolutionary choices about where to get the electrons and hydrogen needed to reduce carbon dioxide. In the sulphur bacteria hydrogen sulphide was being oxidized into sulphur, giving up hydrogen ions and electrons in the process. The transformation of acetate into acetone in non-sulphur bacteria was also a hydrogen-liberating oxidation. If cyanobacteria worked the same way, and if they started off with just water and carbon dioxide, then the water would have to be the thing getting oxidized, thus producing hydrogen or the carbon fixation and oxygen as a waste product.

In the 1930s the idea that microbiology had much to teach the rest of the world of science (as opposed to the world of pathologists and brewers) was still a marginal one. And van Niel was contradicting the widely held assumption that the oxygen plants produce comes from the carbon dioxide they take up. So his idea languished until Hill's experiments provided the evidence needed to revive it by showing that chloroplasts would produce oxygen even if there was hardly a trace of carbon dioxide around.*

The production of oxygen by isolated chloroplasts provided with a suitable electron acceptor became known as the 'Hill reaction', though it was not a name he much cared for, and one which he used seldom and only with self-deprecation. It showed that photosynthesis in chloroplasts – and thus plants – was a redox process like those seen in photosynthetic bacteria, a profound result. Beyond that, as Hill had hoped, it provided a new way for biochemists to tackle their subject. The Hill reaction was a form of photosynthesis independent of living cells, a form that could be analysed and interpreted in the ways biochemists found most enlightening.

Daniel Arnon and the isolated chloroplast

One of the people who took up the Hill reaction was Daniel Arnon, a professor at Berkeley. Born in Warsaw in 1910, Arnon witnessed the famines that followed the Great War; they made him determined to find better ways of growing food. The novels of Jack London, a colleague later remembered him saying, fostered this interest in scientific farming and added a fascination with California, and at the age of eighteen he applied to Berkeley by mail and bought a steamer ticket to New York. Once there he drove a delivery truck to earn the money that would pay his way to California; he had that money stolen from him en route, but eventually got there by hitch-

---

* Working with other scientists at Stanford, Sam Ruben and Martin Kamen went on to confirm the result a completely different way in the early 1940s. When photosynthesizing algae were grown in water enriched with a heavy isotope of oxygen, oxygen-18, the oxygen they produced was rich in oxygen-18, too. When the algae were grown in normal water but with carbon dioxide enriched in oxygen-18, the oxygen they produced contained much less oxygen-18. This was enough to convince all but the hardest core of sceptics.

hiking. He joined Berkeley as an undergraduate in 1930 and made it his academic home for the next sixty years; he was as devoted to it as Hill was to Cambridge.

Though matched in their institutional constancy, the two men were in most ways near-opposites. Where Hill was puckish, Arnon was formal. Where Hill thought freely in abstract terms, Arnon saw things only through experiments. Where Hill was artistic, Arnon was sporty, a swimmer, an oarsman, a keen footballer. Where Hill could barely tell his associates what he was thinking, let alone what he thought they should do, Arnon had a programme that was clearly enunciated and was to be systematically followed. Where Hill grew his plants in a garden, Arnon grew them in hydroponic tanks, their roots fed not by soil but by water to which precisely balanced nutrients had been added.

Those hydroponic tanks were a major part of Arnon's early career as a plant nutritionist. Plants do not need food as a source of energy in the way that animals do – but to build their proteins and other biochemical paraphernalia they need some chemicals beyond the carbon, oxygen and hydrogen they can get from the air and water. In the nineteenth century the great *Organiker* Justus von Liebig showed that they needed nitrogen (in the form of nitrate, which is oxidized, or ammonia, which is reduced) and phosphorus (in the form of oxidized phosphate). He also realized the importance of global cycles of the elements: 'Carbonic acid [carbon dioxide], water and ammonia, contain the elements necessary for the support of animals and vegetables. The same substances are the ultimate products of the chemical processes of decay and putrefaction. All the innumerable products of vitality resume, after death, the original form from which they sprung. And thus death – the complete dissolution of an existing generation – becomes the source of life for a new one.'*

---

* Liebig, 'Chemistry in its application to agriculture and physiology' (1840), quoted in Smil, *Enriching the Earth: Fritz Haber, Carl Bosch and the transformation of world food production* (MIT Press, 2004). Liebig, an argumentative man, was something of a bugbear for Robert Mayer. Mayer felt that the book in which Liebig nailed for good the truth of Lavoisier's idea

Liebig also enunciated the key principle that plants grow at the rate set by the most scarce, and thus limiting, nutrient. In a field well stocked with nitrates but poor in phosphates, crops will grow at a rate set by the phosphate, and no amount of extra nitrate fertilizer will change the matter. This highlights the need, when fertilizing crops or gardens, to provide the necessary nutrient in proportions governed by the needs of the plants. If you miss out one necessary nutrient, it will become limiting, and the addition of anything else will be wasted.

In the 1930s Arnon and his supervisor, Dennis Hoagland, worked tirelessly to assess what trace minerals, in what proportions, were actually vital for plant growth, with the aim of producing properly balanced fertilizers. In later life Arnon used to rail against the concept of marketing anything as 'organic food': plants are organic however they are grown, he argued, and the chemicals that sustain them are inorganic whether they are derived from manure or from a chemical plant. The essential inorganic nutrients Arnon measured over his career included potassium, calcium, magnesium and sulphur (all required in relatively large amounts) and iron, manganese, boron, copper, zinc, molybdenum and chlorine* (required in much smaller amounts). Arnon's work helped give rise to a mixture of essential plant nutrients, 'Hoagland's solution', which is still one of the basic starting points in plant nutrition. His expertise caught the eye of the air force, and during the Second World War Arnon found himself on Ascension Island in the mid-Atlantic growing hydroponic lettuce and tomatoes with which to feed air-crews.

After the war, Arnon increasingly turned his attention away from

---

that respiration was chemically the same as combustion drew on Mayer's observations of lessened need for oxygen in the tropics and the conservation of energy, observations that Liebig had published in his journal *Annalen der Chemie* the previous year.

* Chlorine was the last addition to the list because it was very hard to produce water untainted by it, and it is only by seeing that plants are unable to live without a given chemical that one can say that chemical is essential. The person who finally showed that chlorine was necessary to plant growth was a French biologist called Joseph-Marie Bové, who worked on the problem in Arnon's lab in the 1960s. Bové went on to become an eminent molecular biologist, but has never been anything like as famous as his son José, France's best known campaigner against genetically modified crops and American fast food.

mineral nutrients and towards photosynthesis; his aim was to produce a cell-free chloroplast system capable not just of producing oxygen, as the Hill reaction did, but also fixing carbon dioxide. In 1948 he spent a sabbatical in Cambridge, working with Keilin on electron transfer mechanisms, which Hill's experiments had shown to be as relevant to photosynthesis as to respiration. One of Hill's students, Bob Whatley, joined Arnon in Berkeley in 1953. There Whatley, Arnon and a third colleague, Mary Belle Allen, who had trained with van Niel, took the isolated chloroplast approach a step further.

Their initial focus was on energy. Calvin's group – from which Arnon and his colleagues were quite separate, though Allen had spent time in the Rat House – had at this stage not yet formulated the Calvin-Benson cycle. But they had established the fuels needed to drive it: two smallish organic molecules known by their initials, NADPH and ATP. The NADPH had a specific role as the end of the electron transfer chain (though no one then would have recognized it as such): it fed the Calvin-Benson cycle with the electrons and hydrogen ions that reduced carbon dioxide. The ATP was simply a source of energy, there to keep the cycle turning.

This role for ATP was not specific to photosynthesis; ATP is the primary carrier of energy within the cell. When enzymes catalyse reactions that require an input of energy, by far the most common way for them to get it is to pick up a nearby ATP molecule – the initials stand for adenosine triphosphate – and snap the third of the phosphates off; the energy thus released drives the reaction being catalysed. The molecular machinery that assembles DNA, proteins, fats and just about everything else all runs off ATP. This means that cells need to make ATP in abundance – which they do mostly through respiration and, if properly equipped, photosynthesis. ATP is the chemical form in which the energy released by electrons passing down electron transfer chains is stored. A human body at rest typically creates ATP at a rate of about a billion trillion molecules a second. On an averagely energetic day you will produce three times your own weight in ATP, and use it all up just as quickly.

Creatures with faster metabolisms manage even more extreme feats: some bacteria can produce seven thousand times their weight in ATP each day.

In the 1940s it became widely accepted that respiration created ATP by 'phosphorylating' – sticking a phosphate on to – adenosine diphosphate. How the energy removed from electrons flowing through electron transfer chains was channelled into this reaction was a complete mystery, but the site of the relevant reactions was well known: specialized compartments in animal and plant cells called mitochondria which would produce ATP even when removed from the cells they were normally found in. Calvin and his group assumed that it was the ATP made in these mitochondria which powered the assimilation of carbon dioxide into carbohydrates in the chloroplasts. The NADPH that actually reduced the carbon dioxide was accepted to be native to the chloroplast – but the ATP, they thought, was not. It made sense for cells to have just one ATP-producing system, and the mitochondrion was well established in the role.

Arnon, Whatley and Allen disagreed. Unlike the chemists of the Calvin group, with their uniform flasks of single-celled green algae, Arnon's people were biologists, and they knew a lot about plants. One of the things that they knew was that the leaf cells with the greatest number of chloroplasts – the cells that do the bulk of the photosynthesis – contain remarkably few mitochondria. It just didn't seem feasible that the many chloroplasts in the cells were importing all their ATP, especially since the rate at which the cells produced oxygen through photosynthesis was vastly greater than the rate at which they used it up through respiration. So perhaps they were generating ATP themselves. The Hill reaction showed that there was a flow of electrons in chloroplasts; Hill himself had discovered a new cytochrome that seemed to be involved in it, a cytochrome unique to green plants that was called cytochrome f (f for *frons*, the Latin for leaf). Maybe a flow of electrons could generate ATP in chloroplasts, just it did in mitochondria. The obvious way to prove this was to show that just as mitochondria could produce ATP even when removed from a cell, so could chloroplasts.

The tool with which they looked for ATP production was a radioactive isotope – phosphorus-32 – used as a tracer. They fed it to chloroplasts that Whatley had isolated in a slightly different way from that used by Hill, a way he thought might do a bit less damage to them. And in a series of carefully prepared and endlessly repeated experiments – Arnon always did experiments with as many controls as might conceivably be necessary – they found that their chloroplasts were indeed making ATP that had the characteristic radioactivity of the inorganic phosphate they had added in the beginning. There were, as far as they could tell, no mitochondria in their flasks; the chloroplasts were doing it for themselves.

Arnon called the process 'photosynthetic phosphorylation', soon shortened to photophosphorylation; it was big enough news to get reported in the *New York Times* months before it made it to the pages of *Nature*. The Berkeley team went on to show that chloroplasts prepared according to Whatley's techniques could also reduce carbon dioxide (though not very well). Photosynthesis was not, as had become widely thought, a property unique to whole cells, any more than fermentation or respiration was; it was something which component parts could do just fine on their own. The chloroplast, as Arnon put it in the title of a triumphant paper in *Science*, was 'a complete photosynthetic unit'.

How did it actually work? The answer lies in the redox potentials, and I think the best way to illustrate it is by analogy to a mountain used for skiing. Here is Arnon's view of the mountain, which Hill was later to dispute. The use of sunlight to strip electrons from water – the mechanisms of which were still unknown – gave those electrons the energy they needed to lift them from the redox potential at the bottom of the hill to the one at the top: see it as a ski lift.* When those electrons got to the top of the lift, they had a

---

* Strictly speaking, the analogy is the wrong way round. Just as electrons in electric circuits flow, rather counterintuitively, from the negative electrode to the positive one, so electrons in redox reactions flow from negative redox potentials to positive ones. Light moves the electrons to lower potentials, not higher ones. But this is just a convention, and it seems to

choice of pistes that they could follow. They could go down a short slope to a chalet, where they would stop skiing for the day. Or they could follow a longer slope all the way back to the bottom of the mountain, and then get back on to the ski lift. On the gentle slope the energy given up as they lose altitude corresponded to the energy used to make NADPH, which marks the end of the line for the electrons. On the second slope the energy lost made ATP, and the electrons could pass down this slope as often as they liked. That slope, in Arnon's view, corresponded to a flow of electrons through the two cytochromes that Hill had by this stage discovered in leaves, cytochromes f and b6.

The Z-scheme

In Arnon's lab there were regular meetings in which the future course of the investigations would be discussed in detail before the chilled mortars were let loose on the spinach-leaf-and-sand-filled pestles to release the chloroplasts within. Hill's approach could hardly have been more different. He hated giving voice to ideas he considered obvious, and thus many things which seemed obvious to him but opaque to his co-workers went unspoken. Oblique hints, and displeasure at what he saw as foolishness, were all the guidance he was able to offer.

When he was a schoolboy in charge of his boarding-school dyeing club, Hill wrote to his family that 'the other people are not a bit how I want them to be, and dyeing is such a subtle art ... it is nearly impossible to let people do it merely by explaining. You cannot lay down definite rules for everything. The person must be

---

me that it is easier to think of the electrons as flowing 'downhill'. What matters is that the redox potentials are different at different places along the electron transfer chain, and that's what makes the process work. If you want to talk about this to a chemist, though, probably best to remember that 'higher' here means more negative.

in the right frame of mind to do it.' It seems that he held a similar view about work with the subtler pigments of his adulthood, and his temperament would never allow him to impose that necessary frame of mind. In Hill's infinitely messier sanctum – 'every inch,' his student David Walker was later to recall, 'covered with a plethora of tubes and bottles, a smell of alchemy and total disorder (except that Robin knew where everything was)' – the shy alchemist himself was curiously hard put to tell his apprentices what, exactly, he wanted them to do.* When Walker joined Robin Hill's Cambridge lab as a post-doctoral student towards the end of the 1950s, he soon realized he had no idea what he was meant to be doing.

Hill was also, in some ways, a man increasingly out of his time. New techniques were being brought to bear on photosynthesis, specifically ways of providing precise numerical values for every wavelength of light being absorbed, or emitted, during the process. The results of all this work were not easy to put into context, but the equipment with which the work was done was, for most people, pretty straightforward. As Derek Bendall, another of Hill's students in the 1950s, recalls, 'The first Beckman spectrophotometer . . . was literally a black box with a couple of knobs on, and it seemed simple to learn the sequence of operations that gave the right answer.' Not to Hill. He found it impossible to get on with equipment that didn't display its operating principles to his inspection. He never really got used to electronic weighing devices; he preferred balances where he could see the counterweight. The novelist and physicist C. P. Snow, a Cambridge contemporary Hill admired, wrote of the post-war rise of a previously unknown technocratic class, the 'new men'. Photosynthesis had an ever-growing cohort of new men, many of them moving in from the realms of physics, twiddling the knobs of black boxes assiduously and with insight. Hill was not one of them.

---

* Indeed, though he had lost his intense shyness, self-expression was never Hill's strong point. When rejecting a paper which Hill wrote in retirement about some aspects of thermodynamics, the then editor of *Nature* was moved to ask 'May I also offer what may seem to you to be a gratuitously offensive remark? As an honest outsider, I am perplexed that your manuscript is semantically so perversely confusing.'

Indeed, in some ways Hill was moving backwards, away from new data and towards first principles. In the late 1940s he had developed a strong interest in the basic thermodynamic principles that underlay the processes of photosynthesis and the rest of plant metabolism. So though he was fully aware of all the new work going on in the field, including discoveries about the particular wavelengths of light the photosynthetic apparatus in cells absorbed and used under different conditions, he seems to have been most beguiled by the basic problems of energy, entropy and work. This being so, he was interested in Arnon's research, interested enough that, eventually, Walker was able to infer that it was something he should be working on. The Berkeley group's techniques for getting chloroplasts out of cells were clearly very good, and Walker set about trying to match them. Whatley's innovative deployment of a wisp of cotton wool at a specific point in the process proved crucial.

At the same time, Hill, Bendall and a few others were trying to make sense of what seemed to be an ever-longer and stranger list of factors that the Berkeley team used to tweak – 'poise' was the term of art – the chloroplasts' electron transfer chains so that electrons would actually flow down them. Walker remembers Hill suggesting they try spit, urine and floor-sweepings. 'We shrank from the first two and felt that the third, given Robin's lab, would have been a bit of a foregone conclusion.'

Hill wanted to sort out two problems with Arnon's ideas. One was that adding more of the raw material for making ATP actually encouraged the manufacture of NADPH, suggesting there was more to the story than a choice between two routes. The other was that the redox potentials of the cytochromes involved didn't make sense to him; they didn't seem to make a convincing path back to the redox potential of water, which marks the bottom of the ski lift. So in 1960, in a paper written with Fay Bendall, Hill added a second ski lift. A first light-driven lift took the electrons liberated from water to a first summit, from which they sped down the slopes of cytochromes b6 and f, making ATP as they went. But that path led, not to the bottom of the original ski lift but to the bottom of a second lift which started off higher above the valley floor. That second ski lift then took them to the correct height for making NADPH. To take electrons from water all the way to NADPH thus required both the ski lifts; but electrons didn't have to go straight from the top of the second lift to the NADPH. They could instead return through the cytochromes to the foot of that second lift, producing more ATP. The judicious use of this option allows the system to produce ATP and NADPH in just the ratio the Calvin-Benson cycle requires.*

Hill's explanation quickly became known as the Z-scheme, because when the redox potentials were laid out on a graph the electron flow looked like a zigzagging Z.

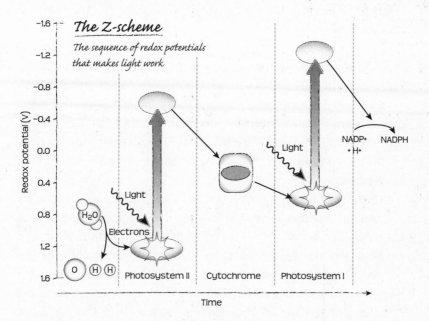

The Z-scheme

The sequence of redox potentials that makes light work

* Quite how much recycling through this mechanism is necessary has been the subject of a great deal of argument. Arnon thought it vital. Others, as the precise mechanism of photophosphorylation has become clear, have seen it as unnecessary. Current opinion is that it is indeed necessary, but at a lower rate than Arnon would have expected; a good recent estimate is that the photosynthetic machinery is set up to recycle one electron in every five that pass through.

It was also pretty much immediately accepted, not least because it made sense of various lines of evidence that had come from the black boxes of the new men and other sources. These had suggested that light was being absorbed at two different key frequencies, one shorter, one longer. To get photosynthesis working properly, you needed to hit both frequencies; doubling the amount of energy provided at one frequency while ignoring the other had little effect. The Z-scheme suggested a natural explanation; the two different frequencies powered the two different ski lifts. Only if both were working well could electrons be moved efficiently all the way from water to the Calvin-Benson cycle.

Within a year or so, the idea that the electron transfer chain contained two different 'photosystems' using slightly different wavelengths of light and tied together by the Z-scheme was the accepted model for the energy transformations in photosynthesis. Because the order in which the photosystems appear in the chain is not the order in which they were named, the story starts with photosystem II, which uses the energy of incoming photons to pull electrons out of water molecules and hurl them into the chain. It continues with photosystem I, which takes the same electrons after they have passed through a couple of cytochromes, re-energizes them with the help of further photons, and then passes them on towards the Calvin-Benson cycle.

For some of the new men in the field, the people whose detailed spectroscopic measurements had shown that chloroplasts, and photosynthetic bacteria, contained different types of chlorophyll, the emphasis laid on the Z-scheme itself was a bit galling. One of the miffed parties was a Dutch physicist, Louis Duysens, the man who named photosystems I and II. (The potentially confusing convention of calling the first ski lift the electrons encounter in the Z-scheme photosystem II and the second photosystem I reflects the order in which Duysens originally worked on them.) In a paper published in 1961 which followed a line of research he'd been pursuing for years, Duysens showed that shorter wavelengths of light would both produce oxygen and pile electrons on to cytochrome f, thus reducing it. Longer wavelengths would oxidize the cytochrome. It was wonderful evidence that the cytochrome sat in between a shorter-wavelength photosystem – photosystem II, the ski lift that took electrons from water – and a longer-wavelength photosystem – photosystem I, the route to NADPH and the Calvin-Benson cycle. Shorter-wavelength light drove photosystem II, driving electrons on to the cytochrome, while longer-wavelength light drove photosystem I, which took them away. But because this work was published after Hill's paper (which made no mention of Duysens' earlier work along similar lines) it was frequently taken as a confirmation of Hill's Z-scheme, rather than a great discovery in its own right. Duysens was not happy about that.

Hill's colleague Derek Bendall sees it a bit differently. Hill's work on the Z-scheme was driven not by an interest in the way light powered the process, but by a concern for thermodynamics and energy flow. The scheme's purpose wasn't to explain how the electrons gained energy – it was to provide a blueprint for how they lost it by passing through the various cytochromes. The fact that Hill didn't mention other work pointing towards the existence of two photosystems in his paper doesn't mean he was unaware of it, or seeking to downplay it. Some of the work which showed the need for two wavelengths of light was by a close friend of his, Robert Emerson, whom we shall meet in the next chapter. But this wasn't at the heart of what he was trying to say. It may, indeed, have been one of those things he thought of as obvious and thus didn't see cause to mention.*

The idea that there might be more than one thing going on when

---

* Reading Duysens' account of the period, it's intriguing to note the degree to which he found the diffident and circumlocutory casting of Hill's arguments irksome in and of itself: 'I have difficulty', Duysens wrote, 'pin-pointing the meaning of their tentative, multi-interpretable and often conditional discussion statements [eg] "We are almost obliged to conclude that these energy relations are significant and not fortuitous".' While taking Duysens' point, there was something of a culture of understatement in Cambridge publications at the time, most memorably demonstrated in the conclusion of Watson's and Crick's paper on the structure of DNA, which came out the year after the paper Duysens was quoting: 'It has not escaped our notice that the specific pairing we have postulated immediately suggests a possible copying mechanism for the genetic material.'

chlorophyll trapped light had been in the air, on and off, for a while; it was mentioned, for example, in the vast compendium on photosynthesis assembled by Emerson's colleague Eugene Rabin-owitch in 1956. Derek Bendall (Fay's husband) and another col-league remember Hill speculating on the possibility of two systems, or even three, earlier in the 1950s. By the end of the decade the idea that light intervened in the photosynthetic process not once but twice had been strongly suggested by a range of data, and the relationship between those two interventions was in the process of dawning, perhaps protractedly, on a number of people. But it was the shy and stammering Robin Hill, owner of the least orderly labor-atory known to man, the lines of his thought obscure even to his closest colleagues, who first put it all together in a clear, succinct, form.

Peter Mitchell and the role of the membrane

The early 1960s marked a cusp in the history of photosynthesis – what might be called the Great Clarification. Before it there was no coherent overview of the process. Afterwards there was. Hill's work on the Z-scheme and Duysens' work on photosystems were two parts of the Great Clarification. The third was to come from Peter Mitchell.

Peter Mitchell was, in his way, as interesting an English eccentric as Hill, though perhaps not as endearing. Long-haired and heavy-faced, shaggy and intense, one of his colleagues remembers him looking 'like the later Beethoven' when he was still in his thirties. In Cambridge in the 1950s – where he, too, was close to David Keilin – he was something of a showman. Independently wealthy, he drove a vintage Rolls-Royce to and from the biochemistry depart-ment, a subject of much comment among his cycling peers, and threw quite impressive parties. After holding positions at Cambridge and then Edinburgh in the 1950s, in the early 1960s he was forced to take a leave of absence due to severe gastric ulcers. While recuperating in Cornwall, he bought a rather grand but very dilapidated country house. With his colleague Jennifer Moyle he turned Glynn House into an independent research institute at which he would both live and work for the rest of his life.

Seeking, as Keilin had, to understand respiration, Mitchell focused on the fact that it took place not just in mitochondria, but in the membranes of mitochondria. Biological membranes are sheets of tightly packed fatty acids that serve to divide things up. Membranes define the edges of cells; in eukaryotic cells they also fence off the nuclei, where the genes are stored, from the rest of the cell. Mitochondria and chloroplasts are set off from the cell by membranes, too – and yet more membranes give them an internal structure.

Using normal microscopes that structure is indiscernible. But one of the welcome intrusions of quantum physics into biology was the development of the electron microscope. Quantum physics showed that light, traditionally thought of as a wave, could also be seen as a particle – a photon. The shorter the wavelength of the light, the greater the energy of the photons it was composed of.* The same mathematics showed that electrons, normally thought of as particles, could also be seen as waves, and thus a beam of electrons, properly focused, could serve as the basis for microscopy. Because the electrons in even a modestly powered beam, such as that in an old-fashioned television's cathode ray tube, have far more energy than the photons of visible light, such microscopes operate at much shorter wavelengths than light microscopes. They can thus make out much smaller details, and peer inside the chloroplast.

Three-dimensional reconstructions based on electron-microscope cross-sections of chloroplasts bring to mind a city of the future

---

* The short-wavelength light that fires photosystem II, for example, carries more energy in each of its photons than the longer-wavelength light which fires photosystem I, which is why photosystem II is more powerful in terms of changing redox potentials. For those who have a smattering of physics, the relationship between the energy of the light and the change in redox potential can be summed up elegantly as $h\nu = eV$, where ν is the frequency of the light, h is Planck's constant, e is the charge on the electron and V is the difference between the two redox potentials.

on the cover of a 1950s science-fiction magazine. Cylindrical sky-scrapers that look like stacks of gambling chips, or anorexic Michelin men, are linked by frets that look like swooping aerial roadways. But while the architecture is complex, and differs widely from species to species, the basic topology is not. A chloroplast is a bag inside a bag. Within its outer boundaries, there is a 'main' space – the volume that contains the chip-stack skyscrapers and their interconnections – and an 'interior' space – the volume within those chip-stacks. The main space is called the stroma, while the interior of the stacks is the thylakoid space. Between them is a membrane. And embedded in this membrane are the cytochromes responsible for electron transport, as well as the proteins which hold together the chlorophyll pigments that capture the sunlight. All the greenness of the plant world is attached to the thylakoid membranes within chloroplasts.

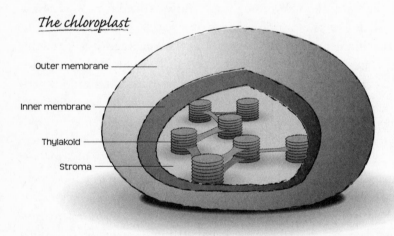

The chloroplast

Outer membrane

Inner membrane

Thylakoid

Stroma

The inside of a mitochondrion looks very different, but the under-lying topology is the same; a bag within a bag, with the key respira-tory cytochromes embedded in the membrane between them. This led Mitchell to think that there was something about the phosphory-lation going on in the two different systems that depended on the membrane. And since a membrane is basically just a way of setting one space off from another, phosphorylation must depend on some sort of difference between the spaces on either side of the membrane, a difference between the inside and the outside.*

Mitchell's hypothesis, published in 1961, rewarded with a Nobel prize in 1978, was that the crucial difference lay in the concentration of hydrogen ions. He argued that the cytochromes embedded in the membranes move hydrogen ions – single protons – from the main space of the stroma into the more confined interior thylakoid space. As a result, the concentration of hydrogen ions inside the thylakoid space is higher than it is outside, and an electric field is also set up across the membrane. Here, again, the Second Law of thermo-dynamics rears its head. Following the dictates of thermodynamics, the ions would rather not be concentrated in one place and excluded from the other. In order to harvest the potential for work represented by this desire to even things out the membrane is studded with portals that allow the hydrogen ions concentrated in the chip-stack thylakoid towers of the chloroplasts to get back into the wide open spaces of the stroma. And as the ions take advantage of these portals they give up energy. The portals use this energy to turn ADP into ATP. As water wheels produce mechanical energy from the flow of water, so these portal-proteins in the membrane produce chemical energy from the flow of hydrogen ions.

In the decades that followed, that analogy was shown to be dis-armingly precise. The proteins that produce ATP in chloroplasts and mitochondria actually spin; each hydrogen ion that passes through one of the channels made of protein that are embedded in the membrane turns the rotor mechanism a bit further round. In human mitochondria, it takes ten hydrogen ions to crank the protein through a full circle and thus produce three molecules of ATP. In plant chloroplasts, it takes fourteen hydrogen ions to get a full turn and produce the three ATPs. The little wheels spin round hundreds of times a second.

---

* This was a view that René Wurmser had foreshadowed back in 1930.

Described like this, the Mitchell hypothesis might sound a little futile. One set of proteins – the cytochromes – pumps hydrogen ions into the interior space; another – the ATP-makers – lets them flow back out. But the crucial thing is to follow the energy. Moving the hydrogen ions into the interior requires energy; releasing them liberates energy. The energy given up by the electrons being passed down the electron transfer chain and through the cytochromes is used to move the hydrogen in; the energy released when the hydrogen moves back out is stored as ATP. So the system takes the energy of an electron flow, which is localized in time and space to a particular stretch of illuminated membrane, and stores it in a chemical form that can be used at any time and any place that the cell has need of it. Shuffling ions back and forth across the membrane translates the energy the cell receives – whether it be from sunlight, in the case of the chloroplast, or from oxidizing food, in the case of the mitochondrion – into the sort of energy that it actually uses.

To mainstream biochemists, especially in America, the Mitchell hypothesis was shocking. The people who studied respiration mostly thought that ATP was made with the help of a less mobile chemical, a 'high-energy intermediate' associated with each cytochrome that stored the energy given up as an electron passed through in a chemical bond. The fact that this intermediate stubbornly refused to be discovered was seen as just one of the inconveniences of science's darkened room. It had to be around somewhere; it was just that no one had found which corner to shine the torch into. Mitchell's idea that the energy was stored, not in a new class of molecule, but in the form of a 'chemiosmotic' difference in the hydrogen ion concentration on the two sides of a membrane was unsettlingly outlandish.

People studying photosynthesis were more welcoming than those studying respiration. One reason, perhaps, was that they had a more physical view of their process than the respiration people did, a view in which potentials over a membrane did not seem quite such an outlandish alternative to the tried and trusted idea of energy stored in chemical bonds. By this stage a lot of the photosynthesis community came from a physics background. They felt that there was more to what they were studying than the proclivities of enzymes, and so they were better able to take Mitchell on board.

It helped, too, that the best early evidence in Mitchell's favour came from a photosynthesis researcher named André Jagendorf. Jagendorf's father, a New York dentist, wanted him to be a farmer. Jagendorf's tastes ran more to science-fiction. As a compromise, he became a plant physiologist. At Johns Hopkins in the early 1960s, Jagendorf set out to find the long-sought 'high-energy intermediate' that took energy from cytochromes and gave it to ATP by seeing what happened to photophosphorylation if the light was suddenly turned off – that is, if the intermediate was no longer being made, and just being used up.

If the high-energy intermediate was in the form of a molecule, and each of these molecules was bound to a cytochrome, then the number of ATP molecules made in the dark, after the electrons stopped flowing, could be no greater than the number of cyto-chromes in the system. But it was – it was dozens of times larger. The energy must be stored in something unprecedented.* Jagendorf had heard one of Mitchell's early lectures on the chemiosmotic idea, and had found it both annoying and incomprehensible. But a British post-doctoral researcher in Jagendorf's lab, Geoffrey Hind, con-vinced him that it was worth looking at. Using an electrode designed to measure the number of hydrogen ions, they watched as thylakoid membranes sucked hydrogen across themselves one way when illuminated and let it back the other way when the lights were turned off, just as Mitchell's idea would lead you to expect.

Hind then showed that if you kept thylakoid sacs in the dark and found some way of taking hydrogen ions out of the medium they were floating in, they would produce ATP with no light at all; the drop in hydrogen ion concentration outside made the ions inside

---

* The same result was found independently by Yunkang Shen, at the Shanghai Institute of Plant Physiology, but the onset of the Cultural Revolution prevented him from following it up; he and Jagendorf were not to meet for more than a decade.

anxious to cross over. All you needed to make ATP with thylakoid membranes was a difference in the concentration of hydrogen ions on the two sides. It didn't matter what the concentration inside was, as long as the concentration outside was considerably lower. Though the battles over chemiosmosis continued for years in the realm of respiration, after Jagendorf's work it became fairly broadly accepted within the photosynthesis community.

Mitchell's hypothesis provided the last link in a chain that stretches far beyond the membranes of the chloroplast, the chain first imagined by Mayer and Boltzmann more than a century earlier, the power cable for the vast engine of Vernadsky's biosphere. It begins with the gravity that pulls every bit of matter in the universe towards all the rest of the matter. Gravity pulls remorselessly on the atoms of the sun, and in so doing subjects those in the star's core to such great pressures that their nuclei, which normally repel each other, are jammed together.* When one of these nuclei hits another just right – after billions of years of doing it not-quite-right – they will merge, and energy will be released; the energy it takes to bind the various protons and neutrons involved into one larger nucleus is less than the energy it takes to hold them together in two smaller nuclei, and the difference is let out as radiation.

The photons produced through fusion percolate through the sun, absorbed and re-emitted countless billions of times as they diffuse from the core to the surface, where they escape into free space. Eight minutes later, a tiny fraction of those photons will hit a chlorophyll molecule sitting in a chloroplast. The chlorophyll absorbs the energy, which sets the chemical bonds that hold its atoms together thrumming. Its neighbours pick up on the vibration and start to vibrate in sympathy, and the energy gets passed from molecule to molecule through this resonance until it reaches the part of the photosystem where one of those chlorophyll molecules, rather than just resonating, actually loses an electron.

---

* Mayer had, in a way, been correct in thinking that the sun's gravity was the ultimate source of its energy, though his eternally pummelling meteors had been wide of the mark.

The electron then slides down the electron transfer chain that is laced through the thylakoid membrane, giving up energy that the cytochromes use to transfer hydrogen ions across that membrane. The chemiosmotic potential created by this pumping is harnessed by the membrane's water-mill hydrogen channels to produce ATP. And at that point, having travelled from the cosmic realm of gravity through the microcosm of nuclear physics, the subtle physical resonances of chlorophyll, the flow of electrons from molecule to molecule and the pressure of hydrogen ions yearning to be free, the energy is trapped in a single chemical bond.

The energy trapped in that ATP will, in all likelihood, be used to power the Calvin-Benson cycle, creating the further chemical bonds that turn carbon dioxide into carbohydrate. And at some point in the future, those bonds will be broken, too, and their energy given up. Perhaps it will be given up within one of the plant's cells, perhaps in one of the cells of something that has eaten that plant; either way, it will probably be in a mitochondrion. There the energy released as the electron liberated from that bond tumbles down another chain of cytochromes will help maintain another chemiosmotic gradient, and produce some more ATP.

It's tempting to see this as a running-down. In fact, thanks to life's Second-Law-skirting ability to make things happen, it is for the most part a story of intensification. Though the sun is very hot – 6000°C at the surface – it is also very rarefied. It takes half a million tonnes of the sun to produce a single kilowatt of energy. Weight for weight, a eukaryotic cell uses energy at 100,000 times the rate that the sun produces it. The processes by which plants and algae capture and concentrate sunlight into chemical bonds are far, far more efficient than the processes by which it is made.

And yet, though the energy density is increased along the way, at every stage some energy is lost to disorder. At every stage entropy is increased. It has to be; the Second Law insists. But that imperfection is part of the point. Without it, the electrons would not flow at all. If energy could be moved from one form to another without a cost, then there would be no reason for it not to move back just as easily. It's the build-up of entropy that makes things irreversible, that gives things direction. It is entropy that gives the electrons flowing through the cytochromes a 'down' to flow to.

And as long as the energy keeps flowing, and keeps ending up in the form of ATP, the build-up of entropy can be dealt with. That's what life does; it accepts the burden of entropy in order to take the benefits of energy. It then exports that entropy while using the energy to arrange brute atoms in vast and intricate molecules, to line up cytochromes in membrane-bound chains, to corral hydrogen ions in tiny reservoirs, to store sugar as starch against the winter, to build leaves and stomachs and eyes and minds and bicycles and gardens and universities. To create interesting, improbable, fragile things, on every scale from the pigment to the planet.

And then to animate them. This work of life never stops, never rests; it must power the muscles as well as build them, swing the leg, pump the heart. All activity requires energy, and all but a tiny fraction of the energy required by life on earth comes ultimately from the sun. This link beyond the earth is the most intimate of our connections to the cosmos. Science writers, especially those of an astronomical bent, are fond of observing that we are made of stardust – that every atom of the carbon, oxygen, nitrogen and other elements which make up our bodies was forged in the hearts of stars that died before the sun was born. The power of this insight is that, at the same time as putting us at one with the cosmos, it emphasizes cosmic scales of distance and space that thwart our imaginations. But there is a complementary power in seeing that while the starry origins of our component atoms may be half a universe away in space and time, the source of the energy that gives the stardust life is as close as the summer which ripened the fruits on our table. The sun's energy, stored by plants, keeps us alive moment by moment, heartbeat by heartbeat, thought by thought. Our bodies are stardust; our lives are sunlight.

CHAPTER THREE

Light

The Thirteenth International Congress
Emerson and Arnold
The photosynthetic unit, and its rejection
The reaction centre
The 46,630 atoms of photosystem II
The exquisite thrill

Are not gross bodies and light convertible to one another? And
may not bodies receive much of their activity from the particles
of light, which enter their composition?

SIR ISAAC NEWTON, *Opticks*, Query 30

The Thirteenth International Congress

Late in the evening of 31 August 2004, an accomplished if slightly jet-lagged blues band is playing up a storm. In a hall set among the rooftops of downtown Montreal, hundreds of people are listening, laughing, drinking and dancing. With a long-legged prowl across the stage and occasional plaintive bursts of harmonica, the singer is giving it all he's got. Over the joyous noise and a couple of beers, a wily veteran whom I met at the bar is filling me in on some of the great photosynthetic characters he has known ('X? Well he was a son-of-a-bitch; Y was a son-of-a-bitch too, but I liked him; he did more with less.'). A little way away from us a dapper Indian man in his early seventies is dancing with his wife and a broad smile; like most of those on the dance floor, they give every sign of delight.

This is the second George Cheniae Blues Concert, the social highlight of the Thirteenth International Congress of Photosynthesis. The congresses take place every three years, a chance for the scientists in the field to swap results and showcase advances, an opportunity for doctoral students and post-docs to look for their next position, an arena for disagreements already rehearsed in print to be thrashed out in person. It's a chance to play football (the Dutch team beat one of the North American teams in the Montreal final) and go out to dinner and, on this particular evening, drink and dance while being entertained by Bill Rutherford, who is not only our blues caller for the night but also a highly respected spectroscopist at the French research council's Département de Biologie Joliot-Curie, near Paris. (The rest of the band are not scientists;

they're the Baskerville Blues Band. Rutherford plays with them at gigs in and around Paris, and the congress has picked up most of the tab for flying them over.) It's a chance to renew friendships and feuds. It's a chance for Govindjee, the Indian-born scientist who, towards the end of a distinguished research career, took on a second role as collator of the field's collective memory, to meet and greet and photograph everyone. He is the congress's honorary president, and gave the first speech; that was him on the dance floor in the first paragraph, and he'd say hello and welcome you to the subject of his devotion with a smile right now if he could.

During the day, the full diversity of photosynthetic studies is on offer. How do plants respond to stress, to too much light, to too little nitrogen, to not enough chloride, or too much copper? What role do forests play in the carbon cycle? What can we learn from the genomes of photosynthetic plankton, and what are photosynthetic bacteria doing inside grains of desert sand? Can rubisco be genetically engineered to make plants grow better? If you drive the basic processes of photosynthesis backwards, can you put water molecules back together? How do the north and south faces of the vines in a Chilean vineyard differ in their sugar production? How does a pine needle work? How much of what's going on in a leaf can you understand by shining a laser on it from afar? How much of what's going on in the biosphere can you measure from orbit?

But though there are people looking at photosynthesis from all sorts of angles and on all sorts of scales, the bulk of the presentations in Montreal are molecular. Martin Kamen used to tell people that the study of photosynthesis was basically parasitic; it fed off whatever techniques and ideas seem most helpful at the time. 'The future of photosynthesis', he wrote in the early 1960s, '[is] the future of whatever science happens to be most relevant in the future.' In Kamen's future and our recent past, the relevant science has been molecular biology, which is what most of the Montreal meeting is focused on.

The idea that the physical nature of molecules might be the key to a deeper understanding of life first began to gain traction in the 1930s, when great physicists such as Niels Bohr and, a little later,

Erwin Schrödinger began to speculate about a revolution in biology comparable to the quantum revolution they had brought about in their own subject. In the 1940s new physical tools began to have an effect. The use of X-rays to discover the structure of DNA, a physical structure which, by suggesting how genetic information could be stored, copied and used, really did open up a deeper understanding of life, finally brought the molecular approach into its own. It has been increasingly dominant ever since.

Photosynthesis sits in a particularly intriguing relationship to molecular biology, because it was yielding up its secrets to similarly physical approaches before molecular biology was properly off the ground. Photosynthesis is a biology of light, and light is a subject for physics. It is not too hard to imagine a world in which the study of photosynthesis might have been the foundation on which molecular biology was based. But molecular biology was from its inception more interested in information than in energy, and that inclination was locked irreversibly in place when Watson and Crick demonstrated the molecular basis of the information contained in genes.

Matters to do with energy, meanwhile, remained in the purview of biochemistry, and while physicists – properly, biophysicists – studied the ways in which the photosynthetic mechanism and the photosynthetic pigments within it responded to light from the 1930s onwards, it was not until comparatively late in the game that they came to apply the approaches of molecular biology to the proteins that hold those pigments in place and make the mechanism work. But that application has now taken place, with spectacular results.

The morning after Bill Rutherford's blues bash, Jim Barber, a professor at Imperial College, London, gave perhaps the best-attended plenary talk of the meeting, reporting a singular triumph in the analysis of how photosynthesis actually works. Barber is a fit-looking sixty-five-year-old Englishman who's easy to imagine with a tennis racket in his hand, a tall man with a longish face that becomes animated when he gets excited, at which point he will also start interrupting himself quite a lot.

Barber's career can be measured out through the history of these triennial congresses. In 1967 he attended the first of them, in Freudenstadt, Germany, as one of Lou Duysens' post-doctoral students. In doing so, he says now, he felt 'ever so proud'; in the Great Clarification of the 1960s Duysens' idea of the photosystem, along with Hill's Z-scheme, had become the subject's all-but-uncontested foundations, and to attend with Duysens was to be in the right place at the right time. At the Tenth International Congress, in Montpellier in 1995, Barber decided to make a radical shift in the tempo and focus of his work. His lab at Imperial College had made valuable contributions to various areas of photosynthesis research, but no breakthroughs. He was somewhat uneasy about the grant from Britain's agricultural research council that had been supporting his work. The link between his basic research and crops in the field was often though not always a tenuous one, and Barber felt his assessors sniffed at him as too much of an ivory-tower man. When they came to the lab 'They always wanted to know where the barley plants were,' he remembers with a touch of pique, seeing in the question an all but unbridgeable difference in outlooks and expectations. So he decided to focus his lab on a single big, dramatic goal: the structure of photosystem II.

Back in the 1960s the late George Cheniae, to whom Bill Rutherford's blues night is dedicated, called photosystem II 'the inner sanctum of photosynthesis'. It is the front end of Hill's Z-scheme, the mechanism that first takes the energy of the sun and transforms it into a change in redox potential – and, mysteriously and vitally, it breaks down water into oxygen in order to do so, providing the wherewithal for almost all the earth's respiration. It is also, by the standards of molecular biology, frighteningly big and gnarly, combining nineteen separate proteins, thirty-six chlorophyll molecules and a number of other bits and pieces into a complex three-dimensional jigsaw. As an audacious target for Barber's refocused lab (and a way of capping off his career in the decade leading up to retirement) the structure of photosystem II would be hard to beat.

Three congresses on, Barber thinks the goal has been achieved (and, in the process, retirement has been deferred). In that morning plenary in Montreal he presented the model of photosystem II that he and his colleagues have put together. In the afternoon he listened as people who have produced less informative models in the past, and who want to produce better ones in the future, criticized his model – which was published some months earlier in the journal *Science* – and his plans for its further elucidation. He didn't let too much of his frustration show. He is satisfied that for the time being he and his team have the best data. Photosynthesis still has its mysteries: but as far as photosystem II is concerned, the size and shape of the puzzle box in which they reside are now known better than ever.

Emerson and Arnold

The path that leads to the structure of photosystem II begins with a classic set of experiments undertaken in the 1930s by Robert Emerson and William Arnold, research which established the key role that physics could play in the study of photosynthesis.

Emerson was part of a distinguished New England family, the great-grandson of Ralph Waldo Emerson's brother, the son of Haven Emerson, the formidable head of the New York Public Health Service; he grew up in an estate on Long Island. Arnold grew up in a house built by his father, a lumber merchant, a little way outside Eugene, Oregon – a house that lacked electricity or indoor plumbing, but which boasted a fine toolshop of which young Bill took full advantage. Emerson studied physiology at Harvard, became interested in plants, and went to one of the greatest laboratories of Europe to pursue his doctorate. Arnold went to Caltech with an eye to being an astronomer, ran short of money, and interrupted his studies to work as a research assistant in the physics department, spending sleepless nights making measurements of the earth's magnetic field too delicate to be undertaken when there were street-cars running on the streets outside. Emerson arrived at Caltech at the point that Arnold felt secure enough to take up his studies again; the new instructor's course on plant physiology was suggested to him as a way of making up his biology requirement. They were a year apart in age and well matched in skills; the patrician easterner and the happy-go-lucky westerner made a great team.

When he'd gone to Europe Emerson had originally planned to study with Richard Willstätter, who had received the Nobel prize for elucidating chlorophyll's molecular structure, and whose book on the subject still sits on Andrew Benson's desk. But he found Willstätter unable to work with him, because the increasingly anti-Semitic university in Munich was blocking the appointment of Victor Goldschmidt, a brilliant Swiss-Norwegian chemist; Willstätter, also Jewish, was in the process of resigning his professorship in protest. Emerson was advised to go to Berlin, instead, and study with Otto Warburg. Warburg was perhaps the leading biochemist of the day, and a strange and redoubtable character – an anglophile homosexual Jewish Prussian cavalry officer with an unyielding in-sistence on additive-free food (his milk was supplied from a special herd at Berlin's School of Agriculture, with the cream separated off in his laboratory centrifuges). He was unremittingly industrious, unforgiving of slights real and imagined, undeniably brilliant, fre-quently wrong, never in doubt, occasionally malicious and incred-ibly stubborn – a seemingly intolerable martinet still capable of inspiring deep pupillary devotion.

He was also, and perhaps above everything, a master of technique. He developed a range of tools and methods for measuring spectra; he invented ways of cutting very fine slices of tissue so that he could consistently produce samples just a few cells deep. Most crucially, he developed instruments for measuring pressure, and the rate at which pressure changed, with great precision. These Warburg ma-nometers became crucial instruments for understanding how much gas was given off or taken up by tissue samples under analysis, and were thus indispensable for studies of both respiration and photosynthesis. In the two years he spent earning his PhD in Warburg's laboratory, eight o'clock to six o'clock, six days a week, Emerson became as skilled in the techniques of manometry as anyone.

He also took on board Warburg's theoretical interests, and would take them back to America with him. Warburg's physicist father, Emil, was a friend of Einstein* and had produced important experimental support for Einstein's 'law of photochemical equivalence'. This held that in photochemistry – chemistry in which the reactions are driven by light, not heat – the number of molecules made is determined by the number of photons absorbed. Otto Warburg sought to follow up his father's work by showing that Einstein's law held in biological systems, too, by precisely relating the number of photons absorbed during photosynthesis with the amount of oxygen given off. He used flasks filled with *Chlorella pyrenoidosa*, a type of single-celled green algae that Warburg found particularly well-suited to experimental investigation, and which subsequently became the photosynthesizer of choice in biochemistry labs around the world.

At Caltech, Emerson told his students about a particular set of experiments aimed at understanding the efficiency of photosynthesis in which he had participated in Berlin. Warburg had compared the rate of photosynthesis under a light that flashed on and off and a light half as bright that was left on all the time. In both situations the number of photons that reached the broth of chlorella was the same – it was just that in one case they came in bursts, and in the other as a steady flow. So Einstein's law suggested that they would set off the same number of chemical reactions. But, puzzlingly, flasks lit with bright flashes produced more oxygen.

Emerson planned to follow up the experiments using a spinning shutter to cut up the light from an incandescent lamp just as Warburg had. Arnold suggested that neon flickering on and off fifty

---

* On the basis of this friendship Einstein wrote to Otto Warburg in 1918 urging him to accept his father's efforts to get him safely removed from active duty with the cavalry. Einstein argued that since Warburg excelled in physiology, an area in which Germany was rather mediocre, it was his patriotic duty to quit the front. It would thus be better, Einstein argued, if 'an average man' were to take Warburg's place in the line of fire.

times a second might do the job better. Emerson was taken with the idea, and Arnold started building the neon system as his coursework. When he graduated, in 1931, he became Emerson's research assistant.

Emerson saw Warburg's discovery, which his own exquisite manometry confirmed at Caltech, as evidence that photosynthesis was not a straightforward photochemical reaction, in which a fixed amount of ingredients, if given a photon each, would produce a fixed amount of product. First, Emerson argued, there was a physical process, the absorption of photons by chlorophyll, which would be effectively instantaneous. Then there were chemical processes in which the energy received from those photons drove biochemical reactions. At this point in the 1930s, no one had the faintest idea what those reactions were; photophosphorylation and the Calvin-Benson cycle were decades away. But Emerson and Arnold were safe in assuming that whatever biochemical shenanigans were going on would be slower than the physical absorption of light. This mismatch in tempos could explain the flash effect: in steady light, energy was being absorbed by the physical front end of the process faster than the biochemical back office could deal with it, and so some was wasted; in flashing light, the periods of darkness gave the chemistry time to catch up with the physics. The system needed photons, but it also needed time to digest them.

To prove their point, Emerson and Arnold devised a new experiment in which the periods of darkness between the flashes could be lengthened. This was no longer a job for simple neon: Arnold built a much more sophisticated light source, one that used a ring of spark plugs and a spinning rotor to provide flashes just twenty millionths of a second long at intervals of as little as three hundredths of a second. They used this to provide flashes to batches of chlorella kept at two different temperatures, 25°C and 1.1°C. The idea was that because more heat means the molecules bang together harder and more often, normal chemical reactions (as opposed to photochemical reactions) are almost always responsive to temperature. So they would be slower in the cold chlorella, and the cold chlorella would, as a result, need longer dark periods between flashes in which to digest all the energy from the photons. And so it proved. When the flashes were only a tenth of a second or so apart, the cold chlorella produced much less oxygen than their warmer colleagues; once the gap between flashes was strung out to almost half a second, though, the yield per flash in the chilled sample was exactly what it was at 25°C. The chemical reactions were getting all the time they needed to digest the energy in the first bunch of photons before the next lot came along.

The experiment took immense technical skill; Emerson's manometry was almost inhumanly precise, picking up changes of less than a thousandth of a millilitre in volume. And when it was done the physiologist and the physicist had shown there was a near-instantaneous physical process at the beginning of photosynthesis which fed energy into a more drawn-out chemical process. Trying to distil the essence of life, Emerson's great-great-uncle once wrote that 'Power . . . resides in the moment of transition from a past to a new state, in the shooting of the gulf, in the darting to an aim'. In their lab at Caltech, Emerson and Arnold had made the first step towards identifying the eternally repeated moment of transition in which life's power shoots the gulf from sunshine to chemistry.

The photosynthetic unit, and its rejection

In histories of twentieth-century physics, crammed as they are with incident, photosynthesis research normally doesn't merit much more than a footnote.* If, as Martin Kamen claimed, photosynthesis research has a tendency to parasitism, in this case the parasite has been largely ignored. And yet there's an odd shadow-of-greatness feel to the story of the decades after Emerson's and Arnold's break-

* See footnote, page 109

through; photosynthesis may have been a minor part of the story, but it was oddly close to the mainstream: a Rosencrantz-and-Guildenstern counterpoint to the great tragic history of physics and the bomb, a glimpse into what was happening in the wings.

Having distinguished physics from chemistry, Emerson and Arnold went on to make the first great discovery about how the physical side of photosynthesis works – a discovery which would be largely disregarded by most of their colleagues for decades. Their aim was to find out whether there was a limit to the rate at which the physical front end of the photosynthetic system could make use of photons. Again there would be flashing lights, their intermittency timed so as to allow the algae ample darkness in which to digest, but this time the flashes would get brighter and brighter. Arnold had to soup up the apparatus to the edge of practicality: higher voltages, bigger capacitors, focusing mirrors, silver linings on the flasks themselves, every means possible was used to maximize the number of photons that hit the chlorella's chlorophyll. At first, brighter flashes meant more oxygen, meaning the increased number of photons was being put to good use. But there came a time at which this ceased to be the case – when the physical front end of the system was working as hard as it possibly could.

The surprise came when they looked at the amounts of chlorophyll involved. If, as Warburg and others had assumed, photosynthesis was a fairly straightforward photochemical reaction, a chlorophyll molecule that absorbed a photon would immediately carry out one of the chemical steps that led to carbon dioxide being reduced and oxygen released: the number of molecules of oxygen produced when the system was working at its maximum efficiency would be pretty close to the number of chlorophyll molecules present. But that wasn't what Emerson and Arnold saw. They found that, when saturated with light, it took a staggering 2480 chlorophyll molecules to produce one molecule of oxygen.

Two explanations were offered for this unexpected finding. One came from James Franck, a physicist who had been a student of Otto Warburg's photochemical father, Emil, and a man of both great influence and great honour. Franck had won the Nobel prize in 1926 for showing that atoms behaved as quantum theorists said they should, with electrons arrayed around the central nucleus at different energy levels. His military service in the Great War and his scientific eminence offered him some protection from the rise of anti-Semitism – the same factors helped keep Otto Warburg safe in Berlin until 1945 – but Franck forfeited that protection when the Nazi employment laws of 1933 required him to fire Jewish subordinates from his physics department at Göttingen University. After searching his conscience and consulting his friends he chose not only to resign his position rather than enforce the law but also to make his reasons for doing so public. Within a few months he left his home country, having made it his business to arrange positions overseas for as many of his junior colleagues as he could.

After some time spent with Niels Bohr, the father-figure of quantum mechanics, in Copenhagen, Franck headed for America,* where a charitable foundation set up a laboratory for him at the University of Chicago on the condition that he apply himself to the problem of photosynthesis. While this may not have been what Franck actually wanted to do – Martin Kamen, who first met him around this time, recalled him being unhappy at the prospect – it was what he ended up doing, taking a traditional photochemical view of the problem as one in which the energy from a photon would cause a chemical reaction in each molecule that absorbed one.

Franck's interpretation of Emerson's and Arnold's results was that photons caused chlorophyll and carbon dioxide to react and form some unstable, short-lived molecule. There was an enzyme which could stabilize this reaction, and make use of the energy trapped in the unstable molecule, but it worked slowly and was in short supply.

---

* To avoid having to bring a significant amount of gold into America, Franck left his Nobel prize medal with Bohr in Copenhagen. When Germany invaded Denmark, the chemist George de Hevesy dissolved Franck's medal, along with that of Max von Laue, in acid. He left the resultant black solution in an unmarked bottle tucked away among others in his laboratory at the Institute. After the war, the gold was precipitated back out of the solution and sent to Sweden, where the Swedish Academy had it recast into new medals.

As a result, he imagined that most of the short-lived chlorophyll-plus-carbon-dioxide amalgams just fell apart uselessly before the enzyme could get to them.

An alternative model was offered by a microbiologist from Otto Warburg's lab named Hans Gaffron and a physicist called Kurt Wohl. Gaffron and Wohl imagined that rather than being distributed randomly throughout the chloroplast, as Franck and everyone else supposed, the chlorophyll molecules were actually assembled into units. Each unit might contain hundreds or thousands of molecules. When a photon was absorbed by one of them, it would excite that molecule's electrons, moving one or more of them to a higher energy level. The laws of quantum mechanics showed that this excited state could be passed directly from one molecule to the next, almost as though it was a discrete object. Gaffron and Wohl argued that these excited states – excitons, as they came to be called – would migrate through the thicket of chlorophyll molecules until they reached a specific point where their energy could be used by an enzyme to do some chemistry. The enzyme didn't go from chlorophyll to chlorophyll to harvest energy, as in Franck's model; it sat still, allowing an entourage of chlorophyll molecules to pass all the energy they absorbed along. Gaffron called this hypothetical lots-of-chlorophyll-surrounding-an-enzyme system a 'photosynthetic unit'.

Gaffron was right; Franck was wrong. The photosynthetic unit is the concept on which the modern idea of a photosystem is based. In a photosystem, large numbers of photon-catching chlorophyll molecules are arranged in and on the membranes in which the electron transfer chain's proteins are embedded. Excitons pass from chlorophyll to chlorophyll until they finally reach one of the membrane-piercing proteins at the heart of the photosystems – the zigs in Robin Hill's zigzagging Z-scheme. There the energy is used to pull an electron loose from a donor and feed it into the electron transfer chain.

Plants, algae and photosynthetic bacteria use various sorts of chlorophyll, and a number of other pigments, to make light-catching antennae of a wide range of shapes and sizes. There are integral antennae built into the fabric of the photosystems, and there are detachable antennae that can move en bloc from one photosystem to another. Scaffoldings of protein lock the three-dimensional arrays of chlorophyll into architectures of propinquity in which excitons can jump from one molecule to the next with the greatest ease. In some systems the architectural response to nature's constraints seems almost random – the arrangement of the ninety-six chlorophyll molecules built into photosystem I, the part of the Z-scheme that pumps electrons up to where they can be used to make NADPH, make no more sense to the eye than the first sight of a building by Frank Gehry or Bernard Tschumi does. The graceful twenty-seven-chlorophyll rings used by purple bacteria have a rich co-ordinated symmetry that would do justice to the Alhambra. In both cases, though, form follows function.

The antenna geometries are complex and varied; the need for them is simple and universal. When you put together the size of a chlorophyll molecule, its capacity to absorb photons and the rate at which the sun rains photons down on the earth, you find that, in bright, direct sunlight, a chlorophyll molecule can expect to absorb ten photons a second. In less than ideal conditions – the late afternoon of a cloudy November day in England comes to mind at the moment of writing – that rate falls to one a second or lower. Given the pace at which enzymes can get things done down at the molecular level, this is abysmally slow. So it makes sense for a number of chlorophylls to pool their resources and shovel all the energy they absorb into a single electron transfer chain.

Unfortunately, from the 1930s onwards Franck was firm in his opposition to Gaffron's idea of the photosynthetic unit, and Franck's views on photosynthesis carried weight. It was not a big field, and he was a big man – a physicist with a Nobel prize, a friend to the giants of the quantum revolution if not quite their peer. He was one of the originators of the ideas that lay behind the exciton, and in work done with Edward Teller at Johns Hopkins he seemed to show that there was no way that they could be used to pass energy through a whole host of chlorophyll molecules. Emerson and Arnold had sympathy for Gaffron's view, and in the late 1930s Arnold and Robert Oppenheimer came up with a different treatment of the problem that allowed the energy to pass more freely, but the war got in the way of their publishing it.

Looking back at this stage of Arnold's career it's impossible not to be struck by the opportunities for meeting historically significant people that seemed to crop up all around him. Part of this was due to luck – Arnold always saw himself as a lucky man. Part of it was doubtless due to his abilities, which must have been earning him a reputation that outstripped his junior status. But a large part of it was down to the relatively modest scale of the academic enterprise at the time – and to the way in which the ripples spreading out from the quantum revolution were rendering the borders between neighbouring subjects oddly porous. It was a good time to be a young physicist with a physiological interest in the fundamentals of life.

Name-dropping on his behalf, and starting from the top, Arnold had met Einstein while still a research assistant; the measurements he was making of the earth's magnetic field during Pasadena's street-car-free nights followed on from measurements that Einstein had made in Berlin, and Einstein stopped by on a visit to Caltech to see how they were going. At Berkeley in the mid-1930s Arnold attended a course by Oppenheimer, with whom he went on to write the lost energy transfer paper. At Stanford he attracted the attention of Warren Weaver at the Rockefeller Foundation, the man who had coined the term 'molecular biology' and funded a large chunk of the discipline's infancy. Weaver had been responsible for bringing to America Max Delbrück, a Göttingen physicist who would become one of molecular biology's founders; when Delbrück arrived in Stanford to take van Niel's general microbiology course, Arnold worked with him on one of the first of the experiments on phage – viruses that attack bacteria – which would be central to Delbrück's research. Weaver gave Arnold a grant to go to the Niels Bohr Institute in Copenhagen, where the physicist Otto Frisch, following the same logic they were employing in the Rad Lab, suggested that radioisotope studies would be more powerful if someone could make a new, long-lived carbon isotope by irradiating nitrogen. Without the power of the Berkeley cyclotrons Kamen had at his disposal Arnold's assault on carbon-14 had little chance of success, but it meant that he was there in the laboratory on the day that Frisch's experiments confirmed a strange suggestion made by Frisch's aunt, Lise Meitner, and her chemist colleague Otto Hahn. It appeared there were circumstances in which uranium atoms could be split in half, releasing a great deal of energy in the process. Arnold later recalled Frisch asking him, in his capacity as a sort-of biologist, 'What do you call the process in which one bacterium divides into two?' Arnold told him the term was fission, and in so doing gave the process of nuclear fission its name.*

The practical implications of that discovery saw Arnold working on the cyclotron-based uranium enrichment systems at Oak Ridge, Tennessee, where Martin Kamen also worked before his fall from grace. He remained at Oak Ridge for the rest of his career; what had been an industrial plant for bomb-making materials grew into one of the new generation of national laboratories under the aegis of the Atomic Energy Commission. James Franck, too, was part of the Manhattan Project, though at a much higher level: he ran the chemistry team at the University of Chicago. He distinguished himself by leading the disregarded call to have the first operational atom bomb detonated on an uninhabited target in front of observers from around the world, rather than in an attack. The call was actually written by Eugene Rabinowitch, another of the young Jews from Göttingen for whom Franck had found positions. Rabinowitch would go on to edit the *Bulletin of the Atomic Scientists*, the journal through which Franck and other Manhattan Project veterans tried to influence, or at least critique, the political uses of their work. He would also become Robert Emerson's closest colleague.

* This is the footnote that photosynthesis sometimes gets in physics histories. Fission also provides a second link between Arnold and Max Delbrück; while Arnold named the process, Delbrück unwittingly managed to delay its discovery by overlooking the possibility when he, Hahn and Meitner were working together in Germany in the early 1930s.

As a physiologist, Emerson had little to offer the Manhattan Project. As a Quaker he had no intention of making any such offering. During the war he acted as a weighty Friend to Caltech's conscientious objectors, including the young Andrew Benson. Angered and ashamed by the internment of Japanese-Americans – a friend once remarked that he loved the underdog as deeply as he loved truth itself – Emerson hit upon the idea of encouraging Japanese-American scientists at the Manzanar camp in Owens Valley to work on the possibility of extracting rubber from a native American shrub, guayule. As a contribution to the war effort, which needed rubber, the guayule project would demonstrate the loyalty of the interned Japanese-Americans. At the same time, Emerson hoped, it would undermine the causes of the war; an America self-sufficient in rubber would have less of a strategic interest in Asia.

Emerson got the Manzanar guayule programme going, with resources from Caltech. He worked on the use of the guayule extract at the American Rubber Company's labs in Los Angeles. The efforts to cultivate and study the shrub in the atrocious conditions of the camp were run by Morganlander Shimpe Nishimura, a physicist who had worked on Rad Lab cyclotrons and then at Caltech before being interned. Nishimura had the sorts of skill the Manhattan Project needed; he was also, like Emerson, an expert gardener. The Manzanar team managed to make progress on the use of guayule that eluded the far larger Emergency Rubber Project run by the Department of Agriculture. The bigger programme showed its appreciation by trying to get the Manzanar research shut down – and at the same time seeking to take credit for its partial success. It was a great disappointment to Emerson that, after the war, the guayule option was not further pursued; he blamed vested interests in the rubber industry.*

After the war Emerson was invited to set up a photosynthesis research group at the University of Illinois, in Urbana. He asked the

---

* Residual interest in the possibilities of guayule persists to this day, not least because the plant can produce a rubber less allergenic than that in everyday use.

university authorities to appoint a physical chemist to work alongside him, so that all aspects of the process could be covered. The chemist they came up with was Franck's amanuensis, Rabinowitch. The two men made a wonderful contrast: Emerson tall, stern, kindly but sometimes terse, poring over his manometers with patient precision; Rabinowitch short, plump, loquacious, surrounded by journals and reprints, doggedly synthesizing other people's results into a magisterial 2000-page monograph that defined the state of the art in 1950s photosynthesis; Emerson a near-teetotal *mens-sana-in-corpore-sano* New Englander devoted to his garden – his family was self-sufficient in its production of vegetables and poultry – whose only evident concession to frivolity was figure skating, at which he was expert and his wife truly excellent; Rabinowitch an asthmatic Russian Jew with a background in journalism and an agility around the pool table that far excelled his prowess at the lab bench, whose wife distilled home-made vodka flavoured with bison-grass (the king of all grasses, at least in this respect). What they shared was science, kindliness, and generosity; they became devoted to each other. In 1959 Emerson, who for a long time resisted flying under any circumstances, died in an aircraft crash. Rabinowitch felt the bereavement deeply.

But while the research group in Urbana provided Emerson with a friendly, collegial setting, his work was in many ways still dominated by ideas and conflicts rooted in the Berlin laboratory in which he had learned his trade. In the late 1930s, Emerson had become mired in a long, wasteful and eventually quite hurtful debate with his one-time boss, the martinet Warburg, about the overall efficiency of photosynthesis. The question turned on counting photons. When the photosynthetic system was running at full efficiency, how many photons did it have to absorb in order to produce one molecule of oxygen and reduce one molecule of carbon dioxide? Warburg's measurements suggested that it took just four photons to reduce a molecule of carbon dioxide. If that were true, it would make photosynthesis extremely efficient; almost all the energy of the photons would end up in the chemical bonds of the carbohydrates.

Other researchers didn't get such impressive results, so in the mid-1930s Emerson took it upon himself to try and make a definitive set of measurements. He found the number of photons required to be between eight and twelve, which was correct. In modern terms, four photons are required by photosystem II, where each is responsible for pulling an electron off a water molecule and injecting it into the electron transfer chain; another four photons are then needed by photosystem I to take those electrons out of the chain's last cytochrome and boost them up to the level where they can be incorporated into two molecules of NADPH, which is what is needed to reduce a single molecule of carbon dioxide. Thus the minimum requirement for producing the necessary amount of NADPH is eight photons. But you also need to produce some extra ATP, which means you need a couple of extra photons in order to send two of the electrons up photosystem I a second time. So ten photons per carbon dioxide is about the best that most systems can manage.

Arnold, in his doctoral thesis, was the first person to show that the higher number of photons was needed, but he never published the result, thinking, correctly, that it would just lead to a bitter fight with Warburg. Emerson, though, took the work forward – and Warburg rejected every result, asserting endlessly that his techniques were superior.

In the late 1940s, Emerson invited Warburg to Illinois; one of the purposes was to try to find him a position in post-war America (Warburg's lab in Dahlem had been closed down under the Allied occupation of Berlin), the other was to see if, by actually sharing a laboratory, they might finally settle the efficiency question once and for all. Given Warburg's fame, the visit was big news; the journal *Science* devoted its cover to a photo of the great man at a lab bench in Illinois. The visit was also an exercise in frustration. Emerson, a dutiful pupil, was open and generous; even by his own standards Warburg was unusually haughty, perhaps because he felt he was in danger of being mistaken for a supplicant. He refused to be bound by agreements about how to interpret the experiments meant to settle their dispute. And even in the midwestern winter he would not allow the lab to be heated. As the visit wore on Emerson became increasingly distressed – 'I can never foresee what his next impossible demand will be', he wrote to Martin Kamen – but also increasingly unwilling to give any ground himself. Eventually Warburg left without troubling to say goodbye, let alone thank you. Cheated of closure, Emerson devoted yet more years to trying to rebut the increasingly unbelievable conclusions Warburg drew from ever more experiments at his now-refurbished institute in Dahlem. The argument with Warburg saddened and frustrated him, and squandered years of his sadly curtailed life.*

His work on all this, though, was not entirely wasted. Emerson would never have seen good work that produced truths as a waste. And some of his research into these matters provided one of the pieces of evidence that led to his friend Robin Hill's Z-scheme. In work on efficiency in the 1940s, Emerson discovered that if he provided photosynthesizing algae with ever-redder light – that is, light of ever-longer wavelengths – there came a point at which the rate of oxygen production dropped fairly precipitously. The chlorophyll was still absorbing photons, but the system as a whole was not responding. This 'red drop' seemed to say that you needed a certain amount of energy per photon to get photosynthesis to work at all; there was a threshold below which the photons just didn't have enough oomph.

When Emerson came back to the topic in the late 1950s, though, he found that the story was a bit more complicated. This time, the algae were illuminated by two independent lights, one in the far-red, one at a shorter wavelength. When just the short-wavelength light was turned on, the algae produced oxygen, though not with immense gusto. When just the far-red light was on their enthusiasm

* Other visits, happily, went better. In the early 1950s Robin Hill came to Urbana and the two men and their families became fond friends. The friendship survived Emerson's return visit to Cambridge, where he found Hill's lab 'filthy' and would come in early to clean it up, something no one else would have dared attempt; the men's gardens were probably a greater source of harmony than their workplaces.

waned a great deal further, as the red-drop experiments predicted. But when both were turned on together, the oxygen production shot up until it was not just higher than it had been for the shorter wavelengths; it was higher than it had been for the shorter wavelengths and the longer wavelengths combined. Something was happening that was more than the sum of its parts.

This 'enhancement effect' showed that light was intervening in the process not once but twice, and that both interventions had to be matched in order for the system to run properly. In the terms Duysens was beginning to use at the same time, the short-wavelength light alone energized photosystem II and its water-splitting powers – but because it left photosystem I lagging, there was no ultimate destination for those electrons. They just sat there clogging up the electron transfer chain in a way that precluded photosystem II from working at full stretch. The addition of long-wavelength light allowed photosystem I to get up to speed, and thus let electrons flow along the whole chain.

Emerson's enhancement effect thus played its role in the final vindication of the work Emerson and Arnold did in the 1930s. Duysens' photosystems were direct descendants of the 'photosynthetic units' invoked to explain why it took so much chlorophyll to keep photosynthesis going. Duysens brought to the field's attention new physical analyses by Thomas Förster, an outstanding German theorist, which said that Gaffron had been right and Franck and Teller wrong about the ease with which energy could pass from chlorophyll to chlorophyll if the molecules were properly arrayed in the chloroplast's membrane. By the time of the Great Clarification of the early 1960s, there was no longer any question about what the chlorophyll molecules did; they funnelled the slow drizzle of sunlight into a stream of excitons intense enough to drive biochemistry at a reasonable rate. Or rather, what doubt there was was restricted to a few holdouts, like Warburg and Franck.* But by then Emerson was dead.

---

* Franck maintained a fierce opposition to the photosynthetic unit until his death, but without the animosity with which Warburg conducted his rearguard defences. Bill Arnold's daughter, Helen Arnold Herron, recalls that 'One of his favourite stories concerned an

The reaction centre

From the flashing neon Bill Arnold arranged for Emerson's chlorella onwards, the great gift that physicists brought to photosynthesis studies was their mastery over light. They understood the general principles behind the production and consumption of photons, and that understanding had a wide range of technological embodiments. The physicists who came to the field in the 1950s knew how to make photometers that counted photons one by one and spectrometers that measured the energy each photon carried. Men like Duysens, Horst Witt in Berlin, Bessel Kok, another Dutchman, and in France Pierre Joliot made their mark by analysing and controlling light with ever-greater precision while measuring the tell-tale details of the ways in which photosynthesizers absorbed it – and also, in some circumstances, produced it. They had 'monochromators' – and in years to come lasers – that could sort out photons by wavelength and flashguns that could space them out in time, filters that polarized them and photomultipliers that amplified their most etiolated effects. Bill Arnold, by this time permanently ensconced at Oak Ridge National Laboratory, claimed to believe that technology's greatest gifts to science were the photomultiplier tube and duct tape.

As well as absorbing photons, the photosynthetic apparatus also emits them; it fluoresces. When a photon hits chlorophyll molecules and raises their electrons to an excited state, that excited state can't last. The electrons feel a thermodynamic desire to fall back down, just as water wants to flow downhill and electric current to flow across a voltage difference. One way for the electrons to calm down is to pass the energy on to a neighbouring molecule in the form of an

---

argument he and James Franck had one day about an idea Bill had mentioned in a seminar. Franck was strenuously pointing out all its faults when his wife came in and scolded him for being so rough on his younger colleague. Later he asked Bill if he had been rude or hurt his feelings. Bill said no. "Well, my wife says I have to apologize. I'm sorry. Now, about this damn fool idea of yours . . ." ' (*Photosynthesis Research* 49, 3–7, 1996)

exciton, a generosity antenna complexes are designed to encourage. Another way is to dissipate the energy as heat, which is a waste, and not encouraged at all. A third way is to just spew the energy back out as a fresh photon; that's fluorescence.

In the 1930s and 1940s, fluorescence was vital to the study of how photosynthetic pigments absorbed light. Its colour made it clear that chlorophyll absorbed in the blue and red parts of the spectrum – we see leaves as green because that is the colour their pigment reflects, rather than absorbs. The colour we associate most with plants is that of the light for which they have least use.* Fluorescence, though, is quite different from reflectance. In fluorescence the light is absorbed according to the rules of quantum mechanics, and then expelled in the same way. The spectrum of fluorescent light thus reveals exactly which wavelengths a pigment can absorb, and how well. In the late 1950s and 1960s, the measuring of fluorescence went beyond the study of specific pigments and became part of the toolkit for analysing whole photosystems.

In accordance with his sense of himself as a lucky man, Arnold discovered one of the most intriguing forms of photosynthetic fluorescence quite by chance. As he told the story, he was approached by a bright young researcher, Bernard Strehler, who asked him, 'How would you like to make one of the fundamental discoveries in plant physiology?' Arnold claimed to have replied, 'OK, if it won't take too long.' Strehler had a way of using proteins from the tails of fireflies to detect the presence of ATP. The firefly proteins used ATP to produce light; that light could be picked up and amplified by a photomultiplier. Strehler's idea was to mix the firefly tail material with some isolated chloroplasts, illuminate it for a minute or so, then put it in the dark and see if it glowed. If it did, it would show that the chloroplasts were producing ATP. Strehler was excited by the possibility because at this point – in 1950, before Arnon's discovery of photophosphorylation – the received wisdom was that

* If our eyes saw more of the spectrum, we would see leaves rather differently, since they reflect very well in the infrared; this is important, because it keeps them cool.

chloroplasts relied on mitochondria for their ATP, and his experiment might prove that received wisdom wrong.

When the experiment was done, the photomultiplier showed that the test tubes containing chloroplasts and firefly extract did indeed glow in the dark. But, much to the experimenters' surprise, so did the test tubes being used as an experimental control, which had no firefly mixture in them at all. The chloroplasts were glowing all on their own.

This utterly unexpected discovery turned out to be intimately related to Arnold's original experiments with Emerson twenty years earlier, those that helped define the border between the near-instantaneous absorption of light by chlorophyll and the slower biochemical processes by which the energy from that light was made use of. Emerson and Arnold had shown that the amount of chlorophyll used by the system was large compared to its capacity for doing chemistry – that the broad physical funnel which caught the sun's energy fed into a fairly narrow biochemical gullet. If energy was coming in faster than it could be squeezed down the electron transfer chain, then sometimes the system would back up, squirting its energy back out into the antenna chlorophyll, from where it would be lost as fluorescence. Delayed light was the photosynthetic equivalent of acid reflux.

This turned out to be a powerful tool for examining the electron transfer chain. By changing the temperature it was possible to change the amount of light that was forthcoming, an effect which allowed the different stages in which the energy was digested to be disentangled – to distinguish an oesophageal belch from something in the pit of the stomach, as it were. Similarly revealing effects could be produced by applying electric fields. Biophysical examinations of delayed light became a commonplace of photosynthesis research. The only drawback to the discovery was that it meant Strehler didn't manage to see the chloroplasts making ATP. A few years later, after Daniel Arnon and his colleagues had shown that the chloroplasts could indeed photophosphorylate ADP to ATP, Strehler tried the experiment again using a filter designed to block the wavelengths of light emitted by fluorescing chlorophyll; sure enough, the dim glow of the fireflies was visible, testament to the ATP-producing powers of the chloroplasts. The serendipitous discovery had masked the sought-after one.

The next great serendipitous discovery at Oak Ridge was made by a biophysicist named Rod Clayton. Clayton had earned a doctorate in biophysics at Caltech with Max Delbrück, and then worked with van Niel, but he had never really found his place. It was at Oak Ridge, with Arnold and the study of photosynthesis, that he felt he at last began to 'feel like a biophysicist. Heretofore, all my preparation had let me be [was] a microbial physiologist who could fix his own amplifier and solve his own differential equations. Now I actually found myself bringing physics, chemistry and biology together in my work.' Clayton and Arnold followed up one of Duysens' insights into the crucial step from physics to chemistry. In purple bacteria, the changes in the way chlorophyll absorbed light at the beginning of photosynthesis showed that some of the chlorophyll* was becoming oxidized – giving up electrons. In an echo of the original Emerson and Arnold experiment, Clayton and Arnold showed that this oxidation was a purely physical effect. They chilled fragments of dried-out bacteria down to just one degree above absolute zero (minus 272°C), a point at which the lack of heat will stop any chemical reaction imaginable, and still the oxidation took place. The oxidation was driven by a photon or an exciton – by physics, not by chemistry.

The changes in absorbance showed that only a small proportion of the chlorophyll was losing electrons. Duysens' interpretation of this was that some of the chlorophyll was, by dint of a particular place in the photosystem, special. Most of the chlorophyll molecules associated with a photosystem would be much of a muchness, and an exciton would be able to bounce from one to another quite freely. Eventually, though, it would reach one of the special chlorophylls,

---

* It's worth pointing out that there are various different sorts of chlorophyll molecules; Clayton's purple bacteria use one called bacteriochlorophyll-b. For simplicity, though, I'm going to use the generic term of chlorophyll except on any occasion where it might be genuinely misleading.

which, because of some subtlety in its position, was particularly prone to losing electrons. The energy from the exciton would then push out an electron, and the biochemical transfer of energy would begin. The special chlorophyll would thus act as a one-way trap that excitons could bounce into easily, but only get back out of with some difficulty – as when, for example, the biochemistry backed up and the chlorophyll started to fluoresce.

Clayton liked the idea that special chlorophylls acted as traps, and with a stroke of luck that would have done Bill Arnold proud he soon found more evidence for it. Coming back to the lab after a two-week vacation, Clayton found that some of his purple bacteria cultures, left in the light while he'd been away, had changed colour from slate-blue to pink. The reason for this 'senescent transvestism', as he called it, was that the chlorophyll they used to absorb the light had lost its magnesium and thus been transformed into a different pigment, called pheophytin, which could not gather up photons in the same way. And yet, under the right conditions, the no-longer-purple bacteria still showed the spectral signature of oxidized chlorophyll that Clayton and Arnold had been investigating. A small amount of chlorophyll was still present, and that remnant was still capable of being oxidized by light and carrying out some photosynthesis. The fact that photosynthesis could still go on this way even when almost all the chlorophyll had been turned into apparently useless pheophytin was the best indication Clayton had seen for the idea there were indeed a few special chlorophylls, tucked away in some central, protected position in the photosynthetic unit, that served as specific traps for the exciton energy, and which could continue to do so even when the antenna chlorophyll was all gone.

Clayton kept studying the transformed bacteria as he moved on from lab to lab over the following years. But it was only after he settled at Cornell as a professor in 1967 that he finally managed to follow up his discovery in the most practical of ways, by using detergents to prise the special chlorophylls and the proteins holding them out of the membrane in which they were normally embedded. This isolated fragment – a few proteins, half a dozen pigments, a couple of small molecules called quinones that could act as shuttles in an electron transfer chain, and some clinging fat from the membrane – was, remarkably, still quite capable of photosynthesis. Stripped of almost all their antenna chlorophyll, they didn't photosynthesize very fast; the mouth of the funnel feeding them energy was not wide. But that same lack meant the stripped-down systems were far easier to study with spectrometers: more signal, less noise. Clayton called them 'reaction centres'.

The discovery of the reaction centre, it could be argued, was what first brought the study of photosynthesis truly into the realm of molecular biology. Reaction centres were just the sort of thing that molecular biology was originally meant to study; the embodiment of a fundamental life process understandable only at the molecular level. Which makes it rather fitting that it was Roderick Clayton who created them, because to look at Clayton's life is to see someone who might have been a key part of the genesis of molecular biology, but went off in a different direction.

Like many of the men who helped to found the field – perhaps most notably Francis Crick – he was a physics student who migrated towards the study of life. And he had the good fortune to do so in a place that was to be a hotbed of molecular biology, Caltech, where new equipment such as ultracentrifuges – the machines with which Clayton's contemporary Sam Wildman first began the study of rubisco, the key protein in the Calvin-Benson cycle – and X-ray diffraction apparatus were allowing proteins to be studied in new ways. On the theoretical front, the great Linus Pauling, Caltech's presiding chemical genius, was working out the patterns into which parts of proteins might fold themselves.

Not only did Clayton have the right interests in the right place to become a molecular biologist, he also had exactly the right mentor. His graduate supervisor was Max Delbrück, one of the founders of the field. Another of the Göttingen physicists, in the 1930s Delbrück was inspired by a lecture on what physics might have to offer the science of life by Niels Bohr, who suggested that there might be an 'elementary fact of biology' analogous to the principle of the quantum, the idea that momentum came in discrete chunks, which Bohr had made the elementary fact of all physics. Delbrück came to harbour a particular interest in photosynthesis; when they were all in Germany, he held meetings on the subject with Hans Gaffron and Eugene Rabinowitch, and it appears he might well have started research in the field had he not known that to do so would be to be forced into confrontation with the obstreperous Otto Warburg. Instead, he tackled genetics, creating mutant fruit flies with X-ray bombardments in a way that let him make the first estimates of the physical size of a gene.*

Delbrück's work on the physical nature of the gene was the direct inspiration for parts of the book *What is Life?* by the great quantum physicist Erwin Schrödinger. Published in 1944, the book divided its eponymous question into two puzzles: the creation of order from disorder; and the creation of order from slightly different order. The first Schrödinger treated as a question about thermodynamics, which he answered in terms that Boltzmann would have approved of: life was about producing entropy in the environment and the capacity for order, 'negative entropy', within its own confines. The influence of Boltzmann here was direct and all but personal; Schrödinger studied with Friedrich Hasenöhrl, who had himself studied with Boltzmann. The second question, the creation of new order from similar order, Schrödinger treated as a question about genetics, suggesting that further physical research like Delbrück's would crack it, and perhaps lead to new kinds of physics as it did so. Schrödinger's book was received sniffily by some traditional biologists, and indeed by some of the people who went on to become molecular biologists, but it undoubtedly inspired many young physicists to consider studying life. It was mind-expanding to learn that a great quantum physicist was even asking such questions.

---

* If one wants to muse about paths not taken and alternative histories of science, it seems to me that this is a particularly interesting decision to consider. If Delbrück had focused on photosynthesis and energy rather than genes and information, how different might the eventual development of molecular biology have been, and how different the course of research into photosynthesis?

By the time *What is Life?* was published and Rod Clayton was his graduate student, Delbrück's experimental work was largely taken up with viruses – 'phage' – that prey on bacteria, a field that would be crucial in molecular biology's further development. But this molecular biology was not the work in which he engaged Clayton. Nor did he lead him into photosynthesis work; Delbrück maintained a personal interest in the questions of light and life to which Bohr had opened his eyes, but he did not pursue research in photosynthesis, and Clayton did not really enter the field until he reached Oak Ridge and Bill Arnold. Instead Clayton was set to studying the ways in which purple bacteria seek light out and home in on chemicals that they need. And yet, his roundabout route led him to something which Delbrück had always sought – a 'hydrogen atom' for biology.

One of the things that Delbrück had taken from Bohr was the need to find a simple system in which the fundamental aspects of life were laid out and easily studied, something analogous to the one-proton one-electron hydrogen atom in physics. That is what led to the phage work: phages were the simplest transmitters of biological information. And that is what the reaction centres represented: the simplest, most tractable systems for the study of photosynthesis, the key to life's energy at the molecular level.

George Feher, a professor of physics at the University of California, San Diego, was one of those who recognized Clayton's reaction centres for the metaphorical hydrogen atoms they were. The career which led Feher to the reaction centre was, like Clayton's, a roundabout one, but it was rather more crammed with incident. In 1941, at the age of seventeen, he left Nazi-occupied Slovakia and made his way overland to Palestine. An inveterate electrical tinkerer and radio ham, he started working for a professor of engineering at the Technion, the technical university in Haifa. One of his tasks was to make an oscilloscope – the sort of display that traces heartbeats in medical dramas – out of scrap. He designed it so the blips moved across the screen from right to left, wiring up the first oscilloscope made in Palestine to be read like a Hebrew text. A Hebrew oscilloscope, though, did not make up for the failed Bible exam that stopped him from entering the Technion as a student. The underground was less fussy, putting him to work developing a system to tap and unscramble the phone line that connected Britain's high commissioner, the governor of the British mandate of Palestine, with Downing Street.*

Feher managed to secure a place at Berkeley, paying for his passage with the proceeds from another electrical sideline, in which he used his skills at growing crystals to make microphones for clubs; demand for entertainment was high, the requisites for its amplification in short supply. The money got him to Berkeley, but didn't pay for much once he arrived – he and his roommate used to eke out their meagre diets by cooking up the legs of the frogs used in physiology classes (things improved once the students moved on to rabbits). In the summer he worked as a fruit-picker, hard work with a culture shock attached. 'Whereas in the kibbutz we discussed ideological issues like the suffering proletariat, while picking fruit at a relatively leisurely pace,' he later wrote, 'here I encountered the suffering proletariat not having time to discuss anything, trying to pick as much fruit as possible since we were paid by the box.'

In Palestine Feher had read Schrödinger's *What is Life?* and been inspired to become a biophysicist; for a man living by growing crystals, Schrödinger's order-from-order idea about life depending on genetic reproductions that were quasi-crystalline was a powerful one. Visas, however, were only granted for more practical avocations, and at Berkeley he studied engineering physics, going on after his doctorate to a research position at AT&T's Bell Labs. It was in its

---

* Another of Feher's underground projects has an intriguing resonance to Emerson's and Arnold's original work with fifty-cycle-a-second neon lamps. Because such lamps flicker on and off, if looked at through a telescope with a shutter set to the same frequency they can look completely dark. Feher and a friend hit upon the idea of changing the frequency on neon signs as a way of transmitting Morse code to people watching with shutter-equipped telescopes, the flickering lights moving in and out of phase with the flickering shutters so as to provide the dots and dashes in a way to which the naked eye was oblivious. The neon signs were installed as Stars of David on the roofs of hospitals and formed a covert communications network throughout the British mandate.

1950s heyday, with semiconductors and solar cells and masers (the precursors of lasers) all under development; Feher worked on new forms of spectroscopy that picked out patterns in the faint radiation given off by wobbling electrons and nuclei within atoms. A 1957 paper he published in Bell's house journal on how best to build these electron paramagnetic resonance (EPR) spectrometers remains, to this day, the most cited of his publications.

In 1960 Feher moved back to the west coast, attracted by Roger Revelle. Revelle had convinced the regents of the University of California to set up a new campus next to his fiefdom at the Scripps Institute of Oceanography: UC La Jolla, soon thereafter to become UC San Diego. He wanted a department that excelled at the sort of physics Bell Labs did so well, the 'solid-state' physics of crystals, semiconductors and the like, and so he courted Feher, outlining the glories of the new campus and showing him a beautiful lot facing the ocean on which he might build a house. (Many recruits were shown and sort-of-promised the same cliff-top lot, Feher recalls; none of them got it.)

Sold on the idea (and turning down a prestigious professorship at Columbia) Feher nevertheless extracted a condition. After he helped set up the solid-state physics shop, he would be allowed to transfer his energies to biophysics; the enthusiasm he had picked up from reading Schrödinger had never left him. In the end, Revelle wasn't able to make good on the deal; he had trodden on too many toes setting up the new campus, and control of it was soon wrested from him. But the establishment he created at UCSD went on to thrive, part of the anticipation of the future that braced La Jolla like the scent of sea salt and eucalyptus hanging on the wind. Behind the university's cliff-top mesa, dreamers at General Atomics were designing nuclear spaceships to fly crews of scientists through Saturn's rings, in thrall to a vision of the space-age that was not to be; at Scripps, down by the shore, Dave Keeling was tabulating the world's carbon-dioxide levels, sensing the breath of the biosphere and setting the twenty-first century's earth-bound agenda.

Harold Urey, the great stable-isotope chemist, now enraptured by the study of the moon and theories of the origin of life, joined the chemistry faculty. So did Martin Kamen, and though not in Revelle's class as a salesman, he made it his business to interest as many of his colleagues as possible in photosynthesis. Kamen had a clever diagram he liked to show at seminars, a timeline that showed the different stages of photosynthesis, from the arrival of the photon to the growth of cells, along with their accompanying scientific special-ities – radiation physics for the light, solid-state physics for the antennae, physical chemistry for the first oxidation, biophysics for the electron transfer chain, biochemistry for the Calvin-Benson cycle, physiology to regulate the biochemistry, botany to put it all into the context of the plant, ecology to match the sun-eating plants to their environment. Kamen would point out the stage in the process where the current 'level of ignorance' peaked, and where there were thus large unanswered questions, promising the audience there were 'tickets to Stockholm' for whoever answered them. Where he actually placed the Nobel-worthy peak in ignorance, of course, depended on the audience he was talking to.

Feher saw through the ruse – among his other skills he is a highly accomplished poker player* – but didn't discount the enthusiasm; Kamen's picture of a biophysics topic shot through with interdisci-plinary connections and with unsolved physical problems at its core appealed to him. In 1967, finally able to make the disciplinary shift Revelle had promised him, he took a summer course in microbiology at Woods Hole, the Massachusetts counterpart to the Scripps Insti-tute of Oceanography, which was taught by Rod Clayton; on that course he was one of the first to hear about isolated reaction centres. Over the next decade Feher's group at UCSD, expert in new spectro-scopic techniques such as the EPR he had pioneered, led the world in the dissection of the purple bacteria's reaction centre.

Clayton's original reaction centres were soon improved by the use of a subtler detergent for breaking up the membranes the proteins

---

* Third place in the America's Cup National Poker Championship, 1992; his preferred game is seven-card stud high/low.

were normally encased in. The newly purified complexes were just a tenth of the size of the first ones, an arrangement of fewer than 10,000 atoms just thirteen billionths of a metre across, yet still fully functional. They were found to consist of three proteins, called L, M and H for light, medium and heavy. Associated with these proteins there were four chlorophyll molecules, two pheophytin molecules, two quinones and an iron atom. By comparing the way that the EPR spectrum of the reaction centres changed when they were illuminated to the spectra of oxidized and reduced pigments and quinones, Feher and his colleagues showed that when an exciton reached the reaction centre it dislodged an electron from a 'special pair' of chlorophylls situated very close to each other, which trapped its energy rather as Duysens and Clayton had expected. That electron jumped to one of the other chlorophylls, then to one of the pheophytins, then to one of the quinones, then to the second quinone.

In living bacteria, the reaction centre and the cytochromes it deals with are embedded in a layer of fatty membrane. The electron lost from the special pair is replaced by one from a nearby cytochrome. The reduced quinone where the electron ends up – fully reduced only when light has sent a second electron to join the first – takes on board hydrogen ions to balance its electrons, and in so doing is released from the reaction centre. The reduced quinone then travels along the membrane to another cytochrome. There the quinone gives up its electrons, and as part of the process also loses the hydrogens it had picked up back at the reaction centre. The clever thing is that the cytochrome is set up so as to eject the hydrogen ions on the opposite side of the membrane from the one on which they were picked up. This is how the system moves hydrogen ions across the membrane and builds up the contrast in concentrations that drives the chemiosmotic production of ATP. The electrons, meanwhile, end up in a part of the cytochrome complex that subsequently detaches itself and floats on through the membrane until it meets another reaction centre. There it will yield up the electrons so that they can reduce the special pair of chlorophylls when next light oxidizes it. And so the cycle goes on.

By the early 1980s, Feher and his team had broken down the sequence of tiny movements in the reaction centre to an almost disconcerting degree of precision. It took just 200 picoseconds (trillionths of a second) for an electron to pass from the pheophytin to the quinone while the transfer from the special pair to the pheophytin took just a couple of picoseconds. The technology for measuring the effects of light in the borderlands between physics and chemistry had come a long way in the fifty years since Emerson's and Arnold's first millisecond flashes. But though the flow had been exquisitely dissected in time, in space it was still a little mysterious. There was no map of the route the electrons took to match against the timetable of their journey; the shape of the system was largely a mystery.

The 46,630 atoms of photosystem II

In September 1985 Jim Barber was giving a talk to the new undergraduates at Imperial College. Among the wonders he showed them in his slides was a complicated and incomprehensible squiggle which some of the freshers, in a sweet surfeit of either eagerness or panic, tried to copy down freehand. They were among the first people in the world to see it, Barber told them. And, he went on, it was going to win a Nobel prize.

That squiggle was the structure of the bacterial reaction centre, laid out atom by atom by means of X-ray crystallography. Molecular biology offers two ways of thinking about a protein – as an ordered list, or as a solid object. The list is as a sequence of component parts called amino acids. Each of these components is stuck to the end of the one in front so as to make a chain hundreds of links long, like one of those necklaces of snap-together beads for little children. But when these chains are put together forces between the side chains on the various beads, and between the beads and nearby water molecules, automatically fold them up into a complex three-dimensional shape defined by the sequence of the beads. It is the shape of this tangle that gives the protein its powers. The shape gives it pockets for holding chlorophylls, or haem groups, or quinones; two proteins that need to stick together will have complementary shapes, like an acorn and its cup, or a couple spooning. Slight changes in the shape can catalyse reactions, bringing together two small molecules – phosphate and the precursor to ATP, for example – in such a way that they react with one another.

The shape is determined by the sequence in which amino acids with different properties are added to the chain; that sequence, in turn, is determined by the DNA that records the relevant genetic information. But, until recently, there was no way to use that genetic information to predict in detail the shape into which the chain would fold up. The shape could only be ascertained by looking, and X-rays – photons with far more energy than light, and thus with wavelengths so short they can interact with the individual atoms within a molecule – are the way to look. Put a crystal in front of a stream of X-rays and the atoms locked in the crystal's regular and repetitive interior structure will spread the X-rays out into a complex pattern unique to the molecules of which that crystal is made.

In the early twentieth century X-ray diffraction was used to work out the structure of simple substances such as diamond and table salt. Victor Goldschmidt, the brilliant geochemist for whose sake Willstätter resigned his post in Munich, used the technique to reveal the structure of minerals in the earth's crust. In the 1920s and 1930s Desmond Bernal, a remarkable Irish communist polymath known almost universally as 'Sage',* took up the application of X-ray crystallography to biological problems. Bernal was interested in everything, but the application of X-ray crystallography was first among equals in the throng of his fascinations. The students to whom he communicated this fascination, Dorothy Crowfoot (later Hodgkin)

* In the 1960s Robin Hill wrote in an article, 'Not so long ago, when he was still with us, the Sage asked me "who said why is grass green and blood red?"' Neither Hill nor the editors of the *Annual Review of Plant Physiology* felt their readers would need to be told to whom he was referring.

at Oxford and Max Perutz in Cambridge, made the technique practical. Perutz, yet another of those to benefit from the kindness of David Keilin, worked with John Kendrew, a chemist who had become enthused about X-rays in wartime Ceylon, where he and Bernal were both advisors to Lord Mountbatten. In the 1950s they worked out the structures of haemoglobin and its relative myoglobin to within a couple of angstroms – an angstrom being a ten-billionth of a metre – which is the scale at which you can assign a place in the structure to more or less every single atom in a protein. Kendrew and Perutz thus earned themselves two tickets to Stockholm.

By the 1970s, when George Feher's group, as well as those at other labs, such as Rod Clayton's, were dissecting the workings of the reaction centre, X-ray crystallography had provided a hundred structures with a resolution of a couple of angstroms. Feher, keen on growing crystals from his boyhood radio-hamming onwards, started to learn the technique in order to ascertain the atomic structure of the reaction centres whose temporal structure he had dissected with such exactitude. But there was a catch. Some proteins were a lot more easily crystallized than others. And the proteins most resistant to crystallization were those which, like those of the reaction centres, had first to be extracted from membranes. Such proteins tend to have scraps of membrane or detergent stuck to them that inhibit the formation of perfect crystals. When Feher applied for a grant to work on crystallizing reaction centres, it was turned down by a reviewer at the National Institutes of Health as 'ill-conceived and futile'. Feher was not to be thwarted. He stuck an Escher print of tessellated wildfowl to his wall and told his team that if Escher could crystallize ducks, they could crystallize reaction centres.

In fact, though, they were pipped to the post. In the mainstream of biology the membrane protein that attracted the most attention was one called bacteriorhodopsin; this pumps protons across membranes in some archaea, and played a key part in the continuing controversies over Peter Mitchell's chemiosmotic theory. A German molecular biologist named Hartmut Michel spent some years trying to crystallize bacteriorhodopsin, and getting called crazy for his pains. In frustration he decided to try his particular mojo on the purple bacteria's photosynthetic reaction centre proteins instead, and it worked; Feher and his colleagues used similar techniques to get their reaction centres to crystallize a year or so later, but the German team was now ahead. In 1985, working with X-ray crystallography specialists at the Max Planck Institute in Martinsried, Michel submitted a full structure of the reaction centre to *Nature*. It was sent out to reviewers, one of whom was Jim Barber; he, in his excitement, showed it to that hall full of first-years.

To practised eyes – and making sense of such structures, which are hard to represent on a slide or a viewgraph, takes practice – the structure was revealing. The L and M proteins were similar in shape and symmetrically arranged. From the side, looking along the plane of the membrane, you can visualize them as two tired sisters dancing, their heads slumped on each other's shoulders. Their ankles and feet protrude from one side of the membrane, the backs of their heads from the other. Seen end on, looking down on the membrane, they are more like two hands with their fingers curled, hooked together in the sort of handclasp bodybuilders use when showing off their arm muscles. The special pair of chlorophylls, which gives up electrons, was held within that grasp. The path of the electrons was immediately quite clear. As had been expected, they moved in a direction perpendicular to the plane of the protein. Leaping from the special pair of chlorophylls at about knee-height to a pheophytin at the level of the L protein's waist, they then travelled up to the 'head' of the M protein, resting on the L's shoulder, and finally across to the L protein's head, which contains the site where the final quinone is reduced and then released.

The reason that Barber was so excited about this, and convinced that it would win Michel and his colleagues tickets to Stockholm, was that he saw its significance for the structure of photosystem II, the 'inner sanctum' of plant photosynthesis. It seemed likely that photosystem II and the purple bacterial reaction centre were related in some way. But few people appreciated, or wanted to appreciate,

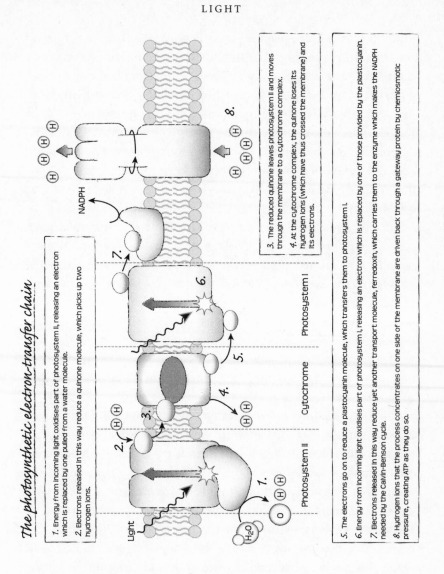

The photosynthetic electron-transfer chain

1. Energy from incoming light oxidises part of photosystem II, releasing an electron which is replaced by one pulled from a water molecule.

2. Electrons released in this way reduce a quinone molecule, which picks up two hydrogen ions.

3. The reduced quinone leaves photosystem II and moves through the membrane to a cytochrome complex.

4. At the cytochrome complex, the quinone loses its hydrogen ions (which have thus crossed the membrane) and its electrons.

5. The electrons go on to reduce a plastocyanin molecule, which transfers them to photosystem I.

6. Energy from incoming light oxidises part of photosystem I, releasing an electron which is replaced by one of those provided by the plastocyanin.

7. Electrons released in this way reduce yet another transport molecule, ferredoxin, which carries them to the enzyme which makes the NADPH needed by the Calvin-Benson cycle.

8. Hydrogen ions that the process concentrates on one side of the membrane are driven back through a gateway protein by chemiosmotic pressure, creating ATP as they do so.

how close they really were. As Barber told me during an afternoon of sifting through old papers and reprints as he prepared to move office in 2004, 'There was a mood that because photosystem II split water it was very different. There was no reason why it would be like a bacterial reaction centre.'

One of the people holding out against that mood was Bill

131

Rutherford, the blues-playing spectroscopist, who knew a very good reason why photosystem II should look like a bacterial reaction centre. As a student in the early 1970s, Rutherford had come across a startling idea being put forward by Lynn Margulis, an American biologist. In the mid-1960s Margulis had become convinced by various lines of evidence, some of them first pointed to decades before, that the chloroplasts responsible for photosynthesis in plants, and also the mitochondria responsible for respiration, were directly descended from bacteria that had colonized complex, eukaryotic cells billions of years ago – that the maintenance of internal symbionts, or endosymbiosis, was the key to what made eukaryotes special. Repeatedly rejected by scientific journals, she promulgated her ideas in mimeographs sent to people she thought might be interested: most weren't, though Bernal was instantly convinced, and congratulated her warmly. She finally had her ideas published in 1967, in a paper now widely seen as a classic. Within a decade, the idea that chloroplasts were descended from bacteria in this way was widely accepted. It was an idea that had been around in less developed form for much of the century, but it was Margulis who convinced the world of it.

But there is a difference between accepting something and having it form part of the heart of your view of the world. For most of those already established in the field, endosymbiosis was moderately interesting but hardly important.* To Rutherford, it was one of the things that first really turned him on to science, and left him with a profound belief that understanding evolutionary stories mattered. Looking back he now says that whenever he has been right and others have been wrong – and it has happened quite a lot in his career – it has been because he takes evolution seriously. He always tries to let ideas about how something might have evolved, and from what, shape his hypotheses about what is going on at the other end

---

* An interesting exception was Bob Whatley, who became very interested in how endosymbiosis might actually have come about. The man who so successfully extracted chloroplasts from plant cells in Arnon's lab put a lot of effort, later on, into working out how the chloroplasts got there in the first place.

of the spectrometer. You might think that this is a commonplace in biology, but it is not; many biophysicists and biochemists give curiously little thought to evolution, seeing life as a puzzle without a past, basing their models on aesthetics or intuitions – both of which can be powerful tools – or just on the first idea to pop into their heads, which is less reliable.

In the early 1980s, as a post-doc at Urbana, Rutherford mounted a series of spectroscopic investigations aimed at showing the similarities between green plants and the much more easily probed reaction centres. The experiments, though, were all done on his own time; the idea that complex plants relied on something just like the bacterial reaction centre for their energy was simply not a popular one to pursue. Rutherford worked without colleagues in the middle of the night; his dance of triumph when the crucial data came in was done in the dark, on his own.

Barber was one of those thinking along similar lines. In the 1980s, he started to bring the tools of modern molecular biology into his lab in order to study the genes responsible for photosynthesis. Though physics, as we have seen, provided molecular biology's first tools, by this stage biology was becoming the toolkit for its own study. The proteins that copy genes were turned into systems for determining the information within them. Antibodies were made in the lab as a way of picking out particular proteins by means of their idiosyncratic shape. Genetic scissors allowed researchers to pick up a gene from one organism and move it to another for more careful study to see what the protein it described might do there.

One of the facts that had led Margulis to expound endosymbiosis was that chloroplasts had their own genes, stored on DNA molecules kept separate from those in the nucleus. Barber, like many others, was interested in the sequences of those genes. Comparing the gene sequences from plants to those from bacteria that were produced in Feher's lab and others showed some striking similarities. The sequences of the L and M proteins looked like those of a protein called D1 that was found in photosystem II and a mysterious near-doppelgänger dubbed D2. D1 had not, until that time, been seen as a particularly important protein, though it was of some interest because various herbicides seemed to bind to it, and it could be induced to break down quite easily. D2's function was unknown – it wasn't even accepted as part of photosystem II. But with the structure of the reaction centre to act as a template, with the genetic sequences that showed which chloroplast protein was analogous to which bacterial one, and with antibodies that showed, in a rough sense, which proteins were where, previous spectroscopy-based ideas about the heart of photosystem II were overturned, and Rutherford's solitary midnight dance was justified. The electrons in plants run up D1 in much the same way as the electrons in purple bacteria run up the reaction centre's L; D2 slumps supportively against its more active partner in much the same way that M embraces L.

As the first structure of a membrane protein – in this case a group of membrane proteins – the bacterial reaction centre would probably have been deemed worthy of a Nobel whatever the proteins did. The fact that the structure threw light on one of the fundamental processes of life on earth and the foundation of all our agriculture clinched the deal. The prize committee agreed; three years on, just after the students Barber bemused graduated, Michel, Johann Deisenhofer and Robert Huber had their tickets to Stockholm.

In the following decade Barber devoted more and more of his time to aspects of photosystem II; finally, in 1995, he made the decision to drop everything else and focus his lab on solving the three-dimensional structure of the entire photosystem – a far more complex undertaking than just studying a stripped-down reaction centre. It would involve laying out in detail the structures and relationships between hundreds of molecules, almost all embedded in a membrane. Knowing this would require someone brilliant at crystallizing proteins which straddle membranes – still something of a dark art – Barber used up favours and trod on toes to open up a professorship at Imperial for So Iwata, a crystallographer then working in Sweden who seemed to have a peculiar knack for growing crystals of membrane proteins. He and his colleagues worked on new ways of isolating whole photosystems and studying them with electron microscopes. They grew films in which photosystem II particles were jammed together, two-dimensional crystals which could be used to scatter electrons as three-dimensional crystals are used to scatter X-rays. And they did it all with the urgency of a team knowing there was someone else ahead of them.

The quarry was Horst Witt, a German physicist of Duysens' generation. Witt had come into photosynthesis studies in the 1950s and had stayed at the forefront of the field ever since. He made his name with a technique that used lots of short flashes of light, thousands to millions of times shorter than Bill Arnold's back at Caltech, to pick up subtle spectroscopic effects. Photosystem II, with its ability to produce oxygen, exerted a particular fascination on him. By the 1980s he, too, was embracing molecular biology in a systematic and well-equipped way. Barber remembers Witt's talks opening with expositions of the Berlin lab's equipment impressive enough to make would-be competitors want to go home and give up.

By 2000, the Berlin team had an X-ray structure of photosystem II accurate to about four angstroms – that is, good enough for some features, but not really precise about where all the atoms are. Jim Barber watched the presentation with a sour feeling in his stomach. It was satisfying that the Berlin structure confirmed some of what he had seen in his lower-resolution electron-crystallography work – specifically, that photosystem II units come in pairs, or 'dimers', each with two D1s and two D2s – but it looked as though, in the end, he was going to be an also-ran. He didn't give up, though. And while the Witt group was used to taking its time, Barber was happy to hurry.

In 2003, the Imperial team took its best crystals to one of the most impressive of all the light sources that physics has yet provided for the study of photosynthesis (or, in this case, of the molecules that make it possible). The European Synchrotron Radiation Facility (ESRF) in Grenoble, like similar facilities in America and Japan, is descended from the cyclotrons that Ernest Lawrence developed for the Rad Lab in the 1930s. But it is the size of a large football stadium. It also works on a different principle. Whereas the point of the Rad

Lab cyclotrons was to put energy into the beam of particles circling within, at ESRF and similar X-ray facilities the idea is to siphon the energy off such a beam. As the stream of electrons is pumped round and round the circuit it gives off X-rays of peculiar radiance and purity like sparks from a Catherine-wheel; it is those X-rays, not the electrons themselves, that the scientists at ring-side make use off. Such synchrotrons are like race-car circuits at which the chequered flag is of no concern but the constant scream of tortured tyres means everything. They generate X-rays about a trillion times more brilliant than those with which Max Perutz and John Kendrew produced the structures of the first small proteins half a century ago.*

In Grenoble, Barber and his colleagues amassed a huge data set from its crystals; back in London, they needed to make sense of it. X-ray diffraction produces a map of the electron density in a crystal, which shows where the atoms all are; but working out a consistent story about which atom fits where is hard work. The Imperial researchers had to thread the amino acid sequences of the proteins they knew to be involved through the electron-density map to try and make sense of all those atomic co-ordinates. When this process works properly, it pulls the data and their interpretation ever more firmly together. At Imperial, it worked well. They produced a 3.5-angstrom structure. It was not all that much more accurate than the Berlin structure, but the edge in resolution was enough to assign a position and an orientation to pretty much every amino acid, chlorophyll molecule and other pigment the system had to offer – nineteen proteins, thirty-six chlorophylls, and various other minor components. This was the structure Barber presented at the 2004 congress in Montreal.

---

* Brilliance in this usage is a technical measure of the rate at which photons are emitted within a beam of a certain cross-section and in a certain energy range. On average, the brilliance of the best X-ray sources available to scientists has been doubling every fifteen months for fifty years. There's less fuss made about it, but X-ray sources get better appreciably faster than microchips, the poster boys for runaway progress. And they've been on the curve for longer.

The part of it in which he takes most pride is the hitherto unseen detail it provides on the little cluster of manganese atoms attached to the D1 protein which actually rips water molecules apart. This 'oxygen evolving complex', it appears, grabs hold of two water molecules, pulls two hydrogen ions (and thus two electrons) from one, and then uses the extraordinarily electron-hungry oxygen atom remaining to pull the hydrogen ions and electrons from the second water molecule, creating a two-atom oxygen molecule in the process. All this is done with a little cluster of manganese ions positioned just so, their redox potential rising and falling in such a way as to pull the water associated with them apart. Understanding how this works could provide the basis of new technologies for splitting water into hydrogen and oxygen using just sunlight: we'll return to those possibilities in the last part of this book.

Already, other groups, as well as Barber's own, are working on surpassing the structure. While 3.5 angstroms is good, a really convincing all-46,630-atoms-in-their-rightful-place structure should probably be accurate to about two angstroms. That will show whether the slightly out-of-focus-looking parts of the Barber structure are right or wrong. Work subsequent to Barber's has suggested that X-rays of the power needed to produce his structure necessarily scramble the picture of highly redox-sensitive parts of proteins, such as the manganese structure in the oxygen evolving complex. But if Barber's structure proves to have been right, then it's possible that he, too, will win a ticket to Stockholm. Like many scientists, Barber is very interested in stories about who the Royal Swedish Academy and its Nobel committee has and hasn't honoured for what. One of the reasons he was encouraged into the physical side of photosynthesis research in the mid-1960s was a belief held by his supervisors that now Melvin Calvin had won a Nobel for the chemistry of photosynthesis, a Nobel for the physics was obviously a possibility. Again like many others, Barber is becomingly reluctant to suggest that he should win, or deserves to win, a modesty that I suspect, in Barber's case, may reflect a nagging insecurity. However, he does think the photosystem II structure is important – that it is the sort of thing that wins prizes. And when the bacterial reaction centre won its prize in the 1980s, many people said a further prize would be in order when the structure of the full photosystem was produced. It may be that Barber lives in hope.*

After the Montreal meeting was over I had the pleasure of an afternoon with a friend who has taken to dividing his time between public-health policy – the field through which we met, at an AIDS conference in Montreal which more than filled the vast convention centre where the cream of the world's photosynthesis scholarship had been accommodated in a small corner – and sculpture. In a converted warehouse he assembles art from twisted pieces of thick wire and rebar that he finds in railway yards, around construction sites, abandoned in gutters. Anchored in brute stone or rough wood, it's art that speaks of both the primitive and the industrial, of the accidental and the intentional; as is the way with found art, the pieces have a meaning that stretches beyond what they are but at the same time is drawn from the fact that, in other circumstances, they would have no meaning at all.

They're not primarily representational, though they're allusive, and they're certainly not intended as comments on protein structure – I'd be quite surprised if anyone else has ever thought of them as such. But after a week of seeing computer-rendered complexities of photosystem II and its like on the screens of lecture halls, they were hard to see as anything else. Structures like Barber's are, in their own way, found art, ways of taking the utterly everyday – if never before observed – and making it visible in a new context and to a new purpose. But they do so in a forbidding way, their visual expression caught between stressing the contorted string of amino acids that makes up the protein's continuous core and the solid volume which the folding of that string defines.†

* Jim Barber's an informal man, but when he gave his plenary in Montreal it was in a blazer and – a rarity, this – a tie. Later, over coffee, he admitted with a grin that the tie, now loosened round his collar, was the one he has by dint of being a foreign member of the Royal Swedish Academy of Sciences. He sees it as his 'lucky tie'.
† Later, on the train heading south from Montreal, it occurred to me that the challenges of this book can be seen in the same sort of way. Like a protein, this book can only fill the

If only, I thought, such representations of proteins could catch the same simple, unmediated substance that this art has, if only they could capture this sense of being things in the world and ideas at the same time. Proteins are, after all, intrinsically sculptural. They are shapes that translate an action into a volume, that give verbs bodies. And they do so in a purer form than any art. The shape of photosystem II doesn't just represent the process by which light is brought into the service of life. The shape is that process; there is nothing to the process but that shape. Make something that shape – as living creatures have been doing for the best part of three billion years, maybe more – and it will do what it does.

Physicists went looking for the secret of life, for a new starting assumption that would allow biology to be rebuilt as physics had been. Instead they found that, down among the molecules, there was no secret. Just atoms in shapes. Simple shapes like the double helix; complex ones like photosystem II.

The exquisite thrill

Late in his life, James Franck was expounding one of his by then entirely outdated theories to Roderick Clayton in a hotel room in Florida. At the end of the exposition, Franck said to the younger man, 'I know I do not have much longer to live, and that in a year or so this will all be nonsense, but just at this moment I have the exquisite thrill of knowing *exactly* how it all works!'

No one knows exactly how it all works. How, for example, does

---

complex space it needs to define with a string – in this case a sequence of words, sentences and stories. To try to do justice to the overall shape of photosynthesis the string curls up on itself, doubles back, folds over, curves round, zigs and zags. Just as amino acids that lie far from each other on the chain may almost touch each other when it is folded up, so various concepts in the book crop up repeatedly, seen from different angles, at very different times. In the end it is the shape that matters. But the string is all I have to work with.

photosystem II live with the problem that the oxygen it produces has a savage tendency to attack the D1 protein that channels the flow of electrons through the membrane? In a well-lit leaf the D1 protein in any given photosystem II will last not much longer than half an hour before being poisoned by the oxygen the photosystem produces – at which point the photosystem opens itself up, disgorges the maimed D1 at its core, assimilates a freshly made replacement, slips itself back into fighting form and gets back to feeding on photons. How do mindless molecules held together by electrostatic forces do something like that? And how do the systems they are part of regulate themselves – how do photosystem cores take on and pass around the light-harvesting complexes that feed them extra excitons? How do the quinones shuttling around the membranes get shared out properly? How do the two photosystems synchronize their work? How can the chloroplasts know which genes need turning on when? How different are the answers to all these questions in different species of plant and algae? All questions asked at Montreal: all areas of research where the detailed structure of photosystem II may be a help, but nothing like a final answer.

I talked to a number of younger researchers in Montreal, including the proud winner of the Robin Hill medal; none of them thought the field was going to wind up and go away for want of more questions to answer. If molecular biology ends up sucked dry of sustenance, some other rising science will be along for the noble parasites to feed on soon enough.

And yet there are times that feel like endings, times when the feeling of knowing it all, although an illusion, has a certain justification. You don't run out of questions – if you have a scientist's mind you never run out of questions – but there do come times when fields of enquiry get wrapped up and retired. The Great Clarification in the early 1960s was such a time; after Hill and Duysens and Mitchell and Arnon and Emerson had made their contributions, the biochemical outline of photosynthesis was set. The spate of X-ray structures that the field has seen in this decade – not just photosystem II but also photosystem I and the cytochrome complex that stands between them in the Z-scheme, as well as the light-harvesting complexes that feed extra energy to the core photosystems – seem likely to mark a similar ending. What used to be suppositions, hunches and arguments have been turned into things. Even when the structures say nothing new about the workings of the molecules in question, they represent what was known before in a new way, accessible and reified.

Since Willstätter and Stoll published their work on the 137-atom chemical structure of chlorophyll in 1915 people have asked how it is that molecules in a cell can take the sun's light and use it to reduce carbon molecules. Today we know. In the span of a single human life – in the time since Andy Benson was born – a fortunate elite has opened the door of the darkened room of photosynthesis, swept it with the beams from their torches, turned on the overhead lights and catalogued the contents. The illuminators don't know everything about how the pieces of this treasure trove work together; but they know the gist, and they know what the pieces look like.

It has happened in the span of a man's life, mixed up with the other events of such lives, dramatic and mundane – with escaping from Nazis and gardening in orchards, with part-time jobs to pay tuition and classified work on weapons, with learning to dive and dying in accidents, with acts of betrayal and acts of affection, with bouts of depression and falling in love and winning at poker. Brilliant people, determined people and lucky people, along with people just doing jobs they were good at, have through their work turned a complete mystery into something visible, something appreciable, something measured. Against the background of the twentieth century they created a new landscape so strange that molecules look like mountains, and a sunlit hour lasts as long as an ice age. And across this microcosm of light and life they have traced a path that links all life on earth to the fires of the sun.

We do not know it all. But no human lifetime has ever taught us more. And in that, there is indeed an exquisite thrill.

PART TWO

In the span of a planet's life

My vegetable love should grow
Vaster than empires, and more slow
ANDREW MARVELL,
'To his coy mistress'

It wouldn't happen in another world
It couldn't have happened to a nicer planet
KINGMAKER, 'Ten Years Asleep'

CHAPTER FOUR

Beginnings

Telescopes and bioscopes
The limits and origins of life
Opening up the world
Oxidizing the earth
Other worlds

Telescopes and bioscopes

At this point, we reverse our stance. We move from the small to the large, from the past to the future, and from a history of science to sciences of history. We look up, not down; out, not in. To put photosynthesis into a planetary perspective, we turn from microscopes to telescopes.

In particular, to a telescope which has not yet been built. We don't know what this telescope will look like or where it will be placed; we don't know whether it will be an 'it' or a 'they'. It might be a souped-up descendant of the Hubble Space Telescope, its mirror all-but-impossibly close to geometric perfection. It might be a small squadron of instruments orbiting in a formation so tight as to be measured to within a fraction of a wavelength of light. It might be a pin-hole camera made of two spacecraft, the pin-hole part half as far from its light-gathering partner as the moon is from the earth. Or it might be on the earth, a vast bowl for catching light built under the clearest of skies – in the high Andes, perhaps, or on Antarctica's thickest ice.

No one knows for sure what it will look like, nor who will build it, nor who will foot the astronomical bill. The photons it is destined to collect and analyse are still many light-years from earth – at least a decade out, more likely two. And until those photons arrive, no one will know what it will find. But for all that, we can be pretty sure that this superlative telescope will be built, because we know what we want to see in whatever number of mirrors it ends up having. We want to see evidence for a second planet like our own – a planet with life. We want to see our living world's reflection.

In the popular imagination, the search for life elsewhere is one of the key motives for space exploration. The life-seeking telescope will be a vast step forward in that search. It will scrutinize planets similar in size to the earth around stars similar in brightness to the sun, analysing the spectrum of the light they reflect as a way of picking up the signs of life that a planet something like ours must have in its atmosphere. Signs which, on earth, are put there by photosynthesis.

These signs are a matter of atmospheric chemistry. If you mix a bunch of gases together, you may see some chemical reactions. Energy will be released, which if the mixture of gases is, say, methane and oxygen in the turbine of a power plant, might be used to do work. At the same time, entropy will be increased. And if the mixture is left alone, in time the reactions will come to an end, leaving the resultant chemicals in a state of equilibrium. If you look at the two most earthlike planets in the solar system, Venus and Mars, you will see that their bulk atmospheres are in a state close to this chemical equilibrium.

The atmosphere which you are breathing, on the other hand, is not. It contains a significant amount of highly-reduced methane and a very large amount of oxygen. On paper, the earth's atmosphere looks like a great big redox reaction just waiting to happen. In fact that reaction is a great big redox reaction that's not even bothering to wait; it's going on all around us. The atmosphere we live in is a slow-burning but never-quenched flame in which methane is being oxidized to carbon dioxide just as surely as in the whirling turbines of a natural-gas power-plant.

In 1965, an English chemist named James Lovelock suggested that this combustibility might be a key indicator of the presence of life on a planet. He was working as a consultant with NASA on chemical instruments for the first generation of planetary space probes, and he was unimpressed by the suggestions for instruments to detect life on missions to Mars. The instruments seemed to him to depend on life elsewhere being very like life on earth, in biochemical terms; whereas what was interesting about the possibility of life elsewhere, in Lovelock's eyes, was that it might be radically different from life on earth.

This was the great dilemma of what had then recently been named 'exobiology', the science of life elsewhere. Because every living thing on earth shares a common ancestor, the study of life on earth cannot be a study of life in general, but just of one particular way of being alive, the way embodied in the use of DNA and ATP and the other biochemical mechanisms everything on earth has in common. A second, unrelated example of life would allow science to escape from this parochialism – hence the idea's allure. But such a second example of life might not be tractable to the techniques of earthly biology. Something different enough to be interesting might be too different to study.

A more universal approach was needed, a science that could recognize life from beyond the confines of earthly biology. And to scientists, there is nothing more universal than thermodynamics. 'It is the only physical theory of universal content,' wrote Einstein, 'which I am convinced, that within the framework of applicability of its basic concepts will never be overthrown.' Schrödinger's *What is Life?* had given Boltzmann's vision of life as a process that reduced entropy a certain currency. To Lovelock, a non-biologist with broad scientific interests, the idea that you could spot life by looking for decreases in entropy seemed a powerful one, anchored in the universal certainties of thermodynamics, rather than the contingencies of earthly biology and biochemistry.

Lovelock came up with a variety of ways of detecting entropy reduction, the best of which was in some ways the simplest. The reduction of entropy innate to living creatures would knock the chemistry of their environment off kilter. If a planet were just a bunch of molecules warmed by the sun, it would be in a chemical equilibrium. But if life were reducing its internal entropy the chemical equilibrium would be disturbed, as it is on earth.

The idea that the earth's atmosphere is evidence of life exporting entropy may, at first, seem confusing. The atmospheres of Mars and Venus are at equilibrium, and as we saw in Chapter Two, a system in equilibrium is in its likeliest state, which is to say that with the highest entropy. The earth's atmosphere, on the other hand, has strikingly low entropy: it is absolutely impossible that you could find this much oxygen and methane mixed together by chance. If life on earth is producing so much entropy, why does the environment hold less of the stuff? The answer lies in remembering that the earth is not a closed system. Just as the cosmos provides the earth with a source of energy, it also provides it with a sink for entropy.

The earth, like everything with a temperature, emits energy by giving off photons. The higher the temperature, the higher the wavelength of those photons: a piece of iron at room temperature emits invisibly in the infrared, but put in front of a furnace and it will go from black to red to orange as it gets hotter and the photons it emits get higher in energy and shorter of wavelength. It is from the white-hot wavelengths of their light that we can tell that the hydrogen and helium at the surface of the sun are at a temperature of 6000°C.

Seen from space, the earth looks like a body with an average temperature of minus 19°C. The chilliness comes from the fact that the infrared photons emitted from the more temperate surface don't make it straight through the atmosphere, but are instead absorbed and re-emitted as they go by carbon dioxide, water vapour and other greenhouse gases. The photons that finally make it out into space thus come from the upper atmosphere, where it is cool, rather than the warmer surface. So if you look at the earth purely in terms of an absorber and emitter of photons, you see a body that takes in high-energy photons travelling in a single direction (sunlight all comes from the sun) and emits low-energy photons travelling in all directions. In terms of energy, the incoming and the outgoing are almost precisely equal:* there are many more of the outgoing

---

* For the past century, the earth has absorbed slightly more energy than it has been able to emit. That imbalance is what we experience as global warming, and forms the theme of the third part of this book.

photons, but each carries proportionately less energy. In terms of entropy, though, the two flows are markedly different. A regimented stream of high-energy photons is much more unlikely and orderly than an omnidirectional wash of low-energy photons. So by turning something orderly into something much closer to equilibrium the earth contributes to the build-up of entropy in the universe at large.

Much of this entropy comes from simply turning hot photons into cooler ones. The physical movement of the winds, the evaporation of surface water into the atmosphere, and the circulation of the oceans also make a contribution; they are effectively great engines driven by the temperature differences between the tropics and the poles, the air and the surface, and in moving billions of tonnes of water and air around day after day they do a great deal of work and produce a great deal of entropy. Similar redistributions of heat in the atmospheres of Mars and Venus produce entropy, too. But the earth has the added ability to produce entropy in large amounts through atmospheric chemistry. The equilibrium atmospheres of the other planets represent a high standing stock of entropy but a low rate of entropy production, since at equilibrium, by definition, there are no net chemical reactions or transfers of energy going on. On the earth, life uses a fraction of the incoming stream of sunlight to make oxygen. Because of that oxygen, the atmosphere is in perpetual disequilibrium, and the subsequent reactions that attempt to restore equilibrium – reactions with oxygen and its by-products that oxidize methane into carbon dioxide – produce a great deal of entropy which is flushed back out into the cosmos in the earth's infrared outwash.

Lovelock did not put things quite like this in the 1960s. But he did argue that the earth's atmosphere showed that there was a lot of entropy being produced. And he noted that the atmosphere of Mars – the planet to which the exobiologists wanted to send their life-seeking instruments – was in chemical equilibrium, which suggested that there was no such production there, and thus that the planet was lifeless. When the exobiologists' instruments got to the great cold desert in the sky on the Viking missions of the mid-1970s, that was indeed the conclusion reached by almost all the scientists participating in the studies: the landing sites, and by inference the rest of the planet, were lifeless.* Lovelock had been right.

By that time, Lovelock had developed his insights into life's effect on the atmosphere further. Once he started seeing the atmosphere as a product of living processes, the constituents of the world begin to change. Living things are systems that keep themselves low in entropy by exporting it, as Schrödinger had argued. But, oddly, their planetary environment was also low in entropy. If you drew a line between the places entropy was reduced and the places it wasn't, that line wouldn't trace the cell-membranes of the bacteria, the skin of the animals, the cuticles of the leaves on the trees. It would trace the edge of the atmosphere. The entropic view suggested the atmosphere itself was part of what was alive on the earth.

What's more, the atmosphere's distance from equilibrium, its chemical instability, seemed to have persisted over hundreds of millions of years; the atmosphere was not just out of balance, it seemed to have stayed out of balance over long swathes of geological time. The atmosphere seemed to Lovelock to be regulated in a similar way to physiological attributes of a human body such as temperature or blood pressure, persisting in a state of great intrinsic instability. Lovelock came to the conclusion that life on earth was not just a collection of zillions of individual organisms – the sum of which scientists call the 'biota' – in a pre-set physical environment. It was a complex system consisting of both the organisms and their environment. He called this system of the earth's organisms and the air and water and rock with which they interacted Gaia. And in the sense that it lowered entropy and regulated itself physiologically, Gaia as a whole was alive.

Since its inception, Lovelock's Gaia has been many things to many people. In the 1970s, Lovelock and Lynn Margulis, the champion of

---

* Far more sensitive studies of the Martian atmosphere mean this conclusion is now open to some possibility of doubt, a subject we will touch on later. But the sort of life currently envisioned as possible on Mars does not necessarily invalidate Lovelock's original position.

the endosymbiotic theory of the origin of chloroplasts, developed it into the 'Gaia hypothesis', which proposed that the living parts of the earth system regulate the inanimate parts – atmosphere, oceans and the like – to their own ends, keeping the environment stable and benign. More recently, Lovelock has preferred a less teleological 'Gaia theory', in which the environmental stability that fascinates him is seen as an inherent property of the system as a whole, rather than as any sort of planned outcome. Gaia has been a *bête noire* to many biologists, a rallying point for greens of various degrees of mysticism, an inspiration to a network of younger researchers, an exercise in the poetics of science and the material for Lovelock's excellent books on the subject, *Gaia: A New Look at Life on Earth* and *The Ages of Gaia: A Biography of Our Living Earth*.

And in all this, there is one thing that can be said about Gaia with certainty. It is photosynthetic. The sublimely ordered oxygen-rich atmosphere that is its calling card is only possible thanks to the photosynthetic use made of the energy from the sun. You can argue about whether it is alive, and in what sense. Changing telescopes for microscopes you could have similar arguments about an isolated chloroplast in Daniel Arnon's lab, or a reaction centre in George Feher's. But you can't argue about what they are all doing. They are eating the sun.

Another thing that can be said surely of Gaia is that it is an instrument. Lovelock is a maker of instruments; it was the reputation that he developed as such while working for the Medical Research Council in Britain that allowed him, at the age of forty, to leave full-time academic employment and work as a freelance for people such as NASA. He has told me that it is in the making of instruments and the design of experiments which use them that his ideas find their surest expression, and the workings of the world become most apparent to him. In this sense Gaia itself is an instrument – a 'bioscope', as he says, 'through which to look at life on earth'.

The real-life equivalents of Lovelock's notional bioscope were the portholes on the Apollo spacecraft, through which the earth was

153

seen as a seeming whole for the first time. Those portholes offered the immanent, photographable 'new look at life on earth' to which Gaia was the theoretical counterpart, a new look that showed its cloud-streaked palette of blues, greens and golds as something extra-ordinarily complicated, its active nature poignantly obvious when contrasted with the dull wastes of the moon. Lovelock invoked the awe engendered by those images in the first paragraph of his first book. One of the Apollo pictures was emblazoned on its cover.

This was not just modishness. Lovelock's ideas had been born out of the study of other planets, and his most enduring claim was that life was first and foremost a planetary phenomenon. Some of its basic properties have to be understood at the level of the planet as a whole, rather than at the level of the organism, or of the molecules that make up the organism. *Gaia: A New Look at Life on Earth* was, in this sense, a bit like Schrödinger's *What is Life?* Schrödinger argued that there had to be an account of life that made sense on the level of the molecule. Lovelock argued that there had to be an account that made sense on the level of the planet. *What is Life?* made its mark by provoking people who thought about physics to turn their minds to the essence of life; *Gaia: A New Look at Life on Earth* invited people thinking about planets to do the same.

The science *What is Life?* helped to inspire was molecular biology. Lovelock's book helped inspire astrobiology, a new discipline created in the 1990s (though the word was used now and then before that). As the births of disciplines often are, astrobiology's origin was in part a genuine intellectual step forward, in part a flag of convenience for practitioners making common cause. In essence, astrobiology is the study of life in its astronomical and planetary contexts. While it has obvious links to, and has indeed subsumed, exobiology, the study of life elsewhere, the stress on context focuses astrobiology on questions of habitability, rather than of life per se. This has the advantage that habitability is more definable, and more discernible at a distance, than life itself is. Another advantage, in terms of actually having something to study, is that astrobiology quite legiti-mately takes the habitability of the earth as part of its subject matter, too, and thus benefits from a living planet which gives it real stuff to talk about, and even opportunities for fieldwork. While the intellectual leadership of astrobiology tends to come from astronomers and planetary scientists, a lot of the work that actually gets done is rebranded microbial ecology.

The great project of astrobiology is to build the telescopes evoked at the beginning of this chapter – life-seeking instruments for which the Gaian bioscope will serve as a sort of conceptual eyepiece. Although there are aspects of Lovelock's thought which are still controversial, and it is certainly not the case that all astrobiologists are Gaians, any more than all molecular biologists were fully convinced by Schrödinger, Lovelock's idea that the composition of atmospheres can be taken as a sign of life is not. It is the basic assumption of those who seek to build the telescopes needed to observe earthlike planets far beyond the reaches of the solar system that, one way or another, they will find ways to pick up enough photons to analyse the planets' spectra and work out their chemistry. And that chemistry, if strongly out of equilibrium, could be used to diagnose life.

Astrobiology's other solution to the apparent lifelessness of our immediate astronomical surroundings is to study planets separated from us in time, rather than in space – to turn the bioscope to periods of the earth's own history when it seemed like another world itself. To look at periods of our planet's past through a lens that reveals the connections between the rocks, the waters, the air, the life and the endless flow of energy from beyond. That is what this and the following three chapters set out to do.

The limits and origins of life

I think I first felt self-consciously Gaian on a walk in the Vercors, a sparsely populated limestone plateau south of Grenoble with hills that roll down to the west and rear up to the east. I was in a high scree-flanked valley beneath the plateau's eastern rim. The white rocks laid down on an ancient seabed; the fusion-bright sun in its bleached blue sky; the resinous pines; the flower-specked pasture; even the camber of the whole landscape, shouldered aside by the uplift of the neighbouring Alps: all seemed to come together as one great but single thing, its rock and air and topography no less alive than the trees and the gentians and the walker suddenly struck by the wonder of it all.

The way in which Gaia calls into question the distinction between that which lives and that which doesn't has a peculiarly pantheistic appeal. But it is far from the only, or the first, twentieth-century idea to raise the problem of life's boundaries. The great problem of biology – which, happily, is of no practical concern to the vast majority of biologists – is that there is no agreed and fulfilling definition of the life that it studies. It is the lack of agreement over life's essential nature that made the exobiological idea of looking for it on other planets so appealing. How could life be different? How different could life be? What distinguishes it from non-life?

In the nineteenth century, many early biologists were convinced that life had an essence, a vital force. The science of biology was in many times and places dedicated to revealing the secret force that animated gross matter and made it into mayflies and marigolds and meerkats. After a century or so of tireless dissection and analysis, though, biologists concluded that the secret was that there was no secret; the eeriness of this haunting was that there was no ghost to explain it. Twentieth-century biochemistry, and later molecular biology, showed that life was made of the same stuff as non-life, and obeyed the same rules. The processes of life did not need to take place in a tissue or a cell, they could all be replicated in a test tube. As far as the electrons are concerned, respiration is no different to rusting.

In the 1930s a truculent Cambridge biochemist, Bill Pirie, gave memorable voice to this idea by arguing that the term 'life' no longer had any scientific meaning. From his infancy on, Pirie boasted a strongly sceptical temperament and a delight in robust debate so marked that his name was changed to match it; called Norman at birth, his opinionated stubbornness led his family to rename him 'Kaiser Bill', and he was Bill for the rest of his life. In the 1920s he found his temperament well catered for in the early days of Hopkins' Cambridge biochemistry department, delighting in its 'gay argumentativeness'. In the 1930s he became one of the first people to crystallize the Tobacco Mosaic Virus (doing a rather better job of it, by some accounts, than the man who went on to win the Nobel prize for the achievement). It was a thought-provoking achievement as well as a technically remarkable one. Something sufficiently alive to infect, subvert and kill the cells of plants was also inert enough to be crystallized like table salt.

In a paper written in Hopkins' honour in the late 1930s, Pirie used an analogy from chemistry to explore the insight that things such as viruses could not be treated as simply 'living' or 'dead'. The words 'acid' and 'alkaline', he pointed out, had once been seen as opposites, but modern chemistry saw them as denoting different levels of a single measurable quantity: acids are rich in hydrogen ions, alkalis poor in them. This raised the possibility that the obviously dead and the obviously alive simply occupied different positions on some sort of continuum of *élan* that was as yet not understood. If so, most of what was interesting to biochemists went on somewhere in the middle of this spectrum, at a point where neither the word 'living' nor the word 'dead' were of much use. 'The words still have a very definite meaning when used by poets, knackers or soldiers,' said Pirie, neatly evoking the professions of life and death, 'but little or none when used to describe the phenomena observed in tissue culture, virus research and kinetic studies on interrelated enzyme systems.'

Pirie's incisiveness did not stop other scientists from using the L-word, and indeed speculating, as Schrödinger did, as to its ultimate meaning – the success of *What is Life?* as a source of inspiration to non-biologists showed that many people thought there should be an answer to such a question. (Pirie thought the book was a waste of time.) In his 1947 lecture on 'The Physical Basis of Life' Pirie's friend J. D. Bernal defined his subject as 'one member of the class of phenomena which are open or continuous reaction systems able to decrease their internal entropy at the expense of free energy taken from the environment and subsequently rejected in degraded form', a thermodynamic view close to Boltzmann's and Schrödinger's – and one that is close to the view that led Lovelock to see the earth itself as alive.* It is not a view that is universally shared, but to many it feels like the basis for a fuller conception we cannot yet grasp.

As Bernal noted, not all the members of this class of phenomena are alive in a commonsense way – flames and hurricanes keep their structures while turning over the set of atoms that make them up in a life-like way, but we do not see them as living. To distinguish life-like from living, Bernal suggested, you needed to supplement physics with history. Life differs from the rest of those 'open reaction systems' because its exemplars share a common ancestor, and thus a common way of doing things. Flames and hurricanes grow from unliving things, from lightning strikes and pressure waves in the tropical atmosphere; life always comes from life. Though Bernal did not put it this way, this common descent depends on life's ability to store and reproduce information, which is what allows life, unlike hurricanes, to evolve through natural selection. Life is a flame with a memory. Pirie derided his friend's ideas as 'vital blarney', charitably allowing, though, that they were 'not as bad as Schrödinger's'.†

The importance of common descent, Bernal argued, was the reason for wanting to compare life as it has evolved on earth with life from somewhere else entirely. Only then could we work out

* Lovelock and Bernal met in the early 1940s, when Lovelock was a part-time student at Birkbeck College in London and Bernal on faculty there, though they didn't ever discuss such issues as the nature of life. Lovelock's clearest memory of their interactions is of arriving at a party of Bernal's with a young lady and being told they were too early and would have to go and wait in the bedroom: Bernal had little time for bourgeois sexual morality.
† Though they have similarities. Bernal's acknowledgement of physics and history is akin to Schrödinger's division of his subject into order from disorder – thermodynamic issues – and order from very similar order – genetic issues.

what properties all living things shared, and what properties, though universal on earth, might not be fundamental. And it was also the reason for trying to understand a particularly vexing specific instance of the general problem of life's definitional boundaries, an instance that intrigued both Bernal and Pirie: life's origin, the starting point of the history that set this particular class of entropy-reducing phenomena apart.

Today, on earth, the key organic compounds of which life is made are always produced either by cells or by chemists – the amino acids from which proteins are made don't just crop up in the environment ready to be strung together. But in the 1920s two Marxist biochemists independently suggested that in the early history of the earth, when it had been a rather different planet, organic compounds would have been an environmental commonplace even in the absence of life. J. B. S. Haldane, Hopkins' first appointment to the biochemistry department at Cambridge, and Aleksandr Ivanovich Oparin, of the Institute of Biochemistry of the USSR Academy of Sciences, argued that in the chemical conditions prevalent in the early solar system organic chemistry might have taken place spontaneously. The earth's earliest atmosphere, Oparin suggested, had been filled with reduced chemicals like hydrogen, methane and ammonia, which would make it an environment that would favour the production of the reduced organic molecules from which life is made. This was not an ad hoc assumption; it was based on astronomical spectroscopy, which showed that the giant planet Jupiter was still wrapped in just such gases today. If all the planets had formed from the same stuff at the same time, it was not unreasonable to think that the biggest, coldest planets preserved most faithfully the conditions that had once been shared by all of them. Reactions in the early earth's reduced atmosphere would have produced organic molecules that would enrich the oceans below to, in Haldane's memorably domestic words, the 'consistency of a hot dilute soup'. Once the necessary organic compounds had been synthesized through these inorganic means, they would have started to react in ways familiar to biochemists, thus producing the 'self-perpetuating pattern of chemical reactions' that, to Haldane, was the hallmark of life.

Oparin and Haldane made it possible to think life was an inevitable part of the universe, a result of common conditions and universal laws of chemistry. This belief, it seems to me, is one of the great legacies to science of the Marxist materialism the two men shared. In the 1950s various chemists, including Melvin Calvin and Harold Urey, tried to reproduce in the lab the primordial soup they had hypothesized about, with various sources of energy (Calvin used the Rad Lab's cyclotrons) and various degrees of success. When exobiology got under way in the 1960s such studies were easily rolled into its remit. Understanding the conditions under which life originated could help guide the search for life elsewhere; studying environments elsewhere similar to those on earth billions of years ago could help guide studies into that origin. In the post-Viking disappointment of a lifeless solar system studies of the origins of life were exobiology's mainstay. Robbed of new worlds, the exobiologists turned to studies of the oldest parts of this one.

One result of this turn towards earth history is that we now have a fair idea of when life began. The earth's 4.5-billion-year history is divided into four great eons. We live in the Phanerozoic, which began 543 million years ago. This is the period in which there are abundant complex life-forms to be found fossilized in rocks. Before it, running from 2.5 billion years ago up to 543 million years ago, comes the Proterozoic. For most of the Proterozoic, almost all the fossils to be found are of single-celled organisms. Before the Proterozoic came the Archaean, and before the Archaean came the Hadean.

We can follow life's traces back to the early part of the Archaean, the best part of four billion years ago. This is not primarily thanks to fossils; things that look like fossils in the most ancient of rocks are notoriously hard to identify with any certainty, and their presence or absence in various samples is a matter of sometimes fairly acrimonious disagreement. But structures in ancient sediments that appear to have been due to the work of microbes – such as fossilized versions of the layered, mat-like communities in which some microbes still live today – reach back well into the Archaean. And evidence from isotopes reveals some of what they might have been doing.

In the 1920s Vernadsky, the Russian who prefigured Lovelock's ideas about Gaia with his vision of the biosphere as a single entity, made the unfounded suggestion that biological processes could make subtle distinctions between isotopes of the same element. This turned out to be true. Living systems have preferences when it comes to the isotopes of nitrogen, sulphur and, most crucially, carbon. When the Calvin-Benson cycle makes carbohydrates it uses carbon dioxide containing carbon-12 more avidly than that containing carbon-13 because its key enzyme, rubisco, has a preference for the lighter molecules.

This means organic matter in an ecosystem where the carbon is fixed by rubisco will be lighter, in carbon isotope terms, than inorganic carbonate rocks such as limestone laid down at the same time. This is true even if the limestone is deliberately precipitated by living things such as shells, or corals – while the carbonate ions in such structures may be being used by life, they don't pass through rubisco's preferential filter. Today, organic matter in sediments contains about three percent less carbon-13 than carbonates do. That is rubisco's signature in the rocks.

The picture is complicated, though, by the fact that other processes can also skew the isotope ratios. While some hold the signature to be discernible as far back as the middle of the Archaean, in the early Archaean things are much less clear. It seems that some process or set of processes was discriminating between light and heavy carbon as far back as 3.8 billion years ago, but that far back any rubisco signature is utterly unintelligible.

Further back than that, we hit a brick wall – or, rather, our planet does. About 3.9 billion years ago, the earth underwent a pummelling with left-over building material from elsewhere in the solar system, a cataclysm called the 'late heavy bombardment' which is beautifully recorded in the pock-marked face of the moon. Objects about as large as the largest asteroids left in the solar system today pelted the planets; some of these impacts would have been big enough to boil the oceans of the earth in their entirety, blanketing the whole planet in live steam and sending a pulse of sterilizing heat down into the rocky depths. This environment may have precluded life of any sort; it would certainly have made persistent life at the earth's surface impossible.

What science has to say about life's origin in or just after the Hadean is as yet far from coherent. There are currently no theories of the origin of life that one need find compelling, or even very plausible – it is a speculative and ad hoc area of science that some observers find remarkably unsatisfying.* It is not even possible to say for sure that life's origin and its first appearance on earth are the same thing: theories that life came from outer space in the form of spores of some sort are more respectable than they sound.

There is, however, a way of thinking about these matters that is at least in keeping with the spirit of this book, and somewhat opposed to the 'primordial soup' picture developed by Haldane and Oparin which dominated early experimental work on the subject. That tradition concentrated on ingredients – can amino acids, or the nucleotides from which DNA is made, be formed spontaneously in the sort of atmosphere expected on the early earth? An alternative, or at least complementary, approach is to look at energy rather than structure, and at process rather than ingredients. Life didn't just form out of things in the world, it formed out of things going on in the world. Energy flowed through the metabolism of the earliest living things, and it's fair to imagine that that flow of energy was derived from flows of energy that were already going on in the non-living world, the sort of flows which will go on anywhere where there are interesting contrasts for the inexorable laws of thermo-dynamics to grind down to equilibrium.

Planets provide such contrasts all the time. Physics forces large planets into layered compositions, and the different layers will have different chemical contents. And the heat trapped inside a large planet – heat left over from the planet's creation in a series of collisions of

---

* See *Life's Solution* by Simon Conway Morris for a bracingly critical take on the field.

smaller bodies, and topped up over its life by the decay of radioactive elements – will mess up that careful layering. On today's earth, a large part of the internal heat is lost through the great convection currents of plate tectonics, which bring hot rock up from the depths and send cool rock back down. By stirring up the chemically strati-fied earth and moving heat from place to place such currents create all sorts of local situations where the chemistry is out of kilter. Today these places offer various forms of life opportunities to earn their livings. In the past, the non-living responses to such disequilibrium could have been the processes from which life developed.

Mike Russell, until recently a geochemist at Glasgow University, offers just such a model for early life at a hydrothermal vent – a place where water heated and stuffed with chemicals deep below the crust bubbles out into cool seawater with a quite different chemical make-up. Highly reduced sulphur in the hot water would react spontaneously with iron in the seawater, creating a spongy lattice of iron sulphide. So far this is simple, and observable, geochemistry. Russell argues that the chambers found in these honeycombs could have acted as tiny chemical flasks. Organic molecules were cooked up and concentrated in them using the energy from the ongoing redox reactions between sulphur and iron. In some of these cells self-catalysing reactions took off, gradually becoming more and more complicated. These reactions could help build more iron sulphide cells by catalysing further redox reactions; they could also harness the spontaneous flow of protons from the more acid ocean to the hot alkaline water in the pores – perhaps using them to fix carbon dioxide into new organic molecules. Nucleic acids – relatives of DNA – then started to take a role, first as more catalysts, then as replicable stores of information. Geochemistry turned to biochemistry.

When in all this does the bubbling water become alive? What part of it becomes alive – the organic molecules, or the metal sul-phide cells that they live in and cause to grow and reproduce, or the whole system? There may be no way to say – and that should not be counted a failure. The reductionism of biochemistry and bio-logy, so powerful in its context, may mislead us by stressing the importance of the small and particular in life, rather than the broad and fuzzily bounded. Russell's argument, whether or not it is right in its particulars, does at least provide us with a somewhat plausible vision in which life did not start in individual things that thus became living, but rather started at large. Life was not an efflorescence, coming from one particular arrangement of matter that worked, but a condensation, a squeezing of the energies of the earth into ever tighter and more productive packages.

Or rather, it was both. The concern over definitions that led Bernal's friend Bill Pirie to reject the word 'living' in the context of biochemistry did not stop him from being interested in the origin of life – quite the reverse. He maintained an interest in the field from his early days at Cambridge with J. B. S. Haldane until his death more than sixty years later. It was only one of many interests: among others were early work on oral contraceptives and a long-running, if rather quixotic, programme for having rubisco mechanically extracted from leaves so it could be used as a protein source in developing countries. (Both lines of research were suggested by his socialist and internationalist ideals.) He self-deprecatingly dismissed his decades of contributions to the subject of life's beginning as 'extremely repetitive'. But he took some pride in the idea he repeated. He was the first to talk of the problem in the plural: of the problem of the origins, rather than just the origin, of life.

Pirie saw the history of life and its antecedents as having the shape of an hourglass. At the top we can imagine a broad range of simple proto-metabolisms and almost-lifes based on all sorts of inorganic geochemistry. As time went on and the hourglass tapered inwards, these simple potentials were gathered together into more complex arrangements until, in the neck of the hourglass, they were condensed down into the thing from which everything now living is descended. Then, below, they spread out into the diversity of our tree of life. Chemical diversity before; biological diversity afterwards. At the crux the starting point of life's history of common descent, the history that sets it apart from all the other open processes it resembles. The point where the flames received a common memory.

Pirie's view of increasingly life-like geochemical processes coalescing seems as apposite a counterpoint to the biochemical approach to life's origins now as it was when he presented it to the first international conference on the origin of life, in Moscow, just a few weeks before the launch of Sputnik. He ended his Moscow talk on life's scattered and diverse origins with some lines from Louis MacNeice's poem 'Snow' which evoke the power of the senses wakened to the everyday. The bolshy but soulful biochemist thought they might also capture the awakening of the earth itself, and I would like to think so too. And if they don't, they certainly convey the temperament needed to imagine such things.

> World is suddener than we fancy it.
> World is crazier and more of it than we think,
> Incorrigibly plural.

Opening up the world

The isotope record assures us that, however they came about, by the early Archaean there were microbes on the earth, their biochemistry capable of distinguishing between light and heavy carbon. While it is impossible to be certain, I suspect that this life was biochemically much the same as life today. Its genes were probably made of DNA, its biochemistry was largely in the hands of proteins composed of the same twenty amino acids that our cells still use, its energy needs were met by ATP, much of which was generated by chemiosmotic potentials across membranes made of still familiar lipids. The reason for thinking this is that these are traits still shared by all things living today.* Everything living discovered on earth fits into one of the

---

* There are some seeming exceptions – microbes that use an extra amino acid, say, or which live almost entirely by fermentation, which does not require chemiosmotic potentials, but these are elaborations on the basic scheme, not alternatives to it.

three kingdoms of life, bacterial, archaeal,* or eukaryotic, and those kingdoms share the basic biochemistry of whatever served as their last common ancestor, right around the neck of Pirie's hourglass.

But in a matter fundamental to this book, life in the early days was probably quite different to life now, because life then was local. The first microbes had to gather energy directly from inorganic chemicals that were out of equilibrium with each other. They had to be in touch, quite literally, with their sources of energy.

Such creatures still exist today. There are microbial communities which live deep in the crust, the creatures at their base reducing carbon dioxide with hydrogen to form methane. Though the numbers are extremely woolly, some estimates suggest that these microbes might outweigh all the surface biota. But even if that is true, under the gaze of the planetary bioscope their sheer mass is largely irrelevant. While there may be a vast number of such microbes, they do remarkably little, reproducing at very slow rates, impacting the environment hardly at all. Their energy supply is limited by the rate at which the earth provides them with chemical imbalances to feed on, and this means they represent only a tiny fraction of the earth's total metabolic activity.

On a young earth still roiling with the internal heat of its creation the possibilities for such life would have been considerably greater than they are today. But they would still have been limited, local and transient. Environmental redox gradients get worn away, just as batteries go flat; life taking advantage of those gradients will always tend to accelerate the process.

It is possible that Mars may offer an example of just such a world. In 2004 measurements far more sensitive than any available to Lovelock in the 1960s picked up what appeared to be traces of methane in its for the most part thoroughly oxidized atmosphere. This implies that there must be a source of methane on the planet,

* There is inevitable confusion in discussions of the role of the archaea, one of the three kingdoms of life, during the Archaean, a period of earth history. To try and moderate the confusion a little, I will use the adjective archaeal when referring to properties of the archaea, as one would use bacterial of bacteria.

because methane in the Martian atmosphere would be constantly destroyed through oxidation (the fact that Mars's atmosphere contains almost no oxygen does not mean it doesn't oxidize things). Though there are various possible sources, a particularly appealing one, from an exobiological point of view, would be microbes living deep below the cold surface, desultorily munching on whatever traces of hydrogen the planet's cooling interior can provide.

When Lovelock took his new look at earth in the late 1960s he realized that the sort of self-regulation he ascribed to Gaia would not be possible without a complex and expansive biosphere, and that the dramatic unregulated environmental swings that would take place in the absence of such Gaian regulation would surely make a planet uninhabitable. Life was therefore, he decided, an all-or-nothing phenomenon. A planet would either possess life in abundance or be so unable to regulate itself that it had no life at all. If there is life on – or more accurately in – dry, cold, dusty Mars, that all-or-nothing notion of life as a planetary phenomenon would be disproved.

But there would still be room for a Gaian distinction between planets where life is an active force and planets on which it is merely a passive passenger, between what one might call living planets and planets that merely carry a little life. In this view biological life – cells and stuff – is a necessary but not sufficient condition for planetary life – a rich biosphere with complex, possibly self-regulating properties. And the extra ingredient needed to get from one to the other is photosynthesis.

The power of photosynthesis is not just that the sun is a far greater source of energy than any hot spring or deep-crust chemistry can be. It is that instead of hastening equilibrium on, as life does when exploiting geochemical gradients, photosynthesis abolishes it. It can build new redox gradients in places that had previously been in chemical equilibrium, thus liberating life from local chemistry. Where respiration, for example, has reduced sulphates to sulphides, photosynthesis can reverse the process. It can fill the biosphere with things that will react together, providing new opportunities for life

– while, in the process, challenging it to regulate the now unbalanced world it finds itself in.

This is why Andrew Watson, a professor at the University of East Anglia who, in the 1970s, was Jim Lovelock's first and only graduate student, argues that the appearance of photosynthesis, not the appearance of life, marks the birth of Gaia. On a planet which never made the transition to photosynthesis – or on which the era of photosynthesis has come to an end, as one day it will on the earth – life would simply lack the power to be a driving force in the environment. My hunch is that Mars is such a planet.

In the past ten years it has become possible to piece together some of the story of how the photosynthesis necessary for a flourishing biosphere evolved on the early earth. This has not been through the discovery of new fossils; fossils or 'pseudofossils' of potential photosynthesizers from the Archaean era have been a subject of heated controversy in recent times, but the debate over them shows no sign of saying anything of interest about the evolution of the process. Instead it has been with the help of the DNA in the plants and bacteria around us today. Experience of the way of all flesh might lead you to think that hard rocks, high mountains and broad continents would be stable and lasting, while the tiny molecules of life were fleeting: biology *brevis*, geology *longa*. In fact, over the history of the planet, the reverse is true. Mountains are worn down to seabeds, continents pulled asunder and ground together; oceans open and close. As a result, only a tiny fraction of the earth's early crust is still available for inspection today. There is for example only one rock formation yet discovered that dates from the Hadean – the Acasta gneiss, in Canada. In the rest of the world the only samples of the Hadean are fragments incorporated in rocks of a younger age. The biggest are the size of small pebbles.

Yet molecules from those shattered days are all around us today, in the form of DNA sequences. Many of our genes are billions of years old; some date back to the universal ancestor itself. While the winds and waves of entropy erode earth's heights, life maintains its inner order across cosmic spans of time.

Not only are the molecules recorded; their patterns of relationship are there to be inspected, too. Genes diverge as the proteins they describe are adapted to different functions in different circumstances; but as they do so, they keep certain similarities. Looking at the sequences of two genes it is possible, up to a point, to work out how they are related, if at all, and which evolved from what. Resemblances like this let us deduce a surprising amount about the development of photosynthesis.

One of the things we can say for sure is in which of the three kingdoms of life the process first arose. In eukaryotes photosynthesis is only possible with the help of chloroplasts, which are suborned bacteria. The archaea, which seen down a microscope are more or less indiscernible from bacteria, but in a biochemistry lab or a gene-sequencing machine look remarkably different, have various non-photosynthetic ways of getting energy. They are the creatures responsible for methane-making in the deep crust, and indeed for reducing some of the sulphur at hot springs; some of them also have ways of using sunlight to produce energy. But they don't have any way of using sunlight to fix carbon, and so can't be said to be true photosynthesizers. The true photosynthesizers are all bacteria.

This should not come as much of a surprise. The bacteria are astonishingly good at finding energy that will let them make a living. More or less everywhere where the earth brings together substances with different redox potentials, there's a bacterium that knows how to take advantage of the situation by passing electrons from one to the other and skimming off some energy as it does so. This vast range of biochemical capabilities makes bacteria vital to the rest of life on earth. The way different bacteria ceaselessly break down and reassemble all sorts of everyday compounds drives the sulphur cycle, the nitrogen cycle and (if we honour their roots by including chloroplasts as bacteria) the carbon cycle. The endless stream of electrons running through these bacterial metabolisms provides the background hum of the earth's habitability.

DNA sequences suggest that the first photosystem was an adaptation of one of the protein way-stations on a bacterial electron transfer chain used in respiration. The sequences of the L and M proteins – the embracing dancers in the purple-bacteria reaction centre studied by Clayton and Feher – are similar to that of a particular cytochrome, and so to some extent is their shape. Some of the hooks with which that cytochrome hangs on to its iron-containing haem groups are in the same positions as the hooks that hold chlorophyll molecules on to reaction centre proteins.

Imagine the gene describing such a cytochrome being duplicated, so that the bacteria in question suddenly found themselves with two copies of it. Such duplication is a powerful factor in evolution; it allows a protein that does two things passably, for example, to evolve into two that both do one thing excellently. Duplication opens up room for evolution to experiment with one copy of the gene while the other keeps things ticking along. In the case of our cytochrome the experimentation might have involved swapping haems for pigments, then making each protein likelier to take a partner, a coupling ancestral to the embracing proteins now seen at the heart of the reaction centre.

For such a transformation to have been possible, it would have had to offer benefits at every stage. Evolution is not going to tolerate half-assembled molecular machines that might one day be reaction centres but so far are just a waste of space. So what might the intermediate attractions of a cytochrome decking itself out with pigments have been? One quite plausible possibility is the absorption of ultraviolet light. The high-energy photons in ultraviolet light can break up water molecules, thus producing 'hydroxyl radicals' which consist of one oxygen and one hydrogen atom. These radicals are extremely chemically active, with a propensity for damaging DNA. Most creatures that live on the planet's surface or in shallow water have evolved defences against ultraviolet light in order to reduce that damage. The need for such defences would have been much greater on the early earth because there was not, as yet, an ozone layer. Ozone is a form of oxygen, and oxygen only got into the atmosphere in large amounts after the development of photosynthesis – more specifically, after the evolution of photosystem II.

So clumps of protein with ultraviolet-absorbing pigments might have been an asset to early bacteria even if they did not, as yet, do anything with the energy they absorbed other than re-radiate it as heat; a warmed-up protein is a lot better than a high level of hydroxyl radicals. If the energy absorbed sometimes oxidized some of the pigments, the electron transfer chain that the original cytochrome sat in might have made use of the loose electron. That would mean that mutant proteins producing more electrons would be favoured; the shapes of the proteins would shift to favour electron movements; the embrace between neighbours would be honed into one like that we see today, with a 'special pair' of chlorophylls at its heart.

This is what evolutionary biologists call a just-so story. It is not necessarily the way that it happened – but it is a broadly plausible way it could have happened. There are other possibilities. Euan Nisbet, of Royal Holloway, a part of the University of London, is a Zimbabwean geologist with an interest in the Archaean era among many, many other things. He suggests that the precursors to reaction centres were sensors rather than sunscreens. For a microbe that lives near a hydrothermal vent, as many early microbes probably did, being able to sense the vent's infrared radiation would be a useful trick. Perhaps the pigment-protein pairings that evolved into reaction centres were originally used to work out what direction would be the warmest to swim in. Some of the chlorophylls used by bacteria are rather good at absorbing the sort of infrared radiation such sensors would need to work with.

Whatever the reason for it, this cytochrome-to-photosystem transition was probably made only once. There are two different photosystems found in bacteria – the sort found in the purple bacteria and some relatives, which is clearly related to photosystem II, and the sort found in other photosynthetic bacteria such as the green sulphur bacteria, which is related to photosystem I – and they look surprisingly different in terms of their gene sequence. But their overall geometry and some details of their proteins are similar enough that it is supererogatory not to imagine that they had a common ancestor. The original cytochrome evolved into an early reaction centre, some of the descendants of which ended up looking like photosystem I and some like photosystem II.

All photosynthetic bacteria have reaction centres of some sort, but most have just one type, either photosystem-I-like or photosystem-II-like: only cyanobacteria have both types, strung together into a Z-scheme. There are two ways in which this could have come to be. There could have been more gene duplication, this time affecting the genes for an entire photosystem, which produced a two-photosystem bacterium in which the photosystems were free to evolve in divergent and complementary ways. Some descendants of this bacterium, as part of an effort to fit into specific niches, might subsequently have lost one photosystem or the other, depending on their particular redox needs, while other descendants kept both and evolved into cyanobacteria. Alternatively, the descendants of the bacteria carrying the original photosystem might have split up into two different lineages, in one of which the photosystem evolved into something like photosystem I and in the other into something like photosystem II. Only later would some sort of merger or recombination have brought the two together. Bacteria have a proclivity for swapping genes around; for one to take up from another all the genes necessary to make a photosystem would be a peculiarly thoroughgoing transfer of genetic material, but it is not necessarily an overdramatic stretch. I have a slight preference for the duplicate-first, split-later account, championed by John Allen of Queen Mary, a college of the University of London, as opposed to the split-first, recombine-later account. But in the grand scheme of things it may make little difference – and with the vast amounts of genome sequencing that will be done over the next decade, the question looks likely to be solved one way or another in the not-too-distant future.

It was within the context of the cyanobacteria, however they evolved, that photosynthesis made its great leap forward, and finally learned to use water as an electron donor. This moved life beyond the potential limits that might be imposed by the scarcity of other electron donors, such as the sulphides and iron used by today's photosynthetic bacteria. On earth, at least, life has a non-negotiable need for water, the substance which gives its membranes shape and make its proteins work. So life can only take place where there is some water around. This meant that once water could be used as the electron source, almost everywhere that was both illuminated and wet enough for life became a potential site for photosynthesis. Water-using, oxygen-producing photosynthesis – the photosynthesis of cyanobacteria and, subsequently, of plants – opened all the habitable parts of the planet up to habitation.*

How photosystem II evolved from a similar but simpler reaction centre closely related to the kind now seen in the purple bacteria is again a matter of conjecture. One possibility is that this, too, had to do with the excessive ultraviolet that was probably a feature of the Archaean. Ultraviolet light makes hydrogen peroxide from water vapour, and peroxide is soluble in water. So the rain of the Archaean would have contained hydrogen peroxide; hardly enough to have bleached your hair, but enough to pose some problems for microbes living in freshwater pools and lakes, since peroxide goes on to produce the dreaded hydroxyl radicals.

Bacteria that face this sort of problem evolve mechanisms to deal with it. One of them is to use an enzyme called catalase to break the peroxide down into water and molecular oxygen, thus defusing it before it gives off hydroxyls. It's quite easy to imagine that at some point a bacterial reaction centre learned to use the catalase process for its own ends, tweaking the system so that the peroxide was turned

---

* The exception lies in the ocean depths, particularly the vibrant communities around some hot vents. But even here, photosynthesis plays a role, as the vast bulk of the oxidants – oxygen, sulphates, nitrates and the like – used by creatures in those depths come from the light-driven surface ecosystems. If the sun were doused, the vent communities would not long outlast the surface ones. They would shrink fairly quickly to something much smaller powered only by geochemical redox gradients such as those that drive the deep biosphere – and in some places, bizarrely, by the light of the hot lava of the vent itself. It has recently been discovered that there are in fact photosynthetic bacteria at depth which never see the sun but make use of this hot glow. While remarkable, this is not geochemically particularly significant, as the flow of energy is tiny. Nor does it speak directly to Euan Nisbet's ideas about the origin of photosynthesis at vents, as these bacteria are modern forms that have adapted to the depths, not primeval ancestors of all.

into molecular oxygen, hydrogen ions and a stream of electrons that the reaction centre could then use for photosynthesis.

Because this is a lot easier to do than splitting water – the peroxide starts off fragile, which is why it is dangerous – it could be done with the relatively weak oxidizing power generated by bacterial reaction centres, rather than requiring the considerably greater oomph generated by full-on photosystem II. But the dismantling of hydrogen peroxide and the splitting of water have obvious chemical similarities which suggest the capability for the second might have evolved from the equipment for the first. Like the part of photosystem II that splits water molecules, catalase uses a carefully crafted arrangement of manganese atoms to prise its molecular prey apart.

Enzymes will sometimes pick up the wrong molecule to work on, all the more so when the molecules are small ones; it's possible that the engineering of early enzymes was a little less well honed in this respect than today's, and that they made such mistakes more often. Catalases will thus sometimes pick up water rather than peroxide. If the reaction centre's oxidizing power were to grow in the course of evolution (as it would have, for example, if the chlorophylls in the special pair were pulled a little further apart) it would thus sometimes be tried out on water, instead of peroxide, and at some point the water would have begun occasionally to yield. The more often that happened, the better for the bacterium, and so catalases more likely to take up water would be chosen over the traditional sort by natural selection.

Again, this is a just-so story, and many in the field would prefer a different scenario. The mystery of how the manganese cluster became incorporated into photosystem II is deeply intriguing and far from solved – as are the origins of other metal-bearing clusters used in fundamental biology. It's possible that in some way they may be descendant remnants of the original geochemical precursors to life in the upper part of Pirie's hourglass. But these details are not necessary to trace the story's broad outlines, revealed as they are in the incontrovertible similarities in protein structure and DNA sequence between the reaction centres of purple sulphur bacteria.

Photosystem II grew out of an earlier photosystem which shared its ancestry with a cytochrome. And when that evolution was done, the earth was opened up to the cosmos, and life on earth had become a truly planetary phenomenon.

Oxidizing the earth

Given its immense importance for the earth's biosphere, you might think that, if you were an alien studying the history of the earth through some far-off bioscope, the invention of photosystem II would be easy to spot. Photosystem II produces oxygen, and it is this oxygen which has pushed our current atmosphere so far and so tellingly away from chemical equilibrium. Oxygen is, in a way, the signature gas of the planet.

The cyanobacteria which first produced oxygen, we can reasonably assume, were an instant success, and would quickly have become the dominant form of photosynthesizer on the planet. Their success would have been in part because they had an effectively limitless supply of electron donors, and in part because having two photosystems wired together through the Z-scheme allowed them to stash away more energy than other bacteria. And yet the advent of the cyanobacteria was not marked by an instant increase of oxygen in the atmosphere. Far from it. For at least 300 million years, and possibly for more than a billion, cyanobacteria managed to dominate the earth's ecology while having no obvious direct effect on its atmosphere. We know this because we can tell when oxygen first became a common component of the atmosphere – and it is well after the first evidence we have for the presence of cyanobacteria.

The appearance of oxygen in the atmosphere – the 'Great Oxidation Event', as it is known – took place about 2.4 billion years ago, shortly after the end of the Archaean. There is a wide range of evidence to back this up, largely to do with the chemistry and mineralogy of soils and sediments. To take one example, there are no rust-red sandstones such as those of the English Devonian, or Arizona's painted desert, in the rocks of the Archaean. That sort of rusting in sediments depends on the presence of oxygen.

A particularly telling way of constraining the oxidation event's date, developed at the end of the 1990s, uses a subtle wrinkle in the chemistry of sulphur. Sulphur comes in four stable isotopes, and the different nuclear configurations in these isotopes is subtly reflected in the behaviour of their electrons. In some reactions involving atmospheric sulphur compounds and ultraviolet light, these slight differences lead to different sulphur compounds ending up enriched in different isotopes.

If there is oxygen in the atmosphere, the compounds that show these tiny but consistent isotopic effects can't last; the sulphur ends up oxidized into sulphate or sulphur dioxide and the evidence for subtle isotopic effects is lost. If there is no atmospheric oxygen, though, the whole range of sulphur compounds can get into the rocks, and the isotopic signatures can be preserved. The spread of sulphur isotopes characteristic of an oxygen-free atmosphere is seen in rocks from throughout the Archaean.

Jim Kasting of Pennsylvania State University has made an impressive career out of applying such models to the early earth and other planets, real and imagined, in order to explore the implications of different atmospheric compositions, and predicted the sulphur isotope effect before it was observed. Even a simple mathematical model, he argues, can serve to move you from arm-waving to science. And according to Kasting's models, the sulphur isotope data from the Archaean mean that the atmospheric oxygen level prior to the Great Oxidation Event can have been no higher than a hundred-thousandth of the oxygen level today, which is twenty percent.

The Archaean has as yet offered up no fossils that can clearly be identified as cyanobacteria by their size and shape. But size and shape are not the only attributes of life that get recorded in rocks. Another of Melvin Calvin's contributions to the further shores of chemistry was a pioneering belief in the value of 'chemical fossils' –

preserved organic molecules that tell you something about the creature they were once part of. Over the intervening eons the heat of the earth will 'cook' the organic molecules into other compounds; this is how oil is made. But if the rocks are not too deeply buried, the cooking can be light enough for the original ingredients to be identified. And there are lightly cooked organic molecules in Archaean rocks that point firmly to the existence of cyanobacteria prior to the Great Oxidation Event.

In 1999 a team of scientists from the Australian Geological Survey and the University of Sydney succeeded in extracting some lightly poached organic molecules from a 2.7-billion-year-old sample of shale brought to the surface from a 700-metre-deep borehole near the town of Wittenoom in northwestern Australia. They found substantial traces of two particularly interesting organic molecules: 2-alpha-methylhopanes and long-chain steranes. The methylhopanes are cooked versions of a sort of lipid that is normally only seen in the membranes of cyanobacteria. The steranes are derived from sterols, which today are mostly made by means of a biochemical pathway that makes use of oxygen drawn from the environment. Finding the two chemical fossils together has been taken as a strong argument for the existence of oxygen-producing cyanobacteria and of other creatures using that oxygen for their own biochemical purposes.

Those shales were laid down about 300 million years before the sulphur isotopes show oxygen bursting into the atmosphere. When we have been throwing billions around playfully, millions of years may not seem like such a big deal. But 300 million years is a long time. It's a gulf more than four times wider than that which separates us from the last dinosaurs to walk the earth; it's enough time for plate tectonics to push all the continents on the earth's surface together and then pull them apart again, or for the sun to make a complete 200,000-light-year circuit of the Milky Way; it's as long as it took for evolution to get from fish to wolverines.

The shales are the oldest known chemical fossils that preserve evidence of cyanobacteria. But there is no reason to think that the

first fossilized cyanobacteria were the first cyanobacteria of any kind
– on the contrary, that would seem extraordinarily unlikely. But
with no earlier chemical fossils, what other evidence is there?

The answer, according to Euan Nisbet, lies yet again in the isotopes of carbon. As we have seen, the distinctive signature of rubisco
– carbonates that have just under three percent more carbon-13 in
them than organic carbon laid down at the same time – persists
quite a long way back into the Archaean. But this is not necessarily,
in itself, evidence for cyanobacteria. Some photosynthetic bacteria
that don't produce oxygen still use rubisco. And one can imagine
other ways of producing organic carbon enriched in the lighter
isotope which might conspire to forge the rubisco signature. The
key, in Nisbet's eyes, is not the difference in the carbon-13 level in
the two types of rock, but the level itself. The absolute level of
carbon-13 in carbonates of a certain age can, if certain assumptions
are made, be used to work out how much organic carbon was being
buried at the time, which may in turn be a proxy for the activity of
the biosphere.

In early 2005, as he explained these isotopic subtleties to me in
his office just west of London, Euan Nisbet rummaged under a relief
map of Cyprus and passed me a chunk of rock roughly the size and
shape of a smallish brick. It was a dirty grey limestone that he'd
collected in Canada, quite finely laminated, a bit heavier than brick
would have been. Its layerings suggested that it had been formed in
something like a reef, by some sort of simple life. Its carbonates,
Euan told me, had the isotope balance of carbonates laid down in
a world where there is about as much light carbon being buried in
organic sediments as there is today. With a few buttressing but
contestable assumptions, that could be indicative of a pretty active
biosphere. Nisbet argues that there is no way that a biosphere similar
in its level of activity to today's could operate except through water-
splitting, oxygen-releasing photosynthesis.

Nisbet told me that he and his colleagues had assigned the rock
and the reef it came from a tentative age of three billion years.

At this point I slightly lost track of the conversation. This unspeakable age was not an abstract concept. It was there in my hand, regular in shape, cool to the touch. In the three dimensions of space, and in every aspect evident to the senses, it was utterly mundane. It might function as an unexceptional paperweight, were it not that the flowing drifts of paper that half-fill Nisbet's office are too far gone for any such discipline – and were it not for the fact that, in its fourth dimension, that rock stretched back two thirds of the age of the earth. Stretched back more than almost a quarter of the way to the Big Bang itself. Most of the visible stars in the night sky are younger. I held its pleasing heft in my hand a good bit longer than was strictly necessary, its age in my mind longer still.

This three-billion-year-old limestone is not the oldest of its sort. A similar isotope signature can be found hundreds of millions of years earlier. To some the idea of a highly productive, photosystem-II-dependent biosphere in the mid or even early Archaean is not surprising. Mike Russell, for example, sees no reason why water-splitting photosynthesis should not have evolved in fairly quick order as soon as there were bacteria in which reaction centres could develop. He once told me that he thinks anyone who believes any aspect of bacterial biochemistry could have taken more than twenty million years to evolve is a closet creationist. But to others it represents a real puzzle. How could there have been an oxygen-producing biosphere spread around the earth for as much as a billion years without that oxygen getting into the atmosphere?

One part of the answer is that you have to look not at the rate of oxygen production by the cyanobacteria, but at the net rate of production by the whole biota – the rate of photosynthesis minus the rate of respiration. While the Archaean biosphere may have been producing oxygen at a rate comparable to today's, if the Archaean environment used that oxygen up promptly then the rate at which it escaped into the atmosphere could have been very low.

For us, it seems necessary that to get from producing plants to respiring animals the oxygen would have to go through the atmosphere; we are used, or should be used, to the idea that every breath we take contains oxygen from the Amazon and the Arctic and everywhere in between, and that the carbon dioxide we exhale will have a similarly global range of destinations. But this is because we think as large animals whose lungs take in their air about a metre and a half above the ground, and who are used to seeing photosynthesizing leaves high above our heads. The microbes of the Archaean lived in a flatter world where the carbon cycle was an intimate and local thing.

On microbial mats found on salt flats and in lagoons today, which may be analogous to the ecosystems that dominated the world back then, life is layered according to its redox preferences. The photosynthesizers sit near the top; below them sit aerobic bacteria, which use up their oxygen; below those live the anaerobes, creatures to whom an appreciable level of oxygen is a poison,* working away on the organic matter and inorganic chemicals produced by the layers above. The oxygen is used up so efficiently that those anaerobes can live happily within centimetres of the oxygen-rich atmosphere. At night, the deoxygenated zone of the mat moves all the way up to the surface as the cyanobacteria stop producing oxygen; indeed, at night in these circumstances the cyanobacteria themselves start to run their metabolisms in an anaerobic way because the mat's oxygen levels become so depleted. Archaean microbial mats could have been parsimonious in the amount of oxygen they leaked into the atmosphere.

But, for all that, they would still have leaked: the carbon-13 ratios tell us so. Even if Nisbet is wrong to argue that the isotopes in Archaean carbonates show organic carbon being buried at a rate comparable to the rate we see in the world today, some was definitely being buried. And that means that the biosphere was a net producer of oxygen. In a closed system, it would be possible for all the carbon reduced into organic matter by photosynthesis to be used up, along with the

---

* Perhaps surprisingly, some of them live in your body. Even in the cells that are laboriously supplied with oxygen by the blood, the oxygen level is less than one percent of what it is in the atmosphere. There are regions of the gut where the life-giving gas is a complete stranger. (This is why the bot-fly larvae David Keilin studied needed so much haemoglobin; they had to be able to hoard the stuff up whenever an air bubble came their way.)

oxygen produced alongside it, by respiration. But if some of the organic carbon is taken out, some of the oxygen is surplus to requirements. Organic carbon being siphoned off into sediments means the biosphere was producing more oxygen than it could use up.

This release-through-burial is a good thing – without it we would never have got any oxygen at all. It is one of the great gifts of plate tectonics. On a planet where the surface just sat there, low rates of sedimentation would make it hard to separate organic molecules from the oxygen produced in their creation. Plate tectonics provides a surface on which new mountains ceaselessly rise and waste away, and plate tectonics creates new ocean depths in which the eroded remains of mountains-gone can be buried. The slow cycles in which the earth's crust is created and destroyed do not add a great deal to the earth's energy flows; the heat they bring up from the interior is totally eclipsed by the energy of the sun. But they add a vital complexity to the physical environment, a way of keeping things sequestered away for millions, even billions of years.

This burial of organic carbon allowed the release of oxygen into the environment. However being released into the environment and persisting in the atmosphere are different things. In the Archaean, the ocean, the atmosphere and the minerals on the earth's surface would all have contained reduced chemicals eager to react with the oxygen, and these chemical 'sinks' could easily have outmatched the biological source. Where local supplies of reducing chemicals were overwhelmed there might have been a little free oxygen; on today's earth, where the situation is reversed, there are bayous rich enough in methane for fiery sprites to play through the swamp gas above them. But in an overwhelmingly reduced Archaean environment life's surplus oxygen would have had no chance of establishing itself on a global scale.

What was needed was a mechanism for making the environment as a whole less reduced – of oxidizing the entire planet. Happily, nature provides one. Earthlike planets will oxidize if left out in the sun as surely as toy cars will rust left out in the rain. At the topmost levels of the atmosphere, sunlight can knock atoms clean off into

181

space, and the atoms it does this to are most likely to be hydrogen atoms, because they are the lightest. And if you remove the hydrogen from something, you oxidize it; remarkably, this simple dictum applies in the case of a planet just as in the case of an organic molecule. Hydrogen escape is almost certainly what has oxidized the surface of Mars to its current rusty red, and it has had a similar effect on Venus. Ultraviolet images of the earth taken from space show a faint but constant stream of hydrogen atoms leaving our planet – a rather beautiful effect known as the geocorona. And as the hydrogen leaves, the earth as a whole becomes more oxidized.

At the moment the geocorona is, while pretty, also pretty negligible: the earth loses fifty tonnes of hydrogen a day from an atmosphere that weighs a hundred trillion times that. This is because, though it is easy to lose hydrogen from the upper atmosphere, it is hard to get it up there in the first place. In today's atmosphere hydrogen released by volcanoes and life (bacterial mats can produce quite a lot) is quickly mopped up by chemical reactions in the lower parts of the atmosphere that tuck it away into water molecules. If these water molecules, or any other water molecules, could get up above the ozone layer, then their hydrogen would be released by ultraviolet light breaking down the water vapour and might make it out into space. But water cannot get into the upper atmosphere; the cold temperatures at the base of the stratosphere freeze it, at which point it falls back down to the surface. Hence the low rate of hydrogen loss through the geocorona.

In the 1990s, a model made by Jim Kasting and colleagues showed that in the Archaean environment, where sources of hydrogen were more powerful and there was no oxygen in the atmosphere to react with it, more hydrogen could get into the upper atmosphere and thus out into space. This hydrogen escape would increase the level of oxidation in the environment as a whole. In 2001 David Catling, Kevin Zahnle and Chris McKay, all then at NASA's Ames Research Center in California, added an extra wrinkle to the idea by bringing methane into the equation.

Archaea capable of making methane were among the first creatures on earth. We know this because the methane-making apparatus, like rubisco, has a distinctive carbon isotope signature. In 2006 it was reported that tiny samples of methane sealed into 3.5-billion-year-old rocks carry that signature, showing that the methane-makers were already active back then. Today, methane in the atmosphere is fairly quickly dispatched by the ever-present, ever-hungry hydroxyl radicals; but in the Archaean, with lower hydroxyl levels, it would have been able to persist far longer. Because methane is light (each molecule has only half the mass of a molecule of oxygen) and has a low freezing point, in the Archaean it would have been able to rise to the top of the atmosphere. There radiation would have torn off its hydrogen atoms and set it off along the geocoronal highway to the stars. The carbon from the methane, now oxidized, would be left behind to fall back like a spent booster rocket. The more methane the earth produced, the more hydrogen it would lose and the more oxidized, on average, it would become, thus paving the way for the eventual emergence of atmospheric oxygen.

The interesting thing about this is that methane production didn't only spur hydrogen loss. It also warmed the planet. Methane is a powerful greenhouse gas – far more powerful than carbon dioxide. In today's atmosphere that power is constrained by the fact that the average methane molecule will only escape oxidation for about a decade. In the Archaean, when methane molecules would have lasted in the atmosphere for hundreds or thousands of years, their greenhouse power would have been greater.

And this was a time when the earth needed warming. Stars of the sun's size get brighter the longer that they burn. In the early Archaean, when the sun was less than a third its current age, it was only about seventy-five percent as bright as it is today. In today's conditions, twenty-five percent less sunlight would leave the earth infeasibly cold. For the Archaean earth to have been something other than a lump of ice it would have needed a strong greenhouse effect. Carbon dioxide was much more abundant in the early atmosphere than it is now, and Kasting's models show that that would have added some warmth, but the early earth would still have been below freezing over most of its surface had there not been some extra greenhousing. Methane is the obvious candidate to provide it. Various geological indicators which suggest that the Archaean climate was for the most part warm, and possibly hotter than today's, provide reasons for believing that a lot of methane was being made back then well before the evidence for it was found in those tiny rock-sealed bubbles.

Before life was widespread, methane levels would have been limited by the rate at which the methane-making archaea could get their hands on hydrogen from the deep earth. But methane-makers can work with organic matter as their starting point, too; the methane that flares up from swamps is made from plant matter decaying in the oxygen-free water. As photosynthesis allowed the earth to increase its total biomass the opportunities for methane-makers to turn some of that organic carbon into methane grew too. This could well have set up what's known as a 'positive feedback' – a process in which the effects amplify the causes. More warmth, more biomass; more biomass, more methane; more methane, more warmth. As Lovelock noted in his 1988 book *The Ages of Gaia*, methane produced by the archaea was probably the first great example of life making the earth more habitable.

But the methane that seems to have warmed the earth also allowed it to lose hydrogen, and by oxidizing the environment in this way it would have set the stage for its own destruction. Catling and Zahnle, working with a student, Mark Claire, have recently pointed out that as the environment oxidized, sulphates would have become more available. The redox potentials involved make oxidizing organic matter by reducing sulphate a more energetically attractive way for a microbe to make a living than reducing carbon dioxide to methane. So the increased availability of sulphate on the ever more oxidized earth would have been to the methanogens' disadvantage. As the rate at which methane was produced dropped, a crucial chemical sink for atmospheric oxygen was reduced, and this may have been what finally let the gas establish itself in the atmosphere worldwide.

Another possibility, which certainly doesn't exclude a role for the sulphate shift, is that tectonic changes increased the rate at which carbon was buried, boosting the oxygen production rate enough to overwhelm the environmental sinks. Once oxygen makes it into the atmosphere to an appreciable extent it changes things to its own advantage by setting up an ozone layer. Lower levels of ultraviolet light in the lower atmosphere mean that it takes a lot longer for oxygen to react with methane. A longer lifetime for the oxygen translates into a higher level in the atmosphere.

The chemical shifts brought about by the Great Oxidation Event just after the end of the Archaean were not just biogeochemically fundamental. They were in all likelihood visible. In the methane greenhouse, the sky would have been thick and hazy, with the planet wrapped in a high-altitude smog of organic molecules produced by reactions between methane and carbon dioxide. The rise of oxygen would have brought with it a host of hydroxyl radicals eager to attack such organic molecules, cutting through the smog like bleach dispelling dye in an advert for toilet-cleaning fluids. For the first time since the dawn of the methane greenhouse – perhaps for the first time since the formation of the planet – the vault of heaven would have turned blue.

Other worlds

As far as photosynthesis is concerned, oxygen is a potentially problematic waste product; but to the biosphere at large it is a great gift. The essence of this gift is its hunger for electrons; in the right circumstances oxygen will pull them off almost anything. This strong redox potential is what makes it such a good thing to have at the end of an electron transfer chain. You can get more energy from such a chain if the electrons are sucked off the end by oxygen rather than by something less voracious, like sulphate or carbon dioxide. As a result aerobic metabolisms can provide much more energy than anaerobic ones – enough energy to power large complex creatures like ferns and flamingos. All large multicellular creatures on earth need the energy levels that only oxygen can provide in order to survive. And so no such creatures were possible before the Great Oxidation Event.

David Catling, now at the University of Bristol, argues that this is not just an earthly restriction – that it is a universal fact of life, the sort of universal fact that exobiology originally went in search of. From what we know of life on earth, there is no way to build complex creatures without an electron acceptor as good as oxygen at the end of the electron transfer chains. And in all the universe there is no other acceptor as good as oxygen. The rules that govern the assembly of atomic nuclei allow for only ninety-two stable chemical elements, and ordain that none of the others, nor any of their simple combinations, is as good a prospect for powering a biosphere.

Oxygen's appetite for electrons, prodigious as it is, does not win it the crown on its own. Fluorine and chlorine have similar appetites; in fact fluorine craves electrons even more than oxygen does. And both fluorine and chlorine share with oxygen the useful trait of being a gas at the sort of temperatures that allow liquid water. While no one has yet shown that water is essential to life in principle, it is hard to imagine living systems remotely like the earth's that don't make use of it. And if life needs liquid water, it also needs an atmosphere that stays gaseous at water-friendly temperatures.

Because chlorine and fluorine are electron-hungry gases, science-fiction by the chemically educated was once full of aliens breathing them. But the chances of such creatures evolving would have been spectacularly low. For one thing, both gases are rare. Oxygen is in plentiful supply throughout the cosmos; the fusion reactions in stars produce oxygen in copious amounts. They produce very little chlorine and fluorine. It is because the universe is rich in oxygen that it is rich in water (though most of it in the form of molecular clouds or ice).

Oxygen not only provides the cosmos with water – it is also

conveniently soluble in the stuff. This means that, all other things being equal, a planet rich in free oxygen will have oceans that offer the same benefits in terms of oxygen availability as the atmosphere. The same solubility also means that oxygen can get in and out of living cells, which are watery places. Fluorine and chlorine, on the other hand, do not dissolve when mixed with water. They react with it to produce viciously powerful acids.

Oxygen's solubility is a special case of a more general helpful attribute: for all its electron-pulling power, it is remarkably stable. When two oxygen atoms bind themselves together to form a molecule – which they will do whenever given a chance – they severely restrict their ability to react with anything else. The bond between them narrows the gullets through which they might swallow extra electrons. As a result, oxygen molecules can mix reasonably harmlessly with all sorts of compounds, wandering the world like muzzled tigers. It is only when some way is found to sneak electrons down those narrowed gullets one by one that oxygen atoms start to become indiscriminately reactive.

The self-denying ordinance of oxygen molecules means that they can mix relatively freely with the biosphere they power while at the same time iron-bearing molecular machinery designed to drip feed them electrons can make full use of their powerful appetites. Chlorine and fluorine molecules are incapable of such subtlety – their atoms pull electrons from more or less anything they touch. Immerse organic matter in fluorine and it explodes.

Oxygen is thus the best tool for excising electrons from other molecules that a planet could be expected to offer, a tool boasting a sharp edge, great ease of use, and a safety guard. It may, as Catling argues, be the only tool that allows life to get its electrons flowing strongly enough to meet the power requirements of large organisms. Life without available oxygen might be doomed to remain small, even single-celled, for ever. Such life is hugely important on a planetary scale – bacteria are still today responsible for almost all the biogeochemical recycling that keeps the earth habitable. But it is not particularly companionable.

This is why it is the possibility of detecting oxygen that most excites the designers of telescopes for studying the planets of other stars. Methane interests them too: methane levels like those suggested for the Archaean might prove detectable, and if out of chemical equilibrium with the rest of the atmosphere would be hard to explain without appeal to life. But oxygen would be easier to detect, and more telling. It would be evidence of a biosphere with enough power to move beyond microbes and into the realms of complexity which creatures like us inhabit – indeed, evidence of what may be the only such type of biosphere that the universe permits.

That said, oxygen is not a fail-safe sign of life. Venus, for example, is thought to have had oxygen in its atmosphere at one point – the point at which it was catastrophically losing the oceans with which it is thought originally to have been endowed. Venus is closer to the sun than the earth, and even the fainter sun of the Archaean delivered forty percent more solar energy to its surface then than the earth receives today. As the sun warmed up, evaporation from any oceans it had would have made its atmosphere ever-richer in water vapour, which is itself a powerful greenhouse gas, and so its increase in the atmosphere would warm the surface even more. Eventually almost no heat from the surface was making it out into space, and the surface temperature climbed far above the boiling point of water, removing the oceans. With that much water vapour in the atmosphere even the high stratosphere became moist; the water was split by ultraviolet light and the hydrogen escaped. As a result oxygen would have been present in the atmosphere. But only for a while: Venus lost all its oceans fairly quickly, and is now as hellishly arid as it is hellishly hot.

So care will have to be taken. But distinguishing living earths from boiling Venuses and other potential confusions should not be too hard in the long run. The odds are with us, too, in that Venus had free oxygen for only a brief period, whereas the earth has boasted the stuff for more than half of its life to date. It should eventually prove possible to weed out any false positives and detect biospheres like earth.

It might even, in time, prove possible to learn more. If we were looking at the earth from a great distance, we would not just see the spectral signs of oxygen – we might well see the spectral signature of chlorophyll. Among chlorophyll's many excellent qualities as a front end for photosynthesis is its high reflectance in the infrared – once wavelengths get appreciably longer than those of the visible they start to bounce right off. If our eyes saw in these infrared wavelengths, forests would glitter as though on fire.

It has been argued that this 'red edge' effect is important to chlorophyll, as it stops it from getting over-energized by absorbing photons too long to be useful, and that it might be a necessary facet of any photosynthetic system. And in the case of the earth, the red edge is a strong enough effect to be seen on the planetary scale. Studies of the spectrum of the 'earthlight' that is reflected from the dark parts of the moon, light which carries the aggregate spectrum of the planet as a whole, show the red edge effect, though not consistently (if the side of the earth that the moon is reflecting is covered in cloud, for example, the spectrum is smoother).

A first-generation planet-studying spectrograph might not be able to see such effects, but further down the line it is not impossible. It might even be possible for such telescopes to see if the photosynthesizers actually use chlorophyll. In the 1970s, the distinguished biochemist George Wald argued that chlorophyll was so well suited to photosynthesis that it was the only pigment that could possibly function in the role, and thus that life elsewhere would have to discover chlorophyll in order to photosynthesize. The current fashion in biology is to think instead that there must be a range of other possibilities that earth-life simply didn't get round to exploring, perhaps because chlorophyll is good enough. But no one will know until they start to study in detail the spectra of other earthlike planets.

If, that is, there are any there to study.

It is this fundamental uncertainty that makes the telescopes in question so popular with astrobiologists; they will tell us how common, or rare, life like that on the earth is. Today we cannot really begin to guess at an answer to that question because we simply do not know how likely the various conditions needed are. It is possible that life gets started only under cosmically rare conditions; it is also possible that it starts on more or less any planet where the energy flows give it a chance. It may be that equivalents to the photosynthetic reaction centre, or the water-splitting complex, are hard things for evolution to come up with; both seem to have evolved only once on the earth.

The likelihood of a system of plate tectonics like that which accounts for carbon burial on earth is similarly unguessable. Any earth-sized planet will start off with a lot of internal heat trying to get out, which is good; but those that start with less water may never develop seas and the erosion-inducing hydrological cycles they allow, while those that start with more water may never develop continents at all. Both low-erosion and all-ocean planets might be hard put to divide organic carbon from oxygen in a way that lets the latter build up.

Which leads to the more general question of oxidation. Plate tectonics may not be necessary for an earthlike biosphere, but free oxygen is, and that requires an oxidized environment. Earth-sized planets that start off with reduced atmospheres will undoubtedly oxidize themselves over time by losing hydrogen, and life that produces methane will speed up that process. But details of the planets' chemistry and size may exert a crucial control. Rather counterintuitively, models suggest that on a planet like the earth but a bit bigger, hydrogen loss will be accelerated. It's only when a planet gets really big, or really cold, that it can hang on to its hydrogen come what may. However, if larger planets oxidize quicker, they will also have greater reservoirs of reduced chemicals in their depths. If the rate at which those chemicals get into the surface environment is high enough, they could still wash away the biosphere's oxygen production. You could imagine a situation where the local equivalent of cyanobacteria went on producing oxygen for billions of years before the gas finally broke through to the atmosphere. You can even imagine the time it takes oxygen to get into the atmosphere exceeding the lifetime of the planet's parent star.*

The questions of how long it takes to oxidize a planet, and how many ways there are to do it, are fiendishly involved, depending on subtleties of chemistry, physics, geology and biology, as well as accidents of planetary composition, such as the nature of the gases stored up in the interior. Teasing them apart is made even harder by the problem that, at the moment, we only have one planetary oxidation event to study, our own. Even there, we are hampered by two billion years in which the prodigiously active earth has tried its best to erode away all the evidence.

But for an astrobiologist in his thirties like David Catling there's every hope that this will change. The first telescopes with which to see the atmospheres of other earths will not be built tomorrow. But, barring a significant collapse in our technological abilities, or a vast change in our outlook on the universe, at some point in the decades to come they will be built, and will be used. The question of how many planets out there are alive – alive in a way that can support complex creatures like us – is answerable by looking for oxygen in the atmospheres of wet, temperate worlds round other stars. In astrobiology – and to some minds in all of science – that is the biggest question that we currently have it in our hands to answer, and that, I think, is why eventually we will choose to build the tools to do so.

The same applies in reverse. We have not yet built a telescope this capable; others, elsewhere, may have done so long ago. If they have, and they have turned it towards us, they will have seen the unmistakeable signs of life. Hundreds of light-years away – maybe thousands, if their telescope-making is a few centuries better than ours, or millions, if they have become like gods – they could be

---

* There are some stars which are much longer-lived than our sun, and would offer ten billion years or more for oxygen to try to find its way into the atmosphere. But these stars tend to have redder spectra than the sun, and thus offer less energy per photon. Without reasonably high-energy photons, it wouldn't be possible to get electrons from oxygen to reduced carbon compounds with a two-jump system; you might need three jumps – a three-photosystem W-scheme rather than a two-photosystem Z-scheme. And that might be very difficult indeed to evolve.

looking at our planet, appreciating the signs of life that it offers, and wondering about its inhabitants, so distant in time and space.

And if they are animals, and not machines, then near them as they wonder there will be photosynthesizers of some sort, opening their world up to the alien energy of its star. And that star will warm their wonder through a slow-burning oxygen sky.

CHAPTER FIVE

Fossils

Driving to Downe
The first snowball
The world of the chloroplasts
The boring billion
Invading the air

The rock garden held her attention for some time. She tried to watch the rocks grow, and failed as always, but did not mind failing in such a beautiful place.

SPIDER ROBINSON, *Night of Power*

Driving to Downe

Towards the end of 1988, in a slightly dilapidated Mini, Bob Spicer, Andy Knoll and I inched our way through the snarled traffic of a southeast London Saturday to Down House, the home of Charles Darwin. Andy, then and now a Harvard professor, had never visited it, and wanted to make a pilgrimage. Bob, then teaching at Goldsmiths College in nearby Deptford, but later to be in charge of a large amount of the earth and planetary science research at Britain's Open University, was being obliging. I, who had met them both at a meeting on extinctions in the fossil record held at the Royal Society over the previous week, had opportunistically wangled an invitation to come along.

That meeting at the Royal Society, Britain's oldest scientific organization, had been an education for me. It had mostly consisted of Britain's older generation of palaeontologists being rude to an eminent American during the presentations, while in the coffee breaks their younger colleagues reassured him that not all Britons were blinkered and boorish. The eminent American was David Raup of the University of Chicago, who had in the mid-1980s suggested that the mass extinctions which had punctuated the fossil record ever since complex animals first appeared – which is to say, for a bit more than half a billion years – could have been caused by comets.

The starting point of Raup's argument was increasingly strong, if contested, evidence that the mass extinction which killed off the dinosaurs at the end of the Cretaceous, sixty-five million years ago, had been caused by the impact of a comet or asteroid. Luis Alvarez, the man who had come ashen-faced into one of Andy Benson's physics lectures to proclaim the news of nuclear fission, and his son Walter, a palaeontologist, had discovered high levels of iridium in sediments laid down right at the relevant time. Since iridium is rare on earth the Alvarezes suggested these deposits were fallout from a massive comet or asteroid impact which had thrown up enough dust to shut down photosynthesis all around the world; half the earth's species then starved in the dark.

Raup and his colleague Jack Sepkoski had taken the idea further, suggesting all mass extinctions were caused by such collisions, and that they come about on a regular basis because of a supply of impacting comets regulated by a slow astronomical cycle. This was all too much for the upper echelons of British palaeontology. They had been schooled to believe that geological explanations should always be gradualist, not catastrophic. The one great paradigm shift they had lived through – the advent of plate tectonics – had rein-forced this view. Even as it provoked a sudden shift in the way the world was seen, plate tectonics denied the need for sudden shifts in the way the world behaved. Continents moving around the globe slowly and sedately since time immemorial could explain pretty much everything. What's more, I realized as the meeting went on, this establishment was dubious of grand theory in all its forms on principle. It suspected that such theorizing involved a lot of input from other sciences – which it seemed somewhat dubious of – and insufficient time spent doing fieldwork in often rainy conditions, an activity which should clearly be the core of the discipline. Thus the hostility to Raup, spiced with such barbs as 'This slide, I should explain for our American colleagues, shows what we call a rock'.

For me, and for many, the first of the supposed flaws in Raup's approach was a large part of its attraction. When he looked at the shell of an ammonite, he saw more than the details of where it was collected and the ways in which it differed from the shells of other ammonites. He was willing to put the study of fossils, often seen as a form of mere stamp-collecting, into the widest of contexts, linking events separated by hundreds of millions of years, tying the deaths

196

of invertebrates on muddy seafloors to cometary encounters on the edge of the solar system. He was convinced that there could be grand truths and overarching patterns to the history of the world, truths that cut across the narrow lines of academic disciplines. And he was convinced that fossils, looked at properly, might reveal them.

This chapter, which takes the photosynthetic history of the earth forward from the Great Oxidation Event about 2.4 billion years ago to the Carboniferous period, 360 million years ago – from cyano-bacteria to trees – is informed by just that way of looking at the earth. It is a story my companions in that traffic-bound Mini, among many others, have helped put together. In its planetary perspective it is, again, an astrobiological story, and though in 1988 Knoll and Spicer would have been seen as a palaeobotanist and a palaeon-tologist, they've both since added astrobiological affiliations to their professional identities. But their main interest, whether on earth or Mars, is fossils.

Knoll and Spicer were admirers of Raup, but hardly disciples – as it happens, Bob Spicer was and remains quite stubbornly sceptical about the effects of the impact at the end of the Cretaceous. But they were like him devoted to the broadest view of what fossils can say about the world. And the fossils in which they were interested were those of photosynthesizers. Because they are so intimately entwined with their surroundings plants are exquisite environmental markers; and they are also primary agents of environmental change. Spicer, in particular, feels that plants get unjustly sidelined in palae-ontology. To counterbalance this, he makes a particular effort to see earth history through the eyes they don't have. A man with a curious mix of enthusiasm and a slightly hang-dog air – a contrast to the equable but reliably up-beat Knoll – he's easily imagined as a sort of shop steward for the vegetable kingdom, asking dubiously of every idea put forward about the planet, 'What's in it for the plants?'

As we drove out to Down House, in the village of Downe, all sorts of outrageous hypotheses and interdisciplinary insights flew around the cramped confines of the car. Spicer was particularly enthused by new statistical techniques he was working with – and works with still – that allowed him to estimate annual averages of temperature and rainfall with remarkable accuracy just by looking at the shapes of fossilized leaves. As we walked along the sandy path where Darwin strolled and thought, the autumn leaf-litter looked like a message to the future.

Knoll, who had spent some of his early career looking at questions of plant morphology, had by this time committed himself largely to the era before plants proper came into being. He wanted to know what fossil photosynthesizers might reveal of the way the world worked during the Proterozoic, the vast tract of time that lasted from the end of the Archaean, 2.5 billion years ago, to the beginning of the Phanerozoic, 543 million years ago. There are no fossil leaves in the Proterozoic. Indeed there are no fossil plants (though there are algae). Until the very end there are no fossils of anything like an animal, either, something which Darwin himself considered rather troubling. But there are plenty of fossils of cyanobacteria and other microbes, if you know how and where to look for them, and by 1988 they had become Knoll's main focus. He thought, and still thinks, that they would let him tell a story of how life and its environment have evolved together, a story much more complex, and more interesting, than a simple succession of fossils – but a story to which that succession forms an indispensable foundation.

The first snowball

The Great Oxidation Event got the Proterozoic off to a dramatic start. It not only changed the chemistry of the atmosphere; it also seems to have changed the climate, suddenly and spectacularly, producing what may have been the first true catastrophe in the history of life on earth. The collapse of the methane greenhouse, which was either a precursor to the rise of oxygen or a part of the process itself, nearly froze the planet solid.

One of the most striking discoveries made about the earth in the past half-century is that it seems to have a natural thermostat. As Harold Urey pointed out in the 1950s, when water and carbon dioxide are both present they turn silicate rocks – of which most of the earth's crust is made – into carbonate rocks, a process known as chemical weathering. This weathering, like most chemistry, intensifies with rises in temperature. Put that temperature-dependence together with the fact that carbon dioxide can drive temperature changes by absorbing the earth's outgoing infrared photons and you have the potential for a negative feedback – a feedback in which the effects counteract the causes.

Imagine a world in which the rate at which carbon dioxide is added to the atmosphere by volcanoes balances the rate at which it is removed by weathering. The carbon-dioxide level will be stable, and so will the temperature. Now add some new volcanoes, as plate tectonics often does. Carbon dioxide will go up, and the planet will get warmer. But when the planet gets warmer, chemical weathering rates will increase, and the rate at which carbon dioxide is removed from the atmosphere will thus also increase. This increase in weathering works against the increase of carbon dioxide, pulling the temperature back down.

Conversely, if the planet cools down for some reason, the weathering rate will slow; carbon dioxide from volcanoes will then accumulate in the atmosphere and warm the world back up. Push the system one way, and it tries to push back – the harder you push, the stronger the resistance. This thermostat, first modelled in the early 1980s, provides at least a partial explanation for the stability that the earth's climate shows over long stretches of time during which the sun has steadily been getting hotter.

The weathering feedback loop is not a terribly tight constraint on the climate. It allows both 'greenhouse' periods in which the planet is all but ice-free and cooler 'icehouse' periods in which there are icecaps at the poles. But in terms of average temperature the differences between these two climate states are pretty small – considerably less than the difference between a winter day and a summer one in many temperate parts of the planet. On Mars and Venus, planets where, for different reasons, the thermostat has either failed or was never properly installed, temperatures seem to have fluctuated much more.*

One of the benefits of this weathering thermostat is that it holds in check another temperature feedback – one of the self-promoting, positive kind. A key control on the temperature of a planet is its 'albedo' – the amount of sunlight it reflects back into space. The higher the albedo, the cooler the planet. Clouds increase albedo; so does the dust from volcanoes. And so do icecaps. If a planet cools down and its icecaps grow, the consequent increase in albedo will, all things being equal, cool it further, leading the icecaps to grow still more. This ice-albedo effect seems to have accounted for a good part of the growth of the icecaps in the ice ages the earth has slipped in and out of over the past few million years.

But as the polar ice spreads across continents, lowering the temperature, it also lowers the rate at which weathering draws down carbon dioxide from the atmosphere. And anywhere where it lowers the temperature below 0°C it stops the process dead, since weathering needs liquid water to work at all. The growth of the icecaps thus leads to a world with more atmospheric carbon dioxide, and thus a stronger greenhouse. Positive and negative feedbacks come into balance.

Without an effective carbon-dioxide greenhouse, though, the icecaps could under some circumstances just keep growing. The larger the icecaps, the stronger the ice-albedo effect. The ice would spread from the poles, through the mid-latitudes, to the tropics and the equator itself. The world would be frozen solid. And in the Archaean, as we have seen, it seems there wasn't an effective carbon-dioxide greenhouse; most of the greenhousing was probably being done by methane. Take away that methane, and the carbon dioxide in the

* On Venus, the thermostat was broken when the planet was baked dry by its runaway greenhouse effect. For Mars the story is probably more complex. An insufficient supply of carbon dioxide may have played an important role, as may acidity in what liquid water there ever was near the surface; carbonates cannot form in water that's too acid.

atmosphere would have been at too low a level to achieve much of anything.

It has been known since the 1960s that various rocks laid down near the beginning of the Proterozoic, about 2.4 billion years ago, showed evidence of being ground down or otherwise altered by glaciers. Since two billion years of plate tectonics have moved the continents around a fair bit since then, in ways still not fully understood, it has never been easy to say exactly where the rocks were when the glaciers rolled over them, and if they were at high latitudes glaciation would not have been that surprising. But an increasing body of evidence suggests that the glaciers reached into the tropics, and down to sea level. In the 1990s Joe Kirschvink, a Caltech geology professor with a taste for dramatic, even pugilistic, theorizing suggested that this could be evidence for a 'snowball earth'. With the methane greenhouse broken, and in the absence of high levels of carbon dioxide, the world froze from pole to pole. Except where volcanoes kept small oases warm, the ice grew thick. The average global surface temperature may have plunged to minus 50°C.

This state, Kirschvink estimates, would have lasted for tens of millions of years. The planet's slow salvation came from its volcanoes and their constant production of carbon dioxide. On the cold, dry snowball this carbon dioxide would not have reacted with the rocks at any appreciable rate; instead it would have built up in the atmosphere. Eventually, this slowly assembled greenhouse would have become strong enough to melt some of the sea-ice, at which point the world flicked from frozen stasis to heated overdrive.

Tens of millions of years of undersea volcanism would have left the ocean more or less saturated with carbon dioxide; when the caps started to retreat, the seas gave up this gas; some have imagined them fizzing like champagne. Carbon-dioxide levels in the atmosphere, already high, were pumped up further still. Water-vapour levels would have leapt up too – and water vapour is also a powerful greenhouse gas. The positive feedbacks would now all be working *against* the ice. As the bright ice retreated ever-greater expanses of dark rock and darker seas warmed themselves in the sunlight as the oceans gave up ever more greenhouse gas, forcing the icecaps further and further back. The temperature difference between the tropics and poles – the factor which controls the amount of work done by the great engines of the climate – would have been enormous. Hurricanes fed by the hot waters of the greenhoused tropics would tear at the retreating fringes of ice like unleashed furies.

Meanwhile the cyanobacteria, having survived, presumably, at volcanic hotspots with thin ice, would greedily set about eating up the most carbon-dioxide-rich atmosphere they had ever seen. Life would spread back around the planet in a titanic, exponential bloom. The reduced iron that had built up in the sealed oceans would be oxidized, precipitating out into great sheets of iron oxide; once the iron was used up, other metals, such as manganese, would suffer a similar fate. This burst of activity would lead to a great deal of organic carbon being buried; the atmosphere and surface oceans would end up more oxygen-rich than ever.*

As is the case for most of what people say about the early earth, the snowball story that Kirschvink and his colleagues have put together is not something for which there is conclusive proof. But there is a lot of circumstantial evidence: there are rocks that were close to the equator which look as though they were shaped by glaciers; in Namibia the largest deposits of manganese ore in the world seem to have been precipitated out of seawater at exactly the same period; the isotope levels in carbonates from that time swing all over the place, suggesting vast upsets in the rate of carbon burial. If we accept Jim Kasting's conclusion that, in the absence of methane, it would have been all but impossible to keep the Archaean tolerably

* A few years ago, when my wife Nancy Hynes and I spent a couple of days with Jim Lovelock and his wife, Sandy, at the home they share with Lovelock's son John on the borders of Devon and Cornwall, I happened to notice a microcosmic recapitulation of the snowball rolling back from the oceans. One morning we took the translucent winter covers off the Lovelocks' swimming pool – a pool kept clean, not by chlorine, but by a system of reedbeds, which, after a winter without use, had walls covered with a film of algae. By the afternoon this algal film had ballooned off the wall at various points, blebbing out like bubble wrap. Uncovered as though by receding ice, the algae had started to grow, to fix carbon, and to pump out oxygen.

warm, and also accept that, not long into the Proterozoic, we can see rocks with chemistries which demonstrate the presence of oxygen in the atmosphere, it is hard to see how the earth could have avoided some sort of climatic convulsion as it moved from one state to another. Or convulsions. There's evidence that the early Proterozoic saw not just one glaciation but a whole series. The ice may have come and gone repeatedly as the system of the earth searched for a new stability.

The earth recovered from this crisis. Other planets may not be so lucky; again, it is possible that Mars provides such a contrasting example. Imagine an early Mars that, like the early earth, was inhabited by methane-makers and something cyanobacteria-like. Because the planet is smaller – just a tenth the volume of the earth – it had less internal heat, and less potential for squirting reduced gases up to the surface through volcanoes. It would thus be much easier to push the atmosphere from a reduced state to an oxidized one, meaning that Mars might have had its Great Oxidation Event much earlier, when the sun, from which Mars is half again as distant as the earth, was considerably cooler. Mars could thus have fallen into a snowball state from which it could never escape. Its photosynthesizers would have eventually died, severing its biosphere's links to the cosmos. Only the methane-makers would survive, deep below the ground, fragments of life buried in a dead planet.

The world of the chloroplasts

There have been three great revolutions in the history of life on earth, one external, two internal. The development of oxygenic photosynthesis was the first internal one, and the Great Oxidation Event was the subsequent external one, making possible the sort of large creatures we see around us today. But in order for life to make use of this new potential it had to undergo a further interior revolution: it needed the eukaryotic cell, much larger and more adaptable than its bacterial or archaeal counterpart. The world of complex shapes, of the genomes that support them and of the behaviours and life-cycles they make possible, is the province of the eukaryotes.

Why could bacteria not evolve into larger creatures? One reason is that despite their spectacular metabolic diversity, the constraints placed on them by rigid walls allow bacteria to grow into only a limited repertoire of shapes. More fundamentally, though, they are constrained to remain small because the membranes across which they build up proton gradients in order to make ATP are membranes that lie between the cell proper and those rigid walls. Thus the amount of ATP a bacterium can generate depends on its surface area. But the amount of ATP a bacterium needs in order to stay alive and reproduce depends on its volume. If bacteria were to grow bigger, their need for energy would increase faster than their ability to provide it. If the length and width of a bacterium were doubled then, all things being equal, it would have a four-times-greater surface area but an eight-times-greater volume. Big bacteria will thus be more energy-constrained, and will have to reproduce slower. Since evolution favours bacteria that reproduce quicker than their neighbours, other things being equal bacteria stay pretty small.*

You might think that, in principle, you do not need large cells to make large creatures; why not just use more smaller cells instead? But small cells mean small genomes, while large creatures need large genomes. You couldn't build a complex organism with genomes of the size favoured by bacteria. Cyanobacteria can afford to be a bit bigger than the average bacterium, because folded photosynthetic membranes give them more energy per unit volume, and they can thus afford genomes complex enough to make different types of cell – a photosynthetic type and a nitrogen-fixing type, which because of the sensitivities of nitrogen fixation has to be kept completely free of oxygen. Some cyanobacteria grow in long multicell strands,

* There are exceptions; they don't concern us.

the nitrogen-fixing cells interspersed between chains of oxygen-producers. This is the height of sophistication for bacteria. But two types of cell is pretty pathetic by the standards of large multicellular creatures.

Eukaryotic cells get round the energy constraints by having discrete organelles within them that produce ATP, and they can have as many of them as they can fit into the space provided. Their capacity to process energy can thus grow in step with their volume. They can be big, if they want to, and can contain big genomes. These ATP-producing organelles are the mitochondria.

Lynn Margulis's great triumph of the late 1960s and early 1970s – the triumph that inspired Bill Rutherford to think about biology from an evolutionary standpoint, and that led Andy Knoll, roughly the same age, to decide that the study of early life was more fascinating than any other – was convincing the world that the mitochondria, and the chloroplasts, were once-free-living bacteria that had been entrained into the eukaryotic cell in two rounds of 'endosymbiosis'. The mitochondrial endosymbiosis came first and was the more profound. A world with no chloroplasts would still be able to rely on cyanobacteria to do the biochemical work that plants do today and keep the rest of the biosphere supplied with oxygen. But a world without mitochondria would lack large creatures able to make use of that oxygen.

Genome analysis suggests that the cell into which the proto-mitochondrial bacteria were incorporated may have been one that made methane, though there is vigorous debate on the matter, and on every other aspect of this original endosymbiosis, including its timing. But to many people – including Andy Knoll, who has thought about this a lot – the fact that there are cells that look too large to be bacterial or archaeal preserved as fossils quite early in the Proterozoic but not in the Archaean suggests that the mitochondrial endosymbiosis took place at a time quite close to the Great Oxidation Event. It thus seems probable that life's great external revolution and its second great internal revolution came to a head at roughly the same time. Whether that is coincidence or not remains to be seen.

Their ability to grow to a large size, and the fact that they had given up the confines of a cell wall for a looser sort of living, meant that the early eukaryotes could make a living by eating bacteria whole, just as amoebae do today. Cyanobacteria would be obvious prey. It seems likely that it was through one such act of predation that the chloroplast was created. As far as we can tell, every chloroplast in every plant across the world today is descended from one particular cyanobacterium that was swallowed up by one particular eukaryote. Not one particular species. One particular creature, at one particular place in space and time.*

The eukaryote in question was not planning a life based on sunlight: evolution doesn't look ahead. It's just that for some reason, on this one particular occasion, it didn't digest what it had swallowed – at least not at first. And the cyanobacterium found that it could reproduce inside the cell. Some sort of steady state was presumably reached in which the host cell digested the cyanobacteria growing inside it at roughly the rate at which those cyanobacteria divided. Eventually a set of proteins was developed that allowed the cell to siphon off sugars from within the living cyanobacteria without killing them – taps like those that let out the maple's sap for syrupmakers. The cyanobacteria were no longer an internalized crop to be eaten at will, but an internalized orchard from which fruit could be harvested continuously.

It's not hard to see the attraction of this arrangement for the eukaryote. Growing food within your own body means never having to go far to get a snack. And it was an advantage that could be passed on. When the cell divided, some at least of its daughter cells would have started off life with cyanobacteria of their own nestled inside their bodies.

---

* There is something very chloroplast-like living inside the cells of *Paulinella*, an amoeba, which seems to be descended from a cyanobacterium that entered into endosymbiosis only recently. The fact that this is a single exception, though, suggests that the original endosymbiosis that led to plants and algae has produced creatures that out-compete almost all newcomers; if they didn't, there would be chloroplasts with all sorts of different ancestries around, and with the exception of *Paulinella*'s there just aren't.

At this point the process was probably still reversible. Symbiotic relationships between things that photosynthesize and things that don't are quite common. Corals are animals that take algae into their bodies in order to reap some of the benefits of photosynthesis; lichens are fungi doing much the same thing (though, for the most part, less prettily). Some free-living animals make use of the same trick – sea squirts and even the occasional snail. But in all of these associations the algae or cyanobacteria involved can be flushed out, or choose to leave.

In algae and plants, though – the creatures that contain true chloroplasts – the contract has been made binding. The chloroplasts can no longer make their way on their own, because they have lost almost all of the cyanobacterial genes they started off with. Some of the genes are gone altogether; some are retained on little loops of DNA within the chloroplasts; but most of those that remain have been transferred to the nucleus, where they sit alongside the genes for the host cell.

This transfer would have started off as a natural side-effect of the peculiarly intimate cohabitation. Though you might expect a little circumspection when it comes to questions of genetic integrity, cells are often surprisingly willing to assimilate strange DNA. When a cyanobacterium sitting in a cell died, it would be unremarkable for some of its DNA to end up in the nucleus. When genetic engineers started putting new genes into the genomes of chloroplasts, they found to their surprise that copies of the gene were quite likely to turn up in the nuclei of some of the engineered plant's descendants a few generations later.

Once a gene had been transferred to the nucleus, its presence in the remaining chloroplasts was not just superfluous; it was damaging. Having to copy the same gene more than once when reproducing represents a cost, and the risk of conflict between co-existing genomes – a fascinating biological topic we won't go into here – makes keeping duplicates potentially dangerous. Proto-algae that didn't get rid of genes in their endosymbiotic cyanobacteria didn't leave descendants, and so the chloroplast genome was gradually whittled away. The extent of this whittling differs in different types of plant and algae. In general, the 'red' algae, which use a slightly different form of chlorophyll, have retained more genes in their chloroplasts than the 'green' algae (from which land plants are descended). In some species of green algae, notably some with just a single chloroplast in each cell, the chloroplast genome is reduced to under 100 genes.

It is common to see this as a form of enslavement, with the genetically hamstrung chloroplast constrained to do the nuclear genome's bidding. But an alternative anthropomorphization is to see it as a radical downshifting which allowed the chloroplasts to lead a simpler life, free of tedious long-term worries. The genes the chloroplast sloughed off still serve it – the proteins they describe are still delivered to the chloroplast as required. But no longer having to worry about such things allows the chloroplast to be devoted to the genes that remain – for the most part, genes that describe and regulate the proteins that get oxidized and reduced as electrons flow through the photosystems and the electron transfer chains, which is to say the genes needed in order to respond quickly to changes in the conditions under which photosynthesis is going on.

The stripped-down genome is a way to live for the now of the sunlit moment, providing all the tools needed to keep the core photosynthetic mechanism working as well as possible, and without the distractions that come from managing any other cellular processes. With nothing else to concern it, the chloroplast can delight in continual adjustments to the disposition of its cytochromes and antenna proteins, as happy in balancing the flow of energy as a skipper pulling on sheets and trimming the course in order to keep every centimetre of his yacht's canvas taut in the wind, lost in the faint rattle of sails beginning to pinch and the shiver of the sea running under the keel. Why not let someone else look after the tedious business of storing and safeguarding the genes that aren't central to your moment-to-moment existence? As long as you can get the relevant genes expressed when you need them, why worry about where they are stored?

And what a magnificent, strange world the chloroplasts sailed off into. Because humans are big creatures, it is natural for us to see the story of this symbiosis as starting with an act of ingestion. But from the cyanobacterial point of view it was much more like a colonization. Photosynthetic bacteria had tackled a number of environments – the open ocean, bacterial mats in tidal flats, lake-bed sediments and many more – before the endosymbiosis. The insides of early eukaryotic cells were just another new environment, one which, if it permitted survival, would be colonized. It was not, to begin with, a very friendly new world, since a good fraction of every generation of colonists were eaten. But some survived. And this new world had advantages. It offered a stable environment, one which, unlike the open ocean, provided a reliable supply of all the chemicals needed for life and growth. What's more, this new environment actually grew as the host cells reproduced; it was like a family home that sprouted new rooms for every generation.

Over time, this ever-expanding home became, like Peake's Gormenghast, an extraordinary world of its own. It stretched all across the planet and billions of years into the future, like a skein of wormholes tangled through the fabric of the universe in a baroque piece of science-fiction. This world of trillions upon trillions of tiny tunnels branching out through space and time is the world chloroplasts still live in today; each time a photosynthetic cell divides a tunnel branches, with some of the chloroplasts taking one route while the rest take the other.

As they have spread out with and through their wormhole world, its chloroplast inhabitants have changed according to the require-ments of the time-tunnels that they find themselves in; evolution applies in there just as it does out here. In some tunnels the chloro-plasts are squadrons of small green capsules that still have the look of bacteria. In others they have evolved into swirling spirals pressed tight against the outer membrane like rifling in the barrel of a gun. Some are red five-pointed stars, one to a cell. But they are all descended from the same original ancestor.

There are many dead ends in the network – hundreds of thousands in every leaf that falls, billions of times as many in every fossil species bereft of descendants today. But the threads as yet unbroken are all but uncountable, and multiply still. All around us, moment by moment, the skein of wormholes pushes itself further into the future.

The chloroplasts inside this network are now responsible for more of the world's net primary productivity than the free-living cyanobacteria that have all the outside world in which to live. Evolution has shaped the plants and algae which they inhabit into beautifully adapted life-support systems providing chloroplasts with the light, water, carbon dioxide and trace nutrients they need. And the chloroplasts, now, can no longer live any other way. They cannot get out of their wormhole world. It is only a fraction of a millimetre away from ours, separated from us by thin translucent membranes. We look into it every day when we see the green of a leaf. But its only entrance is a couple of billion years in the past, and there is no exit.

The boring billion

Today, to be trapped in the intracellular world of plants and algae is no great problem for chloroplasts. According to Knoll there have been periods in the past, though, when it has. Knoll has developed a theory that, for much of the Proterozoic, being a photosynthetic eukaryote was simply not a very good idea.

Of the four periods of earth history, the Hadean, the Archaean, the Proterozoic and the Phanerozoic, the Proterozoic has lasted by far the longest.* Its almost two billion years cover half of what we

---

* To be fair, the Phanerozoic, at 543 million years to date, is not over yet. But as we shall see in the following chapter, we can already predict how, if not exactly when, it will end. This lets us say with some confidence that though the Phanerozoic may outlast the 1.4 billion years of the Archaean, it has no chance of beating the Proterozoic.

take to be life's lease on earth. And it was, for much of its duration, remarkably dull. The Great Oxidation Event and the snowball earth got the Proterozoic off to a dramatic start. Towards its end the earth returned to a similar period of disarray, with more isotopic wildness and what appear to have been several further bouts of global glaciation leading up to the 'Cambrian explosion' at the base of the Phanerozoic, the point where complex animals make a dramatic appearance in the fossil record. In between, though, came a period as uneventful as it was long-drawn-out. The carbon isotope record is as flat as a Sunday afternoon walk round Delft. Though there was some evolutionary innovation, it came at a slow pace. Proterozoic rocks that differ in age by hundreds of millions of years contain essentially identical suites of fossils; for stretches of time that exceed the reign of the dinosaurs, let alone the far shorter ascendancy of the mammals, almost nothing changed. And what changes there were had no apparent effect on the environment. The British palaeontologist Martin Brasier has dubbed the middle of the Proterozoic the 'boring billion', and the name has stuck.

By the early Proterozoic the revolutions apparently needed for life on earth to develop complexity had already taken place. There was oxygen in the atmosphere (though not as much of it as there is today) and there were eukaryotes. Simple fossil algae show that by the time the Proterozoic was halfway through sexual reproduction had evolved, which meant life had a way of evolving large, complex genomes. Yet early simple multicellular creatures such as those algae seem to have just sat there, unchanging, in the same way their microbial cousins did. It was not a good eon for the neophiliac.

It is this stasis that Andy Knoll, now probably the world's most respected authority on Proterozoic life, has been seeking to explain. His arguments, developed with a geochemist named Ariel Anbar, build on work by Don Canfield, an earth scientist working in Denmark who devotes great attention to the study of sulphur isotopes. In the late 1990s, Canfield noted that the Great Oxidation Event had not been quite as great as all that. Canfield argued that the main effect of the atmospheric changes on the oceans was not

211

to fill them with oxygen – the atmospheric level was still too low for that – but instead to change their sulphur chemistry. The oxidized surface of the planet would have provided the oceans with a greatly increased supply of sulphate, which microbes in the oxygen-free depths of the ocean would reduce into sulphides. Something similar can be seen in the poorly aerated waters of today's Black Sea.

Canfield's ideas have changed the way that people see the Proterozoic. Before, it could be treated as an intermediate stage in a fairly straightforward, if long-drawn-out, progression from the conditions of the Archaean to the way things are today. The Canfield ocean gives the Proterozoic a unique geochemical identity, distinct from both the iron-rich Archaean ocean – the sulphides would precipitate out any iron – and from the modern ocean, in which dissolved oxygen is normally available even at depth. It was its own thing, with its own unique attributes.

Knoll and Anbar argue that a significant part of the dullness of the mid-Proterozoic can be put down to the fact that this Canfield ocean was a poor place for eukaryotic algae. For all their sophistication, eukaryotes have some basic deficiencies, the most notable of which is an inability to fix nitrogen. To be useful to life, nitrogen has to be fixed from the gaseous form in the atmosphere into a reduced form, the ammonium ion. No eukaryotes have the energy-intensive electron transfer chains and associated enzymes needed to pump the necessary electrons into the nitrogen; but various bacteria, including some cyanobacteria, do. Until the twentieth century, when the chemical industry took a decisive hand in the matter, two billion years' worth of eukaryotic cells depended almost entirely on nitrogen fixed by bacteria.

The proteins in the electron transfer chains necessary for fixing nitrogen require iron and molybdenum, and in a Canfield ocean both of those would have been pretty scarce. Levels of iron would have been a thousandfold lower than they had been in the Archaean oceans where the nitrogen-fixing machinery evolved. And sulphides in the ocean would have got rid of the soluble molybdenum oxides which provide today's bacteria with their supply of the metal. With iron in short supply and molybdenum in very short supply, nitrogen fixation would have been a tougher proposition back then than it is now, and this would have had its effects on the eukaryotes scavenging for the stuff.

As well as a change in the source of usable nitrogen, there was also more competition for the stuff once it got made. Pretty much everything one sort of bacteria makes, another sort of bacteria will eat. Ammonia is food for bacteria, and some archaea, which can oxidize it into nitrate ions, nitrite ions, nitrogen oxides or plain old nitrogen (nitrogen, like carbon, has many different redox states). The oxygenated water in the shallows of the Canfield ocean would have offered such bugs a far more congenial environment than they ever enjoyed in the Archaean, and there they would have provided newly invigorated competition for ammonia-scavenging eukaryotes.

This would not necessarily have been a problem; eukaryotes can use nitrates and nitrites left over by bacteria, as well as ammonium, as a source of fixed nitrogen. But the machinery that lets them do so depends, again, on molybdenum.* The eukaryotes were thus in a geochemical cleft stick; the more molybdenum they managed to sequester in order to mop up nitrates, the less there would be for the nitrogen-fixing bacteria from which those nitrates were ultimately derived.

So life in general, and eukaryotic algae in particular, were starved of nitrogen. And this, Anbar and Knoll argue, accounts for one of the most obvious oddities of the boring billion: the flat carbon isotope record. In normal times, the carbon-13 level in carbonate rocks moves around a bit, reflecting changes in the overall biological productivity of the marine ecosystem. When plate tectonics builds up big mountains, for example, the productivity of the oceans tends to climb, too, because more mountains means more erosion, and other things being equal more erosion means more phosphates in

---

* It is this that makes molybdenum an essential nutrient for higher plants – a fact discovered by Daniel Arnon during his early work on plant nutrition in the 1930s. Copper can fulfil the same role in some algae, but the chemistry involved means that copper, too, would be in short supply in a Canfield ocean.

the water running off the continents. The productivity of life in the ocean, and elsewhere, is particularly sensitive to the rate at which phosphates are supplied because they are a nutrient that cannot be made from anything else; there's no other soluble form of phosphorus, nor are there any phosphorus-carrying gases in the atmosphere. While life can fix organic carbon from carbon dioxide, and fix atmospheric nitrogen into the ammonia that it needs, when it comes to phosphates life has to make do with what it's given. As Liebig first pointed out in the nineteenth century, a plant's rate of growth will be set by the availability of the vital nutrient that is in shortest supply. In today's marine ecosystems that limiting nutrient is usually phosphate.

Though the middle of the Proterozoic may have been a fairly quiet time tectonically, as well as in almost all other respects, the continents were not entirely quiescent; the supply of phosphates to the sea must have fluctuated over time. If the oceans had been as they are today, that would have sent the amount of carbon fixation bobbing up and down, too, rippling the carbon isotope level. But those ripples are nowhere to be seen. The productivity of the oceans seems not only to have been a bit low, by current standards, but also to have been indifferent to the amount of phosphate fed into the system. Knoll and Anbar interpret this as meaning that, for a billion years, the limiting nutrient was not phosphate but nitrogen. Changes in phosphate availability meant little to a marine ecosystem unable to keep hold of enough fixed nitrogen.

Knoll's argument is an appealing one, backed up by various strands of evidence, but it has not yet proved completely convincing to all his colleagues, and may not stand the test of time. I like it because I have a predisposition towards accounts of the world in which geochemistry and biochemistry, inseparable at the time of life's origins, remain intimately entangled throughout history. The notion that complex life was stymied for a billion years by the subtleties of the sulphur cycle and the chemistry of molybdenum is the sort of idea I warm to.

This taste is not entirely a matter of intellectual style; in my case, at least, it reflects temperament, too. I am not as sensitive as I might be to the subtleties of place. I lack a capacity for the sure recognition and the ready retention of names and distinctive detail. Learning to parse the shapes of leaves or the textures of rocks does not come easily to me, and I have never lived long enough in a non-urban landscape for such things to have seeped in through the capillaries of unattended observation. Given all this, the belief that life's nature needs to be captured at the levels of the molecule and the planet – at levels perceived by the intellect and not the senses – provides me with some succour. It is far more abstracted than traditional ways of feeling close to nature; it is argued more than absorbed. Yet for all that it doesn't grow out of the experience of life in the world, I find that it serves to enrich that experience and to render it more profound. It ties the sky to the seed and the rain to the rock in a way the details of rustic experience cannot. I can see that there is something sad about a oneness with the world that can be felt as easily – sometimes more easily – from the window seat of an hermetically sealed and environmentally damaging passenger jet than when sitting on a riverbank and picking out the trout swimming upstream. But for all that this belief is a creature of the mind, rather than a sentiment grounded in birdsong and summer scents, it has meaning to me that I cannot reduce to analysis and it has the power to move me. And I think it can enhance more traditional forms of empathy with nature. It enriches the way I see trees on a scarp, or grass in the wind, or moss on a cliff, or a star in the sky, even though I can rarely recognize the species, rocks or constellations I may be looking at. A sense of planet can amplify a sense of place.

However, having declared a prejudice for his type of approach to the world, I have to go on to say that an alternative to Knoll's account of the longueurs of the Proterozoic developed by one of his former students makes no use of it at all. Nick Butterfield, now a researcher at Cambridge, argues that the slow pace of evolutionary change in the Proterozoic is due simply to a lack of animals – things with nervous systems and guts and ways of moving around. Today a great deal of natural selection is down to the way animals interact with each other, and with the plants at the base of their food chains. Without those interactions, nature would have less cause to select. And so what seems to us like a glacial Proterozoic evolutionary tempo, says Butterfield, is in fact a perfectly natural tempo for a world in which there were no animals. The rate of evolutionary change in the Proterozoic only seems slow because we are used to the quicker tempo it broke into around the beginning of the Cambrian and which it has kept up ever since – perhaps even accelerating a bit further. The flow of glaciers only seems slow to creatures who are more used to water in its liquid state.

In Butterfield's view of the world, the earliest animals appeared only in the latest stages of the Proterozoic, their presence announced not by fossils of their own bodies but by a sudden increase of diversity in the single-celled creatures on which they fed. Pretty much as soon as they learned the tricks of the animal trade an explosion of diversity such as that seen in the Cambrian was inevitable. The implication is that if evolution had stumbled upon the trick of making animals a billion years earlier, then the Cambrian explosion would have happened then, instead – regardless of the chemistry of the oceans or the availability of molybdenum. As long as someone fixes nitrogen, animals can get at it by eating them.

This flies in the face of a large amount of geological thought which insists there must have been an environmental 'trigger' for the Cambrian explosion. The last few hundred million years of the Proterozoic are a time of spectacular environmental change. There seem to have been at least two more glaciations so severe they could have covered the whole surface in a 'snowball earth'; there are wild variations in the carbon isotope record; there was a great deal of plate tectonic activity; there's even a pretty large asteroid impact. People who want to tie the stories of life and the environment, for example by claiming that these events pumped up the oxygen level from a few percent of the atmosphere to ten or twenty percent, thus have a wealth of material to work with.

Nor do they have to assume that life was entirely passive in the matter. Andrew Watson and his colleague Tim Lenton, a former student of Watson and a disciple of Lovelock, have suggested that the environmental shifts may have been driven by the humble lichen. The first fungi appear in the fossil record during the boring billion, well before the end of the Proterozoic, and there is no obvious reason that some of these early fungi should not have formed symbiotic relationships with cyanobacteria, thus producing lichen.

Before such a symbiosis arose, life on the continents would have been limited to thin bacterial crusts on some rock. Lichen would have been able to spread further and do more, because eukaryotic fungi are much better at leaching nutrients out of the rocks and sands they grow across than bacteria are. Watson and Lenton have suggested that this activity would have increased the rate of chemical weathering on the continents – which is not a particularly contro- versial claim – and thus, through the mechanisms of the great thermostat, drawn down carbon dioxide. More speculatively, they suggest that the lichen might have greatly increased the rate at which phosphates were extracted from the continental rock, and thus increased the flow of phosphate to the oceans, kicking off a new burst of primary production and allowing a greater rate of carbon burial in the ocean depths. More carbon burial, as ever, leads to more left-over oxygen.

Thus the lichen would bring down carbon dioxide and, indirectly, raise oxygen levels. Both of these things might have helped set the stage for glaciations; lower carbon dioxide cools things down, and more oxygen means less methane. The methane levels were by this stage much lower than they were in the Archaean, but would prob- ably still have been high enough to make the gas an important part of the planet's overall greenhouse warming.

It's hard to accept Butterfield's argument that there is no need for such stories, and that the rise of the animals was not intimately connected with the huge environmental upheavals of the late Pro- terozoic. But Butterfield is persuasive. Happily, the ideas do make divergent predictions that can be tested. If the environment really constrained the biosphere in the way Knoll imagines there should be ur-animals further back in the fossil record than Butterfield allows. At the moment it is hard to say for sure. But recent studies have made it increasingly clear that right at the end of the Proterozoic oxygen levels rose dramatically. Large animals and complex algae appeared almost immediately thereafter. And once the algae became this complex – and their evolution came to be driven by animals that could and would eat them – it was only a matter of time before some of them began to colonize the continents.

Invading the air

On the first floor of the east wing of the Natural History Museum in London is a long, wide room, low-ceilinged for its size. The imagination wants museum archives to look antique, but this one merely looks a little old-fashioned, filled with row after row of blank-faced metal-framed cabinets, its walls painted a disconcerting shade of off-green a little too bright to be bland. It's a drab 1970s contrast to the Victorian splendour of the museum proper, where creation's grandest specimens are housed in an iron-vaulted central hall that's half cathedral, half railway terminal. There a skeletal *Diplodocus* dominates the floor while above, where a cathedral would have its rose window, a cross-section from a giant redwood reveals almost two millennia of history.

The wonders in the functional off-green archive are less showy, and so not shown. Inside its cabinets are the museum's palaeobotanical specimens: 250,000 rocks containing fossil plants arranged more or less chronologically, from the oldest to the most recent. Walk down the aisles past these cabinets and you walk through almost 500 million years of the planet's most recent past, assembled through lifetimes of careful collection.

You will not find the first land plants here. The fossil record is not the place to go for firsts. It has refused to yield up the first life, the first cyanobacteria, the first eukaryotes or the first animals. But you will find evidence of when those plants emerged. The Phanero-zoic era is divided into eleven periods,* of which the Cambrian is the first. By the second, the Ordovician, some sediments contain spores almost sure to have come from plants. And these Ordovician spores make obvious one of the differences between the study of fossil plants and animals. Plants leave their mark in the rocks in various different ways. Spores and, later, pollen get everywhere; so do leaves, if they fall into the right sort of muds and silts. But a plant's pollen and leaves won't be found connected to each other, or to any fossil tree trunk you might happen to find. In the vast majority of cases, the specimens that palaeobotanists work with are incomplete fragments.

Plant fossils from the Ordovician and the geological period which followed, the Silurian, are rare in part simply because dry land is not an environment conducive to fossilization; most sediments are laid down in seas, which is why the vast majority of fossil animals are aquatic. On top of that these early plants were small and fragile. When the animals made it to land, rather later on, they brought with them distinctive sizes and shapes and behaviours. Fish dragged themselves across the mud on their stumpy fins rather as they had propelled themselves across the sea floor; arthropods scuttled on their little legs as their aquatic ancestors had done. The plants, though, started more or less from scratch. The algae which sit closest to land plants on the tree of life are small and nothing much to look at. Their immediate descendants probably looked like tiny mosses, and would have been distinctly lacking in the sort of hard parts that stand a chance at fossilization. There were no treelike algae from which trees would descend and shrublike seaweeds to give rise to shrubs. The magnificent range of shapes and sizes in which plants come today was all developed after their common ancestors surfaced from the water.

Within about a hundred million years or so, evolution had

* Twelve if you are an American: what Europeans call the Carboniferous Americans have traditionally divided into two, the Pennsylvanian and the Mississippian.

assembled the toolkit with which that magnificent range would be created. Just after the cabinets containing the earliest spores and fragments in the Natural History Museum you will find that change documented in samples of an extraordinary rock discovered in a field near a Scottish village called Rhynie.

This Rhynie Chert is a dark rock, heavy, hard and tough. It's close to black in places, interrupted here and there by thin veins of something lighter. The surface has a smooth lustre, and if you study it closely you get the feeling that you are seeing just a tiny way into the rock itself. A peculiar mottling of lighter greys seems to sit just below the surface, like dimly seen carp below the surface of a pool. The well-trained eye – an eye such as that of William McKie, a doctor and geologist, who came across a piece of the Chert in a wall while surveying Aberdeenshire a couple of years before the Great War – can tell that this mottling comes from plants preserved within the rock. And when samples of the Chert are ground down into what geologists call 'thin sections' – slices thin enough to shine light through and examine under a microscope – the ancient plants within are revealed to be so well preserved as to beggar belief. The Rhynie fossils are at the same time some of the oldest fossils of complete land plants that geologists have yet found and some of the best-preserved fossil plants of any age at all. It's as though a sound archivist had found a wax cylinder from the days of Edison that everyone had imagined lost and it had turned out to have CD-quality sound on it.

On a grey afternoon in late winter Paul Kenrick, a palaeobotanist at the museum, showed me some of the original thin sections that McKie had made from the Rhynie Chert. Through the microscope, the dark unyielding rock becomes a golden world of tiny translucent beauties, a whole ecosystem captured in minute detail. The remarkable preservation is a fluke of the plants' location. They were growing, it seems, near a geyser or hot spring that submerged them in water saturated with silica. In contact with the plants and animals the silica precipitated out to form a solid mass of tiny crystals of quartz.

Within this clear quartz matrix the ancient plants are preserved down to the level of the very cell – as are the bodies of the wee and quite possibly timorous beasties that lived among them, mites and millipedes and little spider-like things. In one slide you see a stem in cross-section, a circle of closely packed cells in a honeycomb half a centimetre across; in another a prone stem lies across the field of view like a thick-scaled snake. Everything is suffused with a gentle yellow light. Dark-edged cells in the stalks reveal the presence of lignin, the tough, enduring chemical that gives wood its structural strength. Some stems are shot through with the fine tendrils of fungi seeking sustenance. An explosion of distinctively shaped spores seems caught in the act of puffing itself into the wind. A stomate – one of the tiny nostrils through which land plants take in carbon dioxide – is flared open in the cuticle of one of the thicker stems.

It is not the fact that they lived on land that makes these plants so remarkable. In the form of cyanobacterial crusts, and later lichens, there had been photosynthesizers on the continents for a billion years or more. What is remarkable is that, with their stems and their lignin and their windblown spores, they are beginning to leave the surface of the earth and take their first steps into what creatures built from gases might see as their natural home: the air.

Our preconceptions tell us that plants are built from the soil into which they spread their roots. The idea appeals to us: soil is always local, immediate, particular, something with its own feel and smell. We can grind it into the whorls of our fingerprints; seeds having been sown, we can watch as life emerges from its secret places. A patch of soil can belong to us and we, we are told, can belong to it. The conservative and the romantic imaginations both tell us that the soil of our birthplace is second only to our blood in defining us, that the soil houses the spirit of place that makes it special, that makes it worth fighting over, that makes it ours. Rootedness, an essentially soil-bound idea, is our primary metaphor for belonging.

But no soil produces, in itself, special generative powers. Yes, it is vital to the plants we feed on and to those we don't, both as a source of nutrients and as a way of not falling over. Yes, it teems with microbial and other life engaged in subtle processes we as yet only dimly understand. But although the soil is vital and mysterious and distinctive, it is also, at some level, incidental. The particular plot of land in which a tree is rooted is its accident. The tree's purpose is the sky.

Think of a beech tree in winter, its leaves lost, its architecture revealed in dark lines against cold grey cloud. Do what Robin Hill used to urge his children to do to cultivate the artist's eye – take away the tree's established 'common sense' context by turning round, bending over and looking at it upside-down through your legs. Its growth looks less like something pushed from the earth than it does something drawn from the sky. Its limbs, branches and twigs spread into the air like ink into blotting paper or cracks spreading through glass, embodying something between desire and transubstantiation.

The tree's form tells the truth. The tree grows into the air because it grows out of the air. The bulk of the tree is not made from the soil beneath it – indeed, the soil is in large part made by the tree. Both soil and tree are made from carbon drawn from the sky above. Trees are built from sun and wind and rain. The land is just a place to stand.

To look down the microscope into the golden micro-world of the Rhynie Chert is to look at the early days of the plants' ascent into the atmosphere and subsequent spread across the face of the world, an efflorescence of novelty that is called the Devonian explosion after the geological period, from 408 million to 362 million years ago, in which it took place. The tallest of the Rhynie plants reached only to knee-height; most were smaller. They were as yet more or less leafless and had no way of gaining in girth to bolster themselves as they grew higher. They could not stabilize themselves with roots. But they had developed the basic biological mechanisms needed for a life in the sky, as opposed to life in water, and they were poised to make full use of them in order to reap the various benefits of the aerial life – unfiltered sunlight, carbon dioxide flowing freely in and out of the photosynthetic tissue, and winds spreading spores and seeds and pollen as far as possible.

It was for this last purpose, it seems, that stalks first developed. The earliest land plants were in all likelihood low mossy things that kept themselves within the 'boundary layer' that separates free-flowing air from the ground below it. The depth of the boundary layer depends on the roughness of the surface; for the pebbly strands at the edges of a river (a natural place for an early land plant to try and make a living) it will be a few centimetres. Above that, the wind will dry out plant tissues as surely as it dries clothes on a line. Because they had only limited means to deal with such desiccation, the first land plants stuck to the boundary layer, where the still air stayed moist.

There are still plants that live like this today; the bryophytes, which include hornworts, liverworts and mosses, all tend to live low and flat, with only their spore dispersal mechanisms standing proud of the moist surface. Some of them make a pretty good living this way – the peat bogs dominated by sphagnum mosses cover about one percent of the planet's surface, an area similar in extent to the forests of Amazonia, and store an enormous amount of carbon. But over the rest of the earth's surface it has been the plants that took the high road, not the low road, that have prospered most; natural selection smiles on any mechanism that spreads one creature's off-spring to places its neighbours' don't reach.

Stalks that raised themselves above the boundary layer so as to release their spores into wind were favoured by evolution, and over time they came to be equipped with various protections against the arid air, such as waxy cuticles to keep their watery insides moist. Cuticles that keep water in, though, keep carbon dioxide out. If the stems were to be used for photosynthesis, they needed a way to make a trade-off between letting in carbon dioxide and losing water. That trade-off is made by the little orifices called stomata, which sense the amount of photosynthesis going on in green tissue and open or close themselves accordingly. These stomata are, from the point of view of the environment as a whole, one of the most crucial features of plant physiology. They tie the carbon cycle to the water cycle.

Though the stomata limit the loss, opening your tissues up to the air still means losing water, and this is a problem; plants rely on the pressure of water in their cells to keep themselves rigid. But water loss turned out to be something that the plants could use to their advantage. A thin column of water in a tube has a surprising tensile strength; if you pull hard on the water at the top, you will draw up the water from below as surely as if you were pulling on a steel wire. Vascular plants – plants with internal plumbing – learned to use this strength. The loss of water through the stomata, a process called transpiration, provided them with tubes of water pulled up from the soil that reinforced their structure as impressively as cabling would. What was more, the ascending water could be pressed into service to provide the airborne tissue with inorganic nutrients – such as nitrates and phosphates and iron – that the air alone cannot supply. Water loss was an irreversible fact of life for any plants headed skywards. By using that water loss to give strength to their tissues and to pull up a steady stream of nutrients from below, the vascular plants turned the water-loss bug into a feature.

Growing higher also entailed digging deeper. The plants set on the sky needed roots not just to provide water to their airborne tissues but also for structural support – there was not much point in having stems strong enough to take the wind if as a result the whole plant fell over in a stiff breeze. There are already some slender roots in the Rhynie Chert, and fossilized soils from later in the Devonian period show plants learning to grow down as well as up. Crucially, they learned to form chemical contracts with fungi in the soil that could feed them nutrients and with bacteria that could fix nitrogen for them, paying for these services with sugar shipped down from the photosynthetic tissues above. The pervasive nature of these symbiotic relationships is still being unravelled today; their complexity rivals that of the root systems themselves, branching more and more finely through the soil.

Most of the developments that made this Devonian explosion possible were morphological: new genetic programs to govern the building of structures such as stems, stomata and roots. Some of the developments, though, were biochemical – new chemical pathways for making new kinds of molecule. The waxes that give stems and leaves a waterproof coating are one example. The lignins are another. Plant cells have walls made of cellulose and its relatives, polymers made up of chains of glucose and other sugars. Long thin tubes of cellulose make plant stems strong and stiff. To strengthen things further, lignin is added to the cellulose; it cements the tiny fibres together, like the resin in fibreglass, creating hard wood. Reinforcing their cellulose with lignin meant that plants no longer needed to keep their stems engorged with water to stop them from wilting as herbs and flowers do. By turning cellulose into wood, lignin allowed plants to grow higher and higher.

By the end of the Devonian, about forty million years after the Rhynie Chert was laid down, the evolutionary process under way then had provided the planet with its first proper trees. They came in a range of different lineages. Being a tree is not a matter of what family of creature you are descended from, but what sort of lifestyle you have evolved. The basics of treedom – roots, trunk, branches supporting a crown of leaves capable of being lit from the side as well as the top – has evolved in at least seven different lineages of plant. This convergent evolution reflects the fact that being a tree is a really good solution to the problems of photosynthesizing in the air. As a result, though some of the Devonian trees, such as *Archaeopteris*,* are in detail completely unlike anything alive today,

---

* *Archaeopteris*, which made up a fair proportion of the forests which, by the end of the Devonian, had spread across the planet's moister lowlands, demonstrates both an impressive early use of lignin and one of the peculiar problems of palaeobotany. In the early twentieth century a Russian, Mikhail Dmitrievich Zalessky, discovered some interesting tree trunks that looked a little like conifers in the sediments of the Donetz basin. He called them callixylon. It was thought that the impressive callixylon trees – as much as 8.5 metres high and 1.5 metres in diameter – had grown up among fields of a fern that had been discovered fifty years earlier, by a Canadian geologist named Sir J. W. Dawson. Dawson had called his ferns *archaeopteris* for the same reason that, around the same time, the rather more famous discovery of a proto-bird was called *archaeopteryx*; both are Greek names for old feathery things. It was only in the 1960s that an American, Charles Beck, discovered that the assumed association between the two fossils was the wrong way round; the *archaeopteris* leaves belonged at the top of the callixylon trunks, not the bottom. The two 'different' fossils were parts of the same tree.

in broad outline they would easily be recognizable as trees. Bob Spicer goes so far as to suggest that trees may well be a universal feature of planets with dry land and advanced life-forms. On any planet that boasts photosynthetic life-forms capable of taking on complex shapes, some of those life-forms will hit on the possibilities of treedom, and find that they serve a useful purpose. They won't look exactly like our trees, of course. If chlorophyll is not a pigment that evolution seeks out on every planet, their foliage might take on a rather different suite of colours; but they'll be treelike enough for a five-year-old to recognize them for what they are, even if she complains about – or delights in – their oddness. If a few trees from a plant circling some distant star were scattered around our arboreta, many of us would barely give them a second glance. They can hardly look stranger than baobabs.

This points up one of the ways that the Devonian explosion is unlike the Cambrian. The many different body plans tried out in the Cambrian, some of them now very peculiar to our modern eyes, were shaped by interactions between the evolving animals. Some innovations were so useful that it is hard to see life anywhere getting along without them; eyes, legs, guts and fins come to mind. But the packages into which they were assembled took on forms that seem frankly rather arbitrary. The forces shaping plants, on the other hand, were simpler; straightforward needs for air, for light, for moisture, for a sensible balance between surface area and volume. Like treedom, leaves were evolved a number of times, in each case as a response to the same environmental possibilities. Simple computer models that take only straightforward physical constraints into account can produce surprisingly realistic simulacra of early plants. The Devonian explosion was shaped above all else by life's discovery of the basic geometries that come from wanting to photosynthesize in the open air.

The last great innovation of the Devonian was the development of seeds. The sex lives of plants offer little in the way of salaciousness, but make up for this with bewildering complexity, with a range of different strategies making use of different sex cells in different structures at different times. (They clone themselves now and then, too.) The general idea, though, is for each plant to find a way of mixing its genes with as many others as practicable and spreading the offspring thus made far and wide. Animals can do this by wandering off and finding new partners, and encouraging their offspring to do the same. Plants, being stationary, have had to come up with other strategies, and the seed – a throw-away embryo made by one plant with the help of genes sent out as pollen by another – is one of the best.

Seed-making plants, which appear towards the end of the Devonian, went on to displace almost all other vegetation: the vast majority of today's plants set seed. But the seed habit did not give its first practitioners immediate advantages in every environment; the low-lying forests that most of the late Devonian fossils came from were composed mostly of seedless plants and trees. It seems likely that, instead, the seed-setters were establishing themselves in drier and less auspicious surroundings, or places recently disturbed, perhaps by the newly encouraged fires – places that only a blowing seed could end up in, and that were unlikely to provide the conditions in which fossils get laid down. There are thus few fossils of these early settlers of the continental interior. But their pollen, spread out over the world at large, records their presence.

With seeds and stems and roots and leaves (of which more later), with plumbing and stomata, the plants spread from the water's edge to the continental interior and from the ground to the sky. Before the Devonian, plant fossils are few and far between. By the end, in some places, the rocks are so stuffed with dead plants that they are black with solid carbon: coal. The stage has been set for the extraordinary biogeochemical shifts of the Carboniferous – and, indirectly, for the carbon/climate crisis which we are bringing about today as we use carbon fixed by plants in the distant past to fuel our lives and heat our planet.

It is a shame, I think, that the wonders of the Rhynie Chert, the heralds of this new age, are tucked away in the archives of the Natural History Museum. They are a doorway into an extraordinary

227

period in the planet's past – and their microscopic wonders are also straightforwardly beautiful. They deserve to be celebrated and fussed over. Their beautifully preserved spores should be encouraged to settle in the public imagination. But, in general, plants get short shrift in the museum, and in most such museums; for all that plants are exquisite environmental markers, and primary agents of environmental change, animals steal the show. A reconstruction of the Rhynie Chert flora – which would look something like a spindly asparagus patch – would be hard put to earn its keep in a collection where dodos and dinosaurs compete for the visitors' attention.

If you know where to look, though, the museum goes some way to giving plants their due. Some of the pillars in this nineteenth-century cathedral of life have a strange diamond texture to them. The pattern is modelled on the scaly snake-skin bark of a plant just coming to prominence at the Devonian's end: *Lepidodendron*, a telephone-pole of a tree that stood in dense stands in the early coal swamps. The great hall at the heart of the museum, in which the mighty *Diplo-docus* stands facing the doors like a steam locomotive in its engine shed, is held up by the long-dead plants on which the wealth of Victorian Britain was founded, and of whose fossilized remains we still burn millions of tonnes a year.

CHAPTER SIX

Forests and Feedback

The agency of the stationary
Expanding the envelope
The oxygen spike
Life, luck and entropy

There was one picture in particular which bothered him. It had begun with a leaf caught in the wind, and it became a tree; and the tree grew, sending out innumerable branches, and thrusting out the most fantastic roots. Strange birds came and settled on the twigs and had to be attended to. Then all around the Tree, and behind it, through the gaps in the leaves and boughs, a country began to open out; and there were glimpses of a forest marching over the land, and of mountains tipped with snow.

J. R. R. TOLKIEN, 'Leaf by Niggle'

The agency of the stationary

The agency of animals is a visible thing. Their eyes blink, their gills flutter, their hackles rise, their pulses set the rhythm for their lives. They move back and forth, here and there, drawing their histories out behind them like the blur of a cheetah or the slime of a slug. The lines of their lives criss-cross the world, from the gyres of the ocean-circling albatross to the stochastic pinballing of a fly against a windowpane. The whole point of being an animal is trying to get somewhere else. Quite a few – let's hear it for the oysters – have given up on this birthright, and rely on currents and providence to bring them their world. But most of us have not.

Plants, on the other hand, very rarely move themselves around; they just grow, and in almost every case they do so imperceptibly. By and large, the agency of plants is invisible. This is the simplest, and perhaps the most profound, of the differences between those that eat light and those that eat others. It is why plants have a relationship with their environment both more intimate and more abstract than that of any animal. It is why they have no faces and no hearts.

This great difference stems from the fact that sunlight is, at the efficiencies photosynthesis is capable of, a rather dilute source of energy. To appreciate how insufficient it would be to animal needs, imagine the Green Man of forest folklore. Let us assume that his greenness is due to chlorophyll through which he feeds himself. Given the surface area of his skin – and the fact that at any given time some of it will inevitably be averted from the sun – such a

231

green man would have about as much energy on which to run his metabolism as someone restricted to a diet grown in a couple of square metres of garden. All he could eat in a day would be what his little plot could grow in a day. A few leaves for breakfast, maybe a morsel of root for supper: berries for Sunday lunch.

On the sunshine equivalent of this meagre diet, our Jack of the Forest has no energy for moving, or for thinking – nerves and muscles use a lot of energy. He lacks the energy to breathe in or out, or to keep his body any warmer than the outside air. He's not good for much except sitting there repairing the daily entropic wear and tear to his body. Indeed he doesn't really have enough energy for that; quite a lot of him will rot away. The prognosis for the Green Man is vegetable.

Plants simply don't have the energy to rush around like animals, pumping blood and flapping wings and flashing nerve impulses hither and thither along their limbs. So they eschew the compactness of muscle and opt for the looseness of leaf. The lines of their lives do not rush back and forth across the landscape – instead they are recorded in their shoots and limbs and twigs. Plants have shape where we have behaviour; their history is recorded in their form. Where the swoop of the sparrowhawk falling on the pigeon is gone in an instant, the tree's decision to grow its twigs this way or that, depending on the light, is written in wood and lasts the rest of its life – or at least until some rough wind or uncouth animal snaps the relevant limb off.

This passivity makes it hard to see plants as drivers of change. But though they do not walk the world, their slow persistence shapes it, both physically and chemically, to a far greater extent than animals do. Animals have their biogeochemical effects – by processing their leftover food into the first tiny turds, for example, the newly-evolved animals that launched the Cambrian explosion changed the ways nutrients are recycled in the ocean. But their contribution pales into insignificance compared to that of the plants. During the Devonian and the geological eras that followed – the Carboniferous and Permian – the power of the plants rearranged the planet.

232

With stem and trunk the early trees added a new verticality to the world; with canopies they gave it the novelty of shade. Rivers that had once skittered over desert plains like water over glass found themselves newly constrained by root-laced sediments bound together as soil; the riverbank was born. Lightning, which had played without issue over the deserts and oceans for billions of years, found its first tinder; the earth became accustomed to flame. Animals, to which the continents had previously been only a place to die, could now be fed ashore.

New landscapes of shelter and shade. New sounds; a beating of insect wings, the gurgle of a brook, leaves in the wind. A thousand shades of green, from understorey dark to canopy lights, spring green to played-out khaki. All the places where plants grew were changed in such ways. But place was not the whole story.

Despite the fact that we animals, ever anxious for an A and B to move between, tend to see it that way, the world is not just a set of places. It is also a set of processes. It is a set of cycles: cycles of water, carbon, nitrogen, sulphur and more besides. The world of processes, unlike the world of places, has no spaces, no in-betweens. Everything in it is pushed up together – everything affects its neighbour. Each change cannot help but provoke more changes. Sometimes this means the change is amplified; sometimes that it is dampened down.

Land plants changed the world of processes as surely as they changed the world of places. The changes were not as profound as those of the Great Oxidation Event, or as those life may have contributed to, and benefited from, at the end of the Proterozoic. But they are far better documented in rock and fossil and isotope, and so they let us develop much more detailed, testable hypotheses about the underlying principles of such change. Those stories and principles are this chapter's subject matter. Lovelock's Gaia theory, to which we will return, suggests what some of those principles might be – that life would anchor and strengthen the world around it, resisting change. Yet it is possible that the changes set in course in the Devonian led to the greatest extinction the world has ever seen.

Expanding the envelope

Asked about what plants did to their environment in the Devonian, Bob Spicer gives an answer that is, as he points out himself, very Gaian. Life changed the planet in such a way as to make it more to life's liking.

Wherever they spread, the plants raised the atmospheric boundary layer further into the sky. Over the early mosslike plants the boundary layer was just a few centimetres thick; if you'd walked through the plants of the Rhynie Chert it would have come up to your shins. By the end of the Devonian the development of forest canopies, with still and easily moistened air below them, had pushed it up to ten metres. The plants had taken the dampness they thrive on up into the sky with them, increasing the volume in which they could live happily by something like a thousandfold.

Admittedly, the photosynthesizing within this extended boundary layer is a little subdued, because the canopy takes much of the useful radiation for itself. But the plants in the understorey and growing as mosses on the trunks and branches could deal with that lower light, just as their descendants and relations do today – and just as algae in the ocean did before them. Algae grow in stratified communities, some adapted to the dimmer – and redder – light of the depths, some to the brighter, but potentially more damaging, light of the shallows. The forests recapitulated this stratification in the moist air,* offering a wealth of new niches to suit every photosynthetic taste. They were – and are – hollowed-out oceans, the liquid water removed from everywhere except the delicate tracery

---

* Conrad Mullineaux, of Queen Mary, University of London, points out that these low-light tricks had to be learned anew, rather than dredged up from memory: the algae that were the most recent common ancestors to land plants would have been living in freshwater shallows, and their first land plant descendants would have been in the open air, their sunlight entirely undiluted. The need to develop new ways of adapting to low light in the Devonian, Mullineaux has suggested, may explain various interesting differences between the antenna arrangements in plant chloroplasts and the photosynthetic membranes of algae and cyanobacteria.

of stems and stalks and leaves that makes up the world of the chloroplasts, a permanent algal bloom of photosynthesizing cells flushed across their wind-tossed canopies. Fed by dedicated mineral-mining roots instead of the vagaries of ocean upwellings, such blooms can spread for a thousand miles and last for a thousand years, enriching, protecting and moistening the dimmed but fertile world below.

The moistening effects of the new forests, though, were not restricted to their own understoreys. They moistened the sky above, too, changing the planet's hydrological cycle. Deserts are a poor source of water vapour. Rain that falls on deserts either sinks in to become groundwater, drains off in flash floods, or pools in short-lived puddles and playas. Only in the last of those three cases is much of the water likely to evaporate back into the sky whence it fell. Add plants, and things can change dramatically. Their roots make soil, which retains water. Their leaves transpire, returning water to the air through a surface area much greater than that of the ground they cover. The air above them is moistened, and from that moisture comes more rain. A rainforest's rain is not, for the most part, imported from oceans; it is recycled. It's not too poetic a fancy to see the pillars of cloud piled up above a tropical forest in the afternoon as continuations of the towering trunks below, parts of the same cyclic process of water being lifted up and dropped back down. And the same holds true, to a lesser extent, of sparser forms of vegetation.

As plants spread over the Devonian continents, they provided inland sources of water vapour. These new sources moistened the areas downwind of them, which allowed the plants to spread further. As with the boundary layer, there was a positive feedback loop; the more life spread, the more it was able to spread.

This virtuous circle was not, however, the only effect the forests had on the environment. While plants watered themselves, they also began to starve themselves – or at least to diet. By the middle of the Devonian, levels of carbon dioxide in the atmosphere were falling precipitously, and the climate was changing accordingly.

To understand how this came about, you have to appreciate that the earth actually has two different carbon cycles, a biological one and a geological one. The biological carbon cycle is comparatively quick and intense; carbon dioxide from the atmosphere and ocean is fixed by plants in photosynthesis. About a sixth of the carbon dioxide in the atmosphere is fixed this way every year. That which has been photosynthesized is, in fairly short order, returned to the atmosphere through respiration. Carbon may stay in the soil for a few centuries, or dissolved in the ocean depths for millennia, but in a million years all the carbon taking part in this cycle will have gone through its various pathways many times.

The geological carbon cycle, on the other hand, is far, far slower, working on the timescales of plate tectonics. The creation of new crust, which happens at the spreading ridges of volcanoes that run through the oceans like seams, releases carbon dioxide from the mantle that underlies the crust and forms the bulk of the solid earth. Weathering takes carbon dioxide out of the atmosphere and stores it away in the form of solid carbonates. The feedbacks between these two processes, as we have seen, give the earth a thermostat. Many of the carbonates do not last. The tectonic plates on which they sit sink back into the mantle at ocean trenches or grind into each other when continents collide, and these processes let a fair amount of the carbon dioxide stored in carbonates back out. But not all of it. Some carbonates end up where the processes of plate tectonics can't get at them, and as a result the total inventory of carbonate in the earth's crust grows over time.

The geological cycle can be seen as a slowly turning setting for the giddy spinning of the biological cycle. In everyday terms – or indeed in terms of millennia – the biological cycle is what matters. In the long run, though, it is the geological cycle which keeps things moving; without the encompassing armature it provides, the biological cycle would run down. The burial of organic carbon, for example, represents a slow leak from the biological cycle into the geological one, a leak that can only be compensated for by the carbon dioxide springing anew from the earth.

236

The spread of plants across the land in the Devonian represented an intensification of the biological carbon cycle, but that in itself did not change the carbon-dioxide level in the atmosphere. That which was stored by photosynthesis was still respired at a later date, at least at first. However, the physical effects of the spread of plants are thought to have changed the geological carbon cycle, and by so doing reset the earth's thermostat.

It was Jim Lovelock who first saw the connection. In the early 1970s Lovelock had predicted that there must be a Gaian way of stabilizing the earth's climate. When models of the geological carbon cycle which incorporated the carbon-dioxide weathering thermostat came out a decade later, he saw that they could fulfil that prediction. Working with Andrew Watson, he argued that the weathering rate depends not just on temperature and carbon-dioxide levels, but also on the surface area of the rocks being weathered and the moisture available. Life took charge of these extra controls during the Devonian explosion. When roots and their associated fungi break down rock into a porous mass of soil particles they increase its surface area many times over. Just as leaves gave the earth a larger surface area through which to lose water, roots gave the earth a larger surface area over which to absorb carbon dioxide. This new surface area was wet and ready for weathering: the plants brought the rain with them and the soils they made held on to water much better than unbroken rock did. And the rock particles in the soil were exposed to a great deal of carbon dioxide. Plants tend to segregate photosynthesis and respiration: the canopy and the leaves are the sites of photosynthesis while the soil is the site of respiration. It is in the soil that the carbohydrates made by photosynthesis are turned back into carbon dioxide, both by the plants themselves and by the bacteria and fungi that feed off them. By photosynthesizing in the canopy and respiring in their roots, the Devonian forests effectively pumped carbon dioxide out of the sky and into the moist pores of the soil.

Land plants made the interface between the atmosphere and the rocky earth bigger, wetter and richer in carbon dioxide. As a result, weathering rates leapt up. And the increase in weathering rates meant the amount of carbon dioxide in the atmosphere had to fall. So the temperature fell, and it kept falling until the drop in temperature had slowed the weathering rate enough to cancel out the increases due to the creation of soil. Rocks scarred by glaciers show that, by the time the Devonian gave way to the subsequent Carboniferous, 362 million years ago, a sparkling new icecap at the planet's south pole set off the rich greens in which the continents had been lately decked.

This probably wasn't the first time life fiddled with the thermostat. In the late 1980s the geochemist David Schwartzmann and the biologist Tyler Volk – who is one of the handful of scientists to have devoted a significant part of their careers to following up on, and arguing with, Lovelock's ideas – showed how a simple crust of bacteria could have enhanced the weathering of rock as early as the Archaean. As we saw in the past chapter, Andrew Watson and Tim Lenton have since suggested something similar might have been going on in the late Proterozoic, in the time between the stranglehold of the sinister and rather smelly Canfield ocean and the appearance of animals.

The case for increased weathering cooling the planet in the Devonian and Carboniferous is much more robust than the speculation about the Proterozoic, in large part thanks to a geochemist at Yale named Robert Berner. Berner, now in his sixties, was a member of one of the two teams of scientists that independently discovered the power of the thermostat in the 1980s. He is a man with a serious demeanour, an upright carriage and a longish face. Picture him as a pillar of a rather small nineteenth-century community, probably one associated with weights and measures – a respected apothecary, perhaps.

Berner has built the thermostat mechanism into a vast and impressive model called 'Geocarb' which has been evolving for more than a decade. Geocarb takes geological data relevant to all the processes that affect the geological carbon cycle – how much of the continental surface is exposed above the water, how hot the sun is, how many volcanoes are there, are new mountain ranges rising, and so on – for a specific time in the past. It then uses the principles of the thermostat model to work out what the level of atmospheric carbon dioxide should have been at that time.

All this hard work is necessary because the rocks alone tell you little about carbon-dioxide levels. Though there are ways to infer carbon-dioxide levels from various sorts of fossil, they are mostly quite new, developed since Berner first started his work in the 1980s. They are also subject to increasing error the further you go back in time, and often have a tendency to contradict each other.

The figures the latest version of Geocarb provides for levels of carbon dioxide in the past are considerably more than a best guess, but a good bit less than a hard fact. Like most models, it is a way of exploring implications. It is a way of saying 'If the world works the way we think it does, this is what carbon-dioxide levels must have looked like in the past'. If, at some point, the fossil evidence starts to suggest Berner's model is actually getting the wrong answers, then the conclusion will have to be that the world does not work the way we think it does – which is always a useful discovery. So far, though, the evidence suggests that Berner seems to have things about right. And his calculations show a spectacular fall in carbon dioxide starting in the Devonian.

Plant-enhanced weathering is not the only factor behind the fall. Plate tectonics also played a role. In the Devonian, the earth had two major continents, Euramerica and Gondwana. In the subsequent Carboniferous they came together to form a single supercontinent, Pangaea. This had various effects. One was the lifting up of new mountain ranges where the continents collided. Mountain ranges erode quickly, and the debris that flows down their flanks is composed of fresh rock ready for weathering. A spate of mountain-building will tend to draw down some carbon dioxide from the atmosphere regardless of what any plants may be doing, though in practice plants speed the process up. As we shall see, the weathering associated with the rise of the Himalayas has reduced carbon-dioxide levels in the atmosphere a great deal in the more recent past.

Another effect of the continental rearrangement was that the total length of the earth's ocean-ridge system would have decreased. Today the volcanoes of the ridge system are busily creating new crust down the entire length of the Atlantic Ocean, round Africa and up into the Indian Ocean, back down between Antarctica and Australia and on through the southern and eastern Pacific; wherever plates are moving apart, a ridge will be found filling in what would otherwise be the gap between them. It all adds up to about 60,000km of carbon-dioxide-spewing underwater mountain range. When the continents were being assembled into Pangaea, though, the ridge system would have been simpler – the single continent was surrounded by a single ocean, Panthalassia – and shorter. A shorter ridge system means less sea-floor volcanism, and thus less carbon dioxide released to the oceans and, through them, the atmosphere.

Berner's Geocarb model takes both of these plate tectonic factors into account. Neither of them, though, is anything like as powerful as the increase in weathering, which seems to have driven carbon dioxide down by seventy percent to levels lower than any seen before. If a drop of the same proportion took place today, the world would tumble into an ice age far worse than those of the past three million years.

An observation that tends to confirm this picture comes from David Beerling, a young professor at the University of Sheffield who's convinced that plant physiology can tell us a great deal about the workings of the world and ambitious to prove it. Beerling and Berner make quite a sharp contrast in both stature and temperament, but they have become an effective team.

In the 1990s Beerling and his colleagues began to wonder why leaves had taken so long to appear. It's not hard to make a leaf – you just let some webbing grow between some branched stalks. The genes used to control this sort of growth are common to all plants and to many green algae, and so it's fair to infer that the opportunity to evolve leaves was there long before the days of the Rhynie Chert. But leaves do not appear until after the Chert – in the middle of the Devonian. Why so late?

The answer Beerling offered draws on both physiology and environmental history. A side-effect of absorbing a lot of sunlight is absorbing a lot of heat; leaves are prone to heat stress. Transpiration through the stomata will cool the leaves, if there are enough stomata. But in high-carbon-dioxide conditions plants don't need many stomata; just a few holes in the cuticle will let in all the carbon dioxide the chloroplasts require. On the fossilized stems of early Devonian plants the stomata are few and far between by modern standards, reflecting the abundance of carbon dioxide in the atmosphere. If those plants had grown leaves with similarly scarce stomata, the leaves would have been too hot to do much photosynthesizing – rubisco refuses to work at high temperatures. More stomata could have solved the problem, but they would have required a much larger vascular system to keep the leaves supplied with water, and thus much thicker, more complex stems and bigger root systems. Growing a larger number of thin stems made much more sense than investing in the vastly thicker stems that would have been needed to supply leaves with enough cooling water.

As the Devonian went on, though, carbon-dioxide levels fell. Plants needed to pierce their cuticles with more stomata simply to keep their photosynthetic levels up. More stomata made it easier to keep leaves cool – and the need for cooling was dropping, too, as the climate changed. So the relative merits of leaves versus stems started to shift. New branches started to develop flanges of webbing; eventually, familiar-looking leaves appeared, first lacy fern-like affairs, then more solid arrangements.

The answer to the question of why plants spent so long without leaves thus looks similar to the explanation Andy Knoll offers for eukaryotic underachievement in the Proterozoic. In both cases, an aspect of the environment was holding them back; when the constraint was relaxed, the frustrated development took place. The difference is that in this case the plants removed the constraint themselves, by speeding up the rate at which silicate rocks get weathered. They were diminishing their supply of one raw material – carbon dioxide. But in a world with less carbon dioxide it was easier for them to spread their leaves and gather up more of another crucial raw material – sunlight.

As Bob Spicer says, the Devonian does indeed look somewhat Gaian. Living things were making the environment ever more suitable for life. But this may just be a coincidence. The benefits life brings itself are due to feedbacks between the biological and the geological. And to people interested in the issues but sceptical about Gaia, such as Beerling and Berner, the idea that such feedbacks always work in such a way as to stabilize the system, or to favour life, is implausible. There are all sorts of ways for life to destabilize things, too. The new continental ecosystems that evolved through the Devonian and Carboniferous pushed the planet ever-further from the sort of equilibrium that Lovelock's original Gaia hypothesis had envisaged.

The oxygen spike

To understand Jim Lovelock's revolutionary science it's worth remembering that in days gone by he had a bit of a chip on his shoulder. Brought up in south London in somewhat straitened circumstances, and educated in a system that restricted his enthusiasms, Lovelock entered adulthood with a fair amount of self-doubt and a growing suspicion of any authority that depended on unearned deference. Like many of the generation that came of age in the Second World War, he distrusted the old class-based society and dreamed of a new, more egalitarian one, in which origins counted for little and achievements for much.* Lovelock's ideas about Gaia have been moulded by those attitudes. Gaia is a world in which

---

* His dislike for unearned authority blends into a general distrust of received opinion. He won't hold a view simply *pour épater les bourgeois* – but if a few *bourgeois* end up *épatés* when he argues against organic farming, or in favour of nuclear power, he counts it an advantage.

social, or human, values count for little. In *The Ages of Gaia* Lovelock summed the idea up in the phrase 'In Gaia we are just another species'; his friend Peter Horton has shortened this to the acronymic Igwajas, a word he once convinced me to carry around on a card in my wallet as a mantra. In the anti-anthropocentric world of Igwajas creatures are judged by what they contribute, not by what they're descended from; humble bacteria are far more important than the lordly vertebrates, even if they do much of their work in the muck. Both Gaia and Lovelock are bracingly meritocratic. No one gets a free ride – everyone has to pitch in. Gaia is a self-made man's self-made god.

You can see some of these attitudes in the way Lovelock deals with the history of the earth in his first book, *Gaia: A New Look at Life on Earth* – or, rather, by the way in which he chooses not so to deal. 'One of the blind spots in human perception', he wrote in the late 1970s, 'has been an obsession with antecedents' – the sort of obsession with where you came from that had held back a new classless Britain. Textbooks and papers galore, he went on, had been written about the dim and distant past, and it had been accepted that 'this backward view [was] telling us all we needed to know about the Earth's properties and potential'. But looking back in time to the earth's distant past, Lovelock argued, was a poor substitute for looking back from space at its contemporary complexities.

One reason that Lovelock felt able to take such an unhistorical view was that his theory was about balance, not about change or evolution. What most impressed Lovelock about the world was that although it seemed to have an immense propensity to change, it didn't. Where Vernadsky had been impressed by the biosphere's raw power, Lovelock marvelled at its self-control. Gaia was a revolutionary view of a conservative earth.

Lovelock's second book, *The Ages of Gaia: A Biography of Our Living Earth*, proclaimed in its title a new willingness to learn from the past. By this stage Lovelock had moved from the original 'Gaia hypothesis' that he had shaped with Lynn Margulis – the idea that the earth system was regulated 'by and for the biota' – to what he

243

would come to call 'Gaia theory'. In this theory life was no longer acting as if it had a purpose in mind; instead, life was a prerequisite for the emergence of feedback mechanisms that would keep the environment stable. In non-living systems such feedbacks would be the exception; in those blessed with life they would be the rule.

In this new view, the stability on offer was temporary, and subject to *force majeure*. Accordingly, the story in *The Ages of Gaia* was one of stasis and crisis in which the earth's environment remained steady for long periods of time, then changed rapidly and dramatically. Sometimes the change would be just a spike, with life returning its world to the status quo after the interruption; asteroid impacts might be seen in this way. Indeed, Lovelock was keen on David Raup's idea that mass extinctions might in general be caused by comets, not just because it had the drama that he savours in a theory, but also because extraterrestrial attack would absolve Gaia of any blame in the matter. On the face of it mass extinctions seem like the sort of thing a self-regulating biosphere might be expected to avoid. On other occasions, though, a crisis would usher in a fundamentally new regime, as in the case of the Great Oxidation. The influence of the then-fashionable evolutionary idea of 'punctu-ated equilibrium', a purportedly post-Darwinian model of evolution in which long periods of apparent changelessness are set off from each other by crises of innovation, is clear.*

In this framework the climatic consequences of the Devonian explosion could be seen as a movement from one regulated state to another, from a warmer greenhouse world to a colder icehouse one. Indeed the interaction of roots and the weathering thermostat could

---

* Punctuated equilibrium was, like Gaia, controversial, and the two theories had some oppon-ents in common – mainstream Darwinians disliked them both, though for different reasons. Another model of change that may have influenced Lovelock's thought – and indeed that of the proponents of punctuated equilibrium – was Thomas Kuhn's idea that science moved forward in isolated revolutions separated by periods of normalcy. It's tempting to see the revolutionary ideas of the 1970s and 1980s – asteroid strikes as causes of mass extinctions, punctuated equilibrium, shifts between different Gaian regimes – as attempts by would-be revolutionaries to impose something like the imagined structure of their revolution on to the history of the earth itself.

be seen as a tightening of biotic control (Lovelock is keen on the idea that a cooler world may be more productive than a warmer one, and that life might in some way favour the cold). But the spread of the plants across the continents had other consequences far less to Lovelock's liking.

The idea that the atmosphere's oxygen level remains stable over geological time was at the heart of Lovelock's original Gaia hypothesis. The earth's surface is covered with flammable things, such as trees and people, and bathed in oxygen. If the oxygen level went up by twenty-five percent or so, it seemed to Lovelock, the planet would be hard put to avoid burning to a crisp, because wildfires would become much easier to start and harder to douse. If the oxygen level fell by more than twenty-five percent or so, lots of creatures would asphyxiate. It was to explain how the oxygen level could remain safely within these quite tight bounds for millions of years – as apparently it had – that Lovelock first invoked the regulatory powers of the biota. The idea that various loosely evoked feedbacks kept oxygen levels steady was a key claim in his first book, and a strong stance in his later thinking.

In the late 1980s, though, Berner and his student Don Canfield – who was later to plumb the strange sulphidic depths of the Proterozoic ocean – put together a new dataset on the history of atmospheric oxygen through the Phanerozoic. There is no direct fossil evidence for oxygen levels, but by knowing what processes take oxygen out of the atmosphere and bury it in sediments, by laboriously going through geological data to see what types of sediment have been laid down at what times in history, and by trying to estimate what fraction of the sediments originally laid down actually remains, it's possible to make an estimate of what the oxygen levels may have been at a given time in the past.

According to Canfield's and Berner's research, between the Cambrian and the Devonian oxygen levels were more or less stable at a level a bit lower than today's. Then oxygen levels started to climb. They kept rising into the Carboniferous, reaching an extraordinary thirty-five percent. In the subsequent Permian they fell back down, bottoming out a little lower than they had started off, then rebounding a bit.

To build up oxygen in the atmosphere you need a way of burying organic carbon beyond the reach of respiration. The Carboniferous provided just such a sink. The forests became taller, richer, lusher, and often swampier, and a surprising amount of their organic carbon made it deep enough into the sediments to be eventually transformed into coal. Though the Carboniferous represents just two percent of the earth's history, its rocks contain a very significant fraction of the earth's best coal.

This doesn't mean that the forests were being buried in their entirety. Today, 99.9 percent of the carbon fixed by photosynthesis is respired back into the atmosphere in fairly short order; only twelve million tonnes a year are buried. The thin stream of carbon that flows from the biological cycle to the geological cycle through burial represents just one part in a few thousand. In the Carboniferous the proportion of material buried was not much larger. But it was a bit larger, and over tens of millions of years that slight difference was enough to lay down something like ten trillion tonnes of coal.

Why was so much carbon being buried in a reduced form? The most obvious answer is that plants had invented wood. Wood – cellulose and related polymers reinforced with lignin – is designed to be tough and durable. Plants can't break it down themselves; once they have grown a woody limb they're stuck with it. And it seems as if, to begin with, nothing else could break down the lignin, either. Jennifer Robinson, an ecologist who has become a sympathetic but acute critic of various aspects of Gaia theory, pointed out some of the difficulties in the early 1990s. Because lignin molecules are stuck together in a number of ways, you need a wide range of enzymes to get them unstuck – and once this is done the components immediately try and stick themselves back together. They're not soluble in water, they're potentially poisonous – and to add insult to injury they're not even very nutritious.

Despite all these difficulties, a walk through modern woodland shows that there are various fungi that are up to the job of digesting trees, once they are dead and their protective bark is stripped away. But in the Carboniferous the ancestors of these white-rot fungi appear to have had some trouble tucking in. Robinson speculates that they hadn't yet assembled the mixture of strategies and enzymes needed to do the job properly. Another possibility is that the environment may have been against them, too. Breaking down lignin takes a lot of energy, and so has to be done aerobically. But the Carboniferous was swampy, and swamp waters tend to be anaerobic. A tree dropped in a swamp, unlike one left on the forest floor, will not rot. To modern scientists, this preservation is something of a boon; the growth rings in old tree trunks offer a spectacularly well-defined record of past climates, and swamp-preserved trees have allowed that record to be stretched back from the present almost as far as the most recent ice age. In the Carboniferous, though, the timber-preserving properties of swamps would have been part of the problem. Uneaten timber buried in swamps meant oxygen in the air that had not been used up in respiration.

Why was the Carboniferous swampy? One reason may come from plate tectonics. The assembly of Pangaea, as we have seen, meant a shortening of the mid-ocean ridge system. When there are fewer volcanoes in the oceans, the mid-oceans slump down a bit, and this may lead to a drop in sea level. When sea level drops, the broad, flat continental shelves are exposed to the atmosphere. Such seaside flatlands are landscapes predisposed to swampiness.

But as Bob Spicer points out, that is not the whole story. Some of the major plant types – such as *Lepidodendron*, the model for the pillars in London's Natural History Museum – may have been actively keeping things swampy. *Lepidodendron* looked a bit like the Joshua Trees of the American West, with a small crown of strappy leaves. Like those trees it had evolved ways of limiting its rate of transpiration, with a constrained vascular system. But it was living in a swamp, not a desert, and by evolving to use water only sparingly it was trying to keep it that way, ensuring the ground stayed as sodden as possible. The obvious advantage is that the damp is a safeguard against fire. A piece of supporting evidence is that the trees also seem to have grown remarkably fast, reaching their full height of thirty metres – a good size for a modern oak – in as little as a decade or so. A shorter lifespan would be less likely to be curtailed by fire.

It's possible that the swamps of the Carboniferous represent a tangled set of deeply anti-Gaian feedbacks which actively destabilized the oxygen level. The development of wood leads to increased carbon storage, because wood is hard to digest. More carbon storage leads to increased atmospheric oxygen. Higher levels of oxygen increase the risk of fire, a risk to which plants have to adapt.

One way for them to do this is to increase the amount of lignin in their wood because lignin provides fire-resistance. A higher lignin content means more respiration-resistant carbon getting stored, pushing oxygen levels higher still. Another way to resist fire is to make their environment ever more humid and swampy. This adaptation, too, exacerbates the problem. The greater the expanse of swamp, the larger the volume of anaerobic mire in which lignified wood can be preserved. Again, the fire-resistance strategy leads to more carbon burial, and oxygen levels rise.

And fire itself is part of yet another positive feedback loop. This is rather counterintuitive: when wood burns, a lot of reduced organic carbon is turned back into carbon dioxide, using up atmospheric oxygen in the process. But it's not quite that simple. Not everything in a forest fire turns neatly into gas. A significant amount of the biomass is turned into charcoal, wood from which more or less everything except carbon has been baked out by the heat. And charcoal makes lignin look like chicken soup; it is extraordinarily resistant to organic decay, even in aerobic conditions. The fires provoked by higher oxygen levels would, by producing charcoal, have created peculiarly respiration-resistant forms of carbon.

Berner's and Canfield's studies show oxygen levels rising, and the evidence points to various feedback loops that would have driven such a rise. This leaves the original Gaian belief in a stably-regulated atmosphere in some trouble, not least because it reveals a flaw in its foundations.

In the 1970s Lovelock had a visiting professorship in the Department of Cybernetics at the University of Reading, in part so that he would have a base from which to teach graduate students. The only one who ever came along was Andrew Watson, whose thesis was devoted to the key Gaian question of how much oxygen the atmosphere could safely carry. Watson devised a series of experiments in which strips of computer tape were exposed to electric sparks in atmospheres of different water and oxygen content. The results were featured in *Gaia: A New Look at Life on Earth*, and taken to mean that at an oxygen level of twenty-five percent (today's is twenty-one) even damp vegetation would burst into fire when struck by lightning. Twenty-five percent was thus the upper limit beyond which oxygen could not rise.

If you look at the graph of Watson's results that accompanies the discussion in Lovelock's book, though, things don't look quite as clear-cut. At moisture levels of fifty percent, sparks are not sure to light a fire even in an atmosphere of thirty percent oxygen. And in plants moisture levels can rise a lot higher than fifty percent: plants can contain three times their dry mass in water, giving them a moisture level of 300 percent. What's more, as Jennifer Robinson points out, paper is not only too dry, even when soaked, to be a good surrogate for swamps, it has also had all its fire-resistant lignin bleached out of it. In ecosystems full of damp swampy plants laced with lignin, oxygen levels above twenty-five percent would not necessarily make ignition inevitable with every spark – though they would make fires rage far more fiercely once lit.

While Watson, these days, is largely reconciled to such a conclusion, Lovelock remains much less so. But the evidence for oxygen levels above twenty-five percent, maybe even above thirty percent, is quite strong. The coal itself provides some – it frequently contains bits of fossilized charcoal, implying that the coal swamps were no strangers to fire. In today's world, environments like coal swamps rarely burn. Fire's biological equivalent, respiration, provides another line of evidence, of which Bob Berner is particularly fond, and which Watson still resists. The Carboniferous is home to some

Brobdingnagian creepy-crawlies – there are fossils of spiders half a metre across, millipedes a metre long, and, most impressively, dragonflies built on a similar scale. Under current conditions these arthropods – which do not have the benefit of lungs – could not respire fast enough to fly. The simplest explanation for arthropods of such unusual size is that they grew, and flew, in a more oxygen-rich atmosphere.*

At some point, though, the positive feedbacks pushing oxygen levels up must have met resistance from, and eventually been bested by, negative feedbacks – effects of oxygen that led to conditions likely to make less of it, not more. Fire is the most obvious candidate for such feedback. At some point, fire would become so much of a problem that the total amount of photosynthesis would start to drop steeply; oxygen production would thus fall, and so, in time, would atmospheric levels. But there are also subtler negative feedbacks. Ecosystems on land recycle their nutrients, and they do a particularly good job on phosphorus, the nutrient which, unlike nitrogen and organic carbon, can't be fixed out of thin air. Calculations by Tyler Volk suggest that in the modern world phosphorus has a 'terrestrial cycling ratio' of forty-six – meaning that, on average, a given phosphate ion will be used by forty-six different organisms between the time it is first leached out of a mineral by some lichen and the point at which it slips away into the sea. In the 1980s Lee Kump, another notable member of the second generation of Gaians, and now a colleague of Jim Kasting at Penn State, suggested that fires might short-circuit this recycling system, allowing lots of phosphate to blow out to sea in clouds of ash. This would be bad for the land biota, but good for life in the seas. Since the increased carbon burial in coal swamps that was driving the oxygen level was a

---

* Too simple, says Watson, who argues that even doubling the oxygen level would only double the energy available to the insects, while in terms of mass or volume the Carboniferous giants are hundreds of times larger than their equivalents today. An alternative explanation might be a lack of predatory birds and bats. (Incidentally, devotees of the television series *Friends* may have noticed a large dragonfly displayed on the wall of Ross Geller's various apartments: it is a full-sized model of one of these Carboniferous insects.)

continental phenomenon, shifting productivity from the continents to the oceans would counteract the build-up of oxygen.

Subtler still, but possibly more convincing, is a feedback that stems directly from the molecular biology of photosynthesis itself. In his early research on photosynthesis the brilliant and insufferable Otto Warburg discovered that the rate at which plants and algae can fix carbon slows down in the presence of oxygen. This was later found to be because of a quirk in the architecture of rubisco. When it is fixing carbon, rubisco takes hold of a carbon-dioxide molecule – a dumbbell with oxygen atoms at each end and a carbon in the middle – and forces it into the middle of a five-carbon sugar. But rubisco is also quite capable of grabbing on to an oxygen molecule – a shorter dumbbell, still with oxygens at each end, but with no carbon in the middle – and forcing that into the sugar instead.

When rubisco uses carbon dioxide, the result is two molecules of phosphoglycerate, which is good. When rubisco uses oxygen the result is one molecule of phosphoglycerate and one molecule of a two-carbon compound called phosphoglycolate, and this is not good. Five carbon atoms go in and just five come out, so no carbon has been fixed. What's worse, two carbon atoms that were in a useful, receptive form when they were presented to rubisco as part of the original five-carbon sugar are now locked up uselessly in phosphoglycolate. To get those two atoms back into the swing of the Calvin-Benson cycle means shipping them out of the chloroplast and into something called a peroxisome, then into a mitochondrion, then back into the peroxisome and finally back into the chloroplast. This rigmarole is known as photorespiration, because it uses up oxygen and produces carbon dioxide; however unlike proper respiration it consumes ATP rather than producing it.

Today photorespiration reduces the photosynthetic yield in plants quite substantially – in some plants and under some conditions rubisco wastes almost half its time sticking oxygen, rather than carbon dioxide, into the sugars it works with. This sorry state of affairs seems simply to reflect rubisco's origins in a world that no longer exists. In the Archaean, when rubisco first evolved, the world

251

was rich in carbon dioxide and more or less devoid of free oxygen. An inability to tell oxygen dumbbells from carbon-dioxide dumbbells was thus not a problem; rubisco's flaw was invisible. Indeed it was not a flaw, then: it was only after the ratio of oxygen to carbon dioxide in the atmosphere grew roughly a billionfold that photorespiration became a problem. And by that stage the living world was so deeply committed to a system with rubisco at its heart that there was no alternative.

The rate of photorespiration depends mostly on the temperature and on the ratio of carbon dioxide to oxygen in the chloroplast. If oxygen levels rose in the Carboniferous, levels of photorespiration would have risen too – and thus photosynthetic productivity would have dropped. That drop could well have been one of the negative feedbacks which eventually curbed the oxygen spike, though at a far higher level than Lovelock had originally thought acceptable.

Good evidence for this has come from carbon-isotope studies which again combined the skills of Bob Berner and David Beerling. In the 1990s, Berner spent some time studying isotope data to estimate how much isotopically light organic carbon had actually been buried in the Carboniferous. His idea was that a rise in the total amount of carbon buried would offer independent confirmation of the oxygen estimates he had made with Don Canfield in the 1980s. But the carbon isotope data didn't make sense; they suggested considerably more carbon burial, and thus higher oxygen levels, than even Berner thought possible.

Berner realized that there was another possibility. The signal he was looking for stemmed from the fact that organic matter is enriched in lighter carbon. The impossibly large signal might be due to the fact that, in the Carboniferous, the buried organic carbon was even more enriched in light carbon than normal. This sounds absurdly ad hoc: but it is actually an effect that might be reasonably expected to come about as result of increased photorespiration, thanks to a sort of double distillation. The roundabout process that recycles otherwise wasted carbon in photorespiration turns two two-carbon molecules into a three-carbon sugar and a spare carbon dioxide. That carbon dioxide will itself be isotopically light, because all the atoms involved have already been through rubisco once. And that carbon dioxide, already inside the plant, will be the carbon dioxide most likely to be used by rubisco in the future. If you feed rubisco carbon dioxide that is already enriched in carbon-12, the rubisco makes organic matter even more skewed towards light carbon than usual.

At the end of the 1990s Berner, Beerling and their colleagues showed in the lab that photorespiration has exactly this effect; when photorespiration levels are high, organic carbon becomes even more enriched in carbon-12. And when this double enrichment was taken into account, the estimates for carbon burial from Berner's Carbon-iferous studies fell right into line with the predictions based on his estimates of the oxygen surplus. It was a wonderfully consistent piece of work; by taking into account the effects of photorespiration, Berner was able to provide a new line of evidence arguing that photorespiration must have taken place.

It's plausible that some mixture of fire, photorespiration and the oddities of the phosphorus cycle stopped the rise in oxygen levels. Certainly something did. And after the peak came a drop similar in scale and pace to the rise. A significant part of the explanation probably rests on plate tectonics. Once the great continent of Pangaea was assembled its presence changed climate patterns, and the great equatorial coal swamps started to dry out. As they did so carbon burial rates must have fallen; in some places they were reversed as buried carbon was exhumed and eaten up. Thus oxygen levels fell. And though it would not have been the only cause, it seems likely that this fall, even more marked in the oceans than on the land, contributed to the vast extinction that marked the end of the Permian period, the only mass extinction in the fossil record greater than the one at the end of the Cretaceous in which the dinosaurs perished.

Lee Kump, who worked with Lovelock on the idea that phos-phorus flows might regulate oxygen levels, has identified various feedbacks that could have contributed to oxygen levels in the ocean dropping even further than those in the air. Kump thinks that changes in ocean circulation in the Permian could have led to a drop in oxygen levels in the deep oceans, a drop which positive feedbacks to do with the mechanisms of organic carbon burial would have made self-reinforcing. The hydrogen sulphides which dominated the Proterozoic ocean would then have started to build up again. Eventually, in places, the sulphide-rich water would have welled up to the surface, killing vast numbers of animals in the shallow seas and, by bubbling out into the air, poisoning land animals as well.

There is a huge amount of debate over the causes of the end-Permian extinction, and no one thinks that low oxygen and life-driven feedbacks are the whole answer. A vast outpouring of basalt in Siberia, then the most northerly part of Pangaea, was probably important, pumping up the greenhouse effect, possibly damaging the ozone layer and simply poisoning many animals. A related fall in sea levels harmed things too. Some people think an asteroid played a part, as at the end of the Cretaceous. But recent evidence goes some way to bearing out Kump's account of the apocalypse. Distinctive chemical fossils from shales laid down in the sea at the same time reveal something truly startling. It appears that right at the end of the Permian there were thriving communities of anaerobic photosynthetic bacteria in the open seas, in which case there must have been wide areas in which oxygen-free sulphide-rich waters were close enough to the surface that they were bathed in sunlight.

The extinction at the end of the Permian was the most unpleasant episode to which animals and land plants have ever been subjected. The recovery which followed was far slower than that following any other extinction event in the fossil record – life took some twenty million years to get back up to speed. And that extinction may well, in part, have been driven by feedbacks between life and its environment.

Life, luck and entropy

In the summer of 2002 almost everyone mentioned in this chapter met in Valencia, Spain, for the second conference on Gaia organized by the American Geophysical Union. The first such conference, held in San Diego in 1988, is widely seen as having marked a crucial step towards respectability for Gaian ideas. To Lovelock, though, it had been a bitter disappointment. An American physicist, James Kirchner, presented a paper in which he parsed writings about Gaia by Lovelock and Margulis with a logician's precision in order to distinguish between a range of different allied concepts they discussed under the term Gaia. He argued that all such uses could in effect be classed either as 'weak Gaia' formulations that were simply restatements of the obvious or 'strong Gaia' formulations that were unacceptably teleological. Lovelock saw the criticism more as sophistry than science, but felt unable to respond – an inspiring writer, he is not a confident debater. He left the meeting despondent.

Despite Lovelock's gloom, though, the years between that first AGU meeting in San Diego and second one in Spain were good ones for Gaian ideas, even if they did not always go into battle under the Gaian banner. Lovelock himself started to talk about the science of Gaia as 'geophysiology': as physiology deals with the processes, as opposed to the anatomies, of living things, so geophysiology would look at the processes of living planets. It's a nice term, but though I like it it's never really caught on, in part because Lovelock himself soon abandoned it. Instead, the term 'earth-system science' started to develop as a catch-all category for areas where climate, life, biogeochemistry and geological processes come together. Researchers and those who give them funds came increasingly to realize that these intersections between disciplines are crucial to an understanding of forthcoming global change.

In his opening address at the Valencia meeting Lovelock pointed to the 'Amsterdam declaration' on earth-system science drafted by the International Geosphere Biosphere Program, the World Climate Research Program, and kindred organizations in 2001. These are not fringe groups; they represent the state of the art in the earth sciences. And their declaration began, 'The Earth System behaves as a single, self-regulating system comprised of physical, chemical, biological and human components.' In a world where such a declaration commanded widespread assent, Lovelock argued, his ideas had won, whatever the name they did so under. Having declared victory Lovelock then got out, leaving the meeting early on because of prior commitments back in England. He was obviously happy to do so, his memories of San Diego still painful.

Lovelock's declaration of victory is understandable. The earth-systems approach, which stresses the feedback loops tying together chemistry, physics and biology, is indispensable to an understanding of the combined changes in the carbon cycle and the climate that are currently under way. The idea that life is content passively adapting to environments over which it has no sway – which really was the dominant paradigm just forty years ago – has gone for good, and Lovelock played a defining role in its demise. Studies of the ice ages, among other things, make it clear that the earth system exhibits some degree of self-regulation. That said, another reason for a stress on self-regulation such as that in the Amsterdam declaration is as a contrast to the unregulated changes to the system humanity is currently imposing. An understandable desire to stress the scope and potential for harm of human interference may lead some scientific opinion to over-stress the self-regulating prowess the planet might lose as a result.

For all the successes of a broadly Gaian, earth-systems view, though, what can now be said about stability, which is still at the heart of Lovelock's ideas? Berner and Beerling, among others, argue that life is much more likely to adapt to the changes it sets in motion than to counteract them. And yet there really does seem to be a tendency towards stability in many environmental factors over many timescales. Within the general realm of earth-system science, is there still a distinctively Gaian explanation for these environmental stabilities?

The strongest proponent of such an explanation, in Valencia and elsewhere, is Tim Lenton. When he was a disillusioned undergraduate chemist in the early 1990s, Lenton's father sent him copies of Lovelock's books, and he was bowled over. He wrote to Lovelock asking what he should do in order to be able to work on Gaia. Lovelock suggested he talk to Andy Watson, by then a professor at the University of East Anglia – and that he visit Lovelock himself down in Cornwall. The deep affection that sprang from that meeting changed the lives of both men. Lenton had found his inspiration. Lovelock had found an heir in whose hands he felt he might safely leave his ideas.

Lenton has worked on many different aspects of biogeochemistry, but one of his overarching concerns has been to make Gaia acceptable to biologists by showing how stability can be a naturally emerging property of the way the world works, requiring no intention on the part of life or the system it is part of.

At the heart of these Gaian arguments is the unproblematic assertion that living things have preferences. For any given environmental variable – temperature, salinity, oxygen level, whatever – a given life-form will be able to get by over a range of values. But at the top end of the range it will do better if things decrease, and at the bottom end it would rather see an increase. The direction of the creature's response to a change in the level of temperature, or salinity, or oxygen will thus depend on the level. Any feedback loops involving such a creature will thus be context-specific; at one extreme they will go in one direction, at the other they will reverse themselves. This ability of a creature or class of creatures to change the environment in different ways under different circumstances makes feedbacks involving biology fundamentally different from, and more stabilizing than, those which would be seen in a lifeless world.

Think of a simple model that combines plant growth, temperature and carbon dioxide. We start in a hot world in which the plants are stressed and to get things rolling we drop the temperature a little. The plants grow a little better; the carbon-dioxide level drops; the temperature drops; the plants grow better still. A positive feedback is set up. At some point, though, the temperature will drop below the level that the plants like best. Now further decreases in temperature will decrease plant growth, and thus the rate at which carbon dioxide is taken up will slow. Eventually there comes a point at which the temperature will stabilize: if you raise the temperature you will increase the growth rate, which will draw down carbon dioxide and thus bring the temperature down; if you lower the growth rate you will raise the carbon-dioxide level (because plants will be dying) which will increase the temperature and thus restore the growth rate.

This equilibrium point may be a long way from the point the plants actually prefer. But the important thing about there being a preferred point is not that the system will aim for it. It is that a point of preference divides the zone of tolerance into a part where lower temperatures are good and a part where lower temperatures are bad. Any positive feedback that starts in one side of the zone will run into trouble when it crosses over to the other side, and life begins to resist what previously it encouraged. This stops positive feedbacks in which life is implicated from running away with themselves in the way of purely physical and chemical effects like the ice-albedo feedback on a snowball earth or the runaway water-vapour greenhouse which boiled Venus's oceans. And it means that since the positive feedbacks limit themselves in this way, life's relationship with the environment is likely to be characterized by negative feedbacks, which bring stability.

This does not mean life is going to be able to get things entirely to its own liking – the regulation 'by and for the biota' Lovelock and Margulis once talked of is long gone. The system may get quite a long way into the other side of the zone before it stabilizes itself; the point where the feedbacks balance will depend on the chemical and physical and historical factors that govern their strengths. And the system will still be vulnerable to upsets due to circumstances beyond its control. When continents bang together or split apart, or when evolution comes up with radical new inventions like cyano-bacteria or coal swamps, the rules will change. The system may then wander around aimlessly for quite some time until positive feedbacks push it to somewhere where negative feedbacks will stabil-ize it. But eventually that new stability will be found.

Providing, that is, that life is firmly enough embedded in the relevant processes to do its thing. A positive feedback in which life played no role could, in principle, push the earth into inhabitability. This means that, for Gaia to work, life must insert itself into as many of the relevant processes taking place on the earth as possible. And it is here, Lenton would argue, that natural selection comes to life's aid. Anything that can be tried will be tried, and anything that works will be persisted in. Life will spread, and that spread, through its new effects on the environment, will produce the possibility for more feedback loops. Those may in the short term destabilize things – but in the long term, the greater the influence of life, the lesser the risk of catastrophic runaway events.

Lenton's assiduous work on these issues has gone some way to making Gaia a bit more biologically respectable, attracting the inter-est of the two greatest British evolutionary biologists of the second half of the twentieth century. The late John Maynard Smith, one of J. B. S. Haldane's last students, was instrumental in getting a paper in which Lenton outlined some of his arguments accepted by *Nature* in the late 1990s. The late Bill Hamilton went so far as to write a paper with Lenton. Neither grand old man was convinced by all the arguments, but they both thought Gaia was interesting enough to be worth taking seriously. Hamilton suggested that history might come to see Lovelock as a figure similar to Copernicus. Copernicus will be remembered for ever as the man who suggested that the earth moved, but it took more than a century for Newton to explain why it moved in the way it did. Lovelock's insight – that living systems stabilized their global environment – struck Hamilton as a similarly profound and plausible suggestion without, as yet, a compelling explanation.

It may be, though, that there is no compelling explanation to be found. Andrew Watson has become increasingly drawn to the idea that Gaia might be a matter of chance. Watson has always been interested in space research – before meeting Lovelock, he intended to take a doctorate in astronomy, and his first post-doctoral work was on the atmosphere of Venus – and he has found himself increasingly intrigued by the 'anthropic principle', an idea which has been almost as contentious in cosmological circles as Gaia has in biological ones. The anthropic principle comes in various strengths and flavours – shades of Kirchner on Gaia here – with some of them off-putting to most tastes. But one of its least exceptionable, almost banal, formulations is that the universe we observe must be a universe which permits the existence of intelligent observers. We could not observe a universe in which the periodic table never got further than hydrogen and helium, or where stars burned out in centuries, and so the laws of nature must allow long-lived stars and a wide range of atomic nuclei.

In a similar vein Watson argues that, for any given set of intelligent observers to exist, there must be a planet which has stayed relatively stable long enough for those observers to evolve on it. If we are typical of such observers then it takes a long time for such observers to evolve, and all the planets on which they do evolve will by definition have remained habitable throughout that time.

In billions of years, though, a planet like the earth will have had to deal with some fairly profound changes in its circumstances. Its sun will have warmed up, as ours has. Its atmosphere and surface environment will have been oxidized by hydrogen escape. Yet despite these challenges, it will not have ended up locked into an unending snowball, as Mars seems to have done, or have lost its ocean to space as Venus did. Feedbacks of some sort will have saved it from all the worst things that can happen, and will have allowed it to maintain a rich and varied biosphere, one capable of eventually producing intelligence.

This means, Watson argues, that a species like ours looking back at its planet's past must see an environment that appears to have been regulated. If it didn't, then the observing species simply wouldn't be there. But that does not necessarily mean that the ability to self-regulate is an innate feature of all biospheres. There may be no coherent reason why regulation should take place and there may be no common cause for the different sorts of regulation that have taken place on different planets. The history of life and its earthly environment could be no more than the opposite of the writings of Lemony Snicket – a series of fortunate events.

Rather nervously, Watson laid out his ideas in a keynote address on the first day of that conference in Valencia. Knowing Lovelock as well as he did, he knew that by suggesting a 'lucky Gaia' – and implying that Lovelock's discovery might be a fluke – he was courting disapproval. He duly received it; relations between the two men became strained to near breaking point, though they have now improved again. It may seem strange that ideas as abstract as the anthropic principle can drive a wedge between men who were once very close. But then, that closeness had itself been built on ideas. And the sensitivity to the interpenetrating intimacy of life and its environment that distinguished Lovelock's first work on Gaia is a reflection of his broader tendency to mix categories. His strength lies in not making the distinctions that other people make between analysing and building, living and not living. A difficulty in seeing the distinction between criticism and antagonism, between disagreement and disrespect, seems to me to be part of the same characteristic temperament.

Lenton and Watson were not the only people trying out new takes and variations on Gaian thought at Valencia. So were various other old hands such as Tyler Volk, of New York University, who has carved out a niche as, crudely, a Gaian who doesn't believe in Gaia which particularly irritates Lovelock. Their collective ruminations were an inspiration to a young German ecologist named Axel Kleidon, a man with the sort of broad face and joyful smile that one might expect to find immortalized in some minor masterpiece of mediaeval carving, the grain of the wood catching the curls of his beard. Kleidon was a post-doc at Stanford at the time, where he was modelling the relationship between root depth and climate. Steve Schneider, the Stanford professor who was one of the organizers of the Valencia meeting, invited Kleidon to come along.

At one point, in the Botanic Gardens, he heard Tyler Volk and his friend David Schwartzmann discussing entropy, one of those conversations, he says now, that you're not really sure you're taking in at the time, but which lives on in your memory. When Kleidon had been a PhD student, he and his colleagues had deliberately skipped the section on entropy in their textbook on the physics of climate on the basis that entropy, as everyone knows, is weird and complex and confusing. After his Gaian initiation in Valencia, Kleidon went back to that skipped chapter and, through it, discovered a largely forgotten line of research. It seemed that simple climate models matched the world best when they allowed the flow of energy through the world to produce as much entropy as it possibly could.

The classic example of this came from a model of the climate as a heat engine produced in the 1970s by an Australian physicist named Garth Paltridge. The heat gradient that drives the earth's great engine runs from the tropics (hot because the sun shines straight down) to the poles (cool because the sun shines on them askew). But the temperatures at the two ends of the gradient are not a given. If the planet's heat transport systems were very efficient – something which would correspond to a system in which winds and ocean currents were extremely smooth – the poles would soon end up at the same temperature as the tropics. There'd be a lot of heat flowing through the engine – the sun would still heat the tropics more – but because the temperature gradient was so flat no work would be done, and no entropy produced. Nothing irreversible would be going on, no disorder created. At the opposite extreme, if the processes that move heat – the winds and currents and so on – were very bad at their job, then the poles would be cold and the tropics hot, because heat won't be able to get from A to B. Now the gradient will be steep – but nothing will flow down it, because the heat transport processes don't work. Again, no work and no entropy.

A minimal difference in temperature and an overpowering heat flow minimized entropy production; so did a maximum difference and no heat flow. The difference in temperature that's observed on the earth today, though, falls right in between – and corresponds to an arrangement that maximizes entropy production. Paltridge thought that this was an intriguing result, but not one it was easy to take further. It didn't seem to tell anyone anything useful that they didn't know before; climate modellers knew the temperature at the poles and the equator. What they were interested in was the working of the great climate system that lies between them. Knowing that this system was producing entropy as fast as it could didn't help you understand how the entropy-producing processes actually worked. It was a new way of describing the system, not a new way of explaining it.

In this, though, Kleidon saw a similarity with Gaia. Gaia was a way of describing – a bioscope through which to see – as much as it was a way of explaining mechanisms. Maybe Gaia, too, could be described in the language of maximum entropy production. Biological processes produce entropy; it's an unavoidable by-product of having a metabolism. So the more productive the creatures of the biosphere were, the more entropy would be produced. If the planet's climate was arranged so as to maximize entropy, why shouldn't the biosphere as a whole do so too?

Kleidon's interest in maximum entropy approaches brought him to the attention of Ralph Lorenz, a British-born planetary scientist working in Arizona who had also been bitten by the maximum entropy bug. Computer models of the atmosphere of Titan, Saturn's largest moon and the long-standing object of Lorenz's fascination, weren't able to reproduce the workings of the system in a way that matched reality. They didn't get the temperature difference between the poles and the equator right. But Lorenz had found that if you just asked how the atmosphere could maximize its entropy production, then the correct temperatures came out straightaway. And the Martian atmosphere, too, seemed to be maximizing its entropy, though here the process was a little more complex, due to the fact that a significant amount of the Martian atmosphere freezes to whichever pole is enduring its winter at any given time.

By expanding the range of maximum entropy approaches from the earth's climate to the climates of other planets, and from physical climatology to biosystem modelling, Kleidon and Lorenz reinvigorated the field. Researchers who had been working on the ideas in complete isolation were thrilled to discover that there were other people interested in the same stuff. A far-flung network started to take shape.

In 2003 one of this little band of brothers, a mathematician named Roderick Dewar, published a general analysis of open systems – systems into which and out of which energy is flowing – that are far from equilibrium. He found that if they had sufficient 'degrees of freedom', meaning that there were lots of different ways in which energy could flow through them, and if they didn't have fixed boundary conditions, meaning that there was no external control over their temperature and other salient details, then there was an overwhelming statistical likelihood that such systems would maximize their production of entropy.

If Dewar's proof means what it seems to mean, maximum entropy production is something like a law of nature – 'a codicil to the second law of thermodynamics', as Garth Paltridge puts it. Open systems running stably far from equilibrium would have no choice but to fall in line. And as Kleidon points out, this would mean that they would tend to be stable, in the way that Gaia theory suggests the world should be. Because the system naturally tends towards the state that maximizes entropy production, anything which pushes it away from that state will tend to run into negative feedbacks that push right back.

This sort of thinking doesn't only apply at the scale of the planet. If it is as general as Dewar's proof seems to say it applies at all other scales, too. Dewar is currently applying the ideas to models of the stomata, seeing their opening and closing as ways of changing the entropy production of the system. Davor Juretić and Paško Županović, two theorists in Croatia, recently applied the approach to the processes that drive hydrogen ion transport across the membranes in chloroplasts. The basic principle is like that seen in heat transport from the tropics to the poles. If it's really easy to move hydrogen across the membrane then you can't get any work out of the system. If it's really difficult, the hydrogen won't flow. The sweet spot is in between – at the point where the entropy production is maximized. One can't help thinking that it's an idea that Robin Hill would have savoured as he watched the seasons turn in the garden at Vatches Farm.

What Kleidon takes from Dewar's work is that life provides new degrees of freedom, new paths for the flow of energy. The most notable of these would be photosynthesis, which produces entropy at a startling rate.

Kleidon is now exploring a range of ways in which the biosphere might help to maximize the earth's entropy production – by providing a surface roughness that optimizes the turbulence of the winds, by opening its stomata in such a way as to optimize the transport of heat through evaporation, and so on. In both cases, optimization would imply stability; in both cases, Gaia would have a preference. The hard thing, he admits, is to find out how relevant the models he produces are to the real world.

Roughly speaking, the earth's plants produce entropy at the same rate as its winds do, though with a great deal less fuss. And there's some evidence to suggest that, within today's constraints, the earth is photosynthesizing as hard as it can. There's some reason to believe that the current dispostion of climate and vegetation could be predicted simply on the basis of the system's boundary conditions – no alternative history with different plant distributions could do any better.

The idea that there is a 'codicil to the Second Law' that applies to everything from the Z-scheme to the regulation of the earth system is a heady one. And it takes us back to Bernal's 1947 definition of life as 'one member of the class of phenomena which are open to continuous reaction systems able to decrease their internal entropy at the expense of free energy taken from the environment and subsequently rejected in degraded form'. These are exactly the same sorts of thing that the maximum entropy principle would apply to.

Bernal, though, went on to distinguish life from hurricanes and flames by appealing to genetics, to the historical connection which is passed on from life-form to life-form, which cannot cross the boundaries of the inanimate. In that sense, Gaia is not alive, because it does not stand in a chain of historical descent. But it has such chains within it, and they have given it what flames and hurricanes lack – an ever-expanding repertoire of behaviours, ever more ways to generate entropy.

The earth is not brute matter. It is matter with a genome, a genome spread across all the species it contains. This genome does not evolve as those of organisms do; its interaction with the world does not of itself generate new proteins and organelles and me-tabolisms and morphologies. But as natural selection inscribes the potential for such novelties into the genes of specific species, so they are added to Gaia's capabilities. The planet's genome is not pared down for efficiency, as a bacterium's is, nor elaborated into the sort of program that allows cells and organisms to change, develop and specialize in the way a eukaryote's does. It is a lesser thing – merely an accumulation.

But still it makes a difference. In flames and hurricanes, whatever might happen next can depend only on what is happening now. For the planet as a whole, whatever happens next depends on everything that happened in the past, because the record of that past, and the means to reproduce its processes, are locked away in molecules that may be accessed and used at any time. Extinctions may remove species and shapes and behaviours, but they do little if anything to biochemical possibilities. For as long as they have a use, the planet will never forget the workings of photosystem II, or the fixation of nitrogen, or the trick of being a tree, the sense of when to relax a leaf's stomata. Unlike a simple flame, the planet can be tomorrow what it is not today, and the day after change yet again, while all the time remaining itself. The planet is not, in itself, alive in the same way as a tree. But thanks to life, it is a flame with a memory.

CHAPTER SEVEN

Grass

On the Downs
From chalky seas to burning savannas
The ice ages
The fertile millennia
The end of plants
The long walk

You came, and looked, and loved the view
Long known and loved by me
Green Sussex fading into blue
With one grey glimpse of sea.

TENNYSON

On the Downs

To get a sense of the current role that plants play in the system of the world, just take a walk. A walk anywhere might do. The intricate relationships between life and its environment permeate the world as roots do a soil – wherever you go there's evidence of what plants have been up to, and how it has affected the planet. Every sort of local reflects the global to some extent. Every landscape plays some part in the story.

This is certainly true of the landscape under whose influence I have written most of this book: the stretch of the South Downs that surrounds the town of Lewes, about a hundred kilometres from London, a little way inland from the cliffs of the Sussex coast. I have written most of this book in Lewes, taking many walks across the Downs, low, smooth chalk hills that rise steeply from the Weald below, for inspiration or distraction. It is here, at the moment, that the connections which tie the world together seem densest and at the same time most clearly discerned, here where the naked back of the Downs rests on the rumpled sheets of the fields and meadows below. So though a walk anywhere might do, it's a walk over the Downs I'm going to take you on.

Leaving aside the fact that this landscape, like all the English countryside, is a human in(ter)vention, the Downs are unlike any-thing that could have been seen in the Devonian, Carboniferous and Permian world we have just visited. Almost all the 700 or so species of native plant that grow round here have anatomies, reproductive strategies and relationships to the animals around them

271

that have evolved since the Permian. What's more the rocks over which they grow are made of a substance unknown in the days of the Permian.

In the aftermath of the oxygen-free ocean's brief return at the end of the Permian new types of single-celled algae – which, when living in the open ocean, are called phytoplankton – came to prominence. The two major families of phytoplankton that rose to greatness at this time, the dinoflagellates and the coccolithophores, had different ways of armouring themselves against the world around them. In the case of the coccolithophores, the armour is made from calcium carbonate precipitated from the seawater around them. The bosses and spokes of this armour are the principal constituents of chalk, giving it a soft whiteness that no previous form of limestone has ever matched.

The coccolithophores from whose armour the chalk of the South Downs is derived lived about a hundred million years ago, in the Cretaceous period. Pangaea was broken up and most of Europe was submerged; the warm seas over what is now England stretched east for hundreds of kilometres. Coccolithophores flourished in the shallow continental seas – as did the predatory plankton that, despite the precautionary armour, had evolved to feed on them. The process of being eaten compressed the coccoliths into pellets dense enough to sink to the sea floor, and over millions of years a thick layer of carbonates built up on the seabed. Later these sediments were buried, compressed and transformed from ooze to chalk. This bit of the Downs, I am told, once boasted the largest concentration of chalk quarries in England, and the white half-moon bites they left in the hillsides can still be seen today. In the winter, I could see the one on the spur of Offham Hill from the window of my flat. I cannot see it now, though, because spring has washed up the scarp of the Downs like a wave riding up the rim of the world, thickening the air with leaves.

A walk west along the bottom of the scarp, below that old quarry, takes us through gentle farmland, some arable, some dairy. The fields are watered by a line of springs; at the foot of the scarp rain which has soaked down through the porous chalk comes up against the impermeable clay beneath it, and is thus squeezed sideways to the surface. Between the fields and streams there are small stands of oak and thorn and ash – the trees of ancient English memory that Kipling celebrated in his wonderful *Puck of Pook's Hill*, set in similar country a day or two's walk east of here.

Everywhere there are flowers; white hawthorn blossom in the hedges, bluebells under the trees, bright dandelions in the pasture, whole fields of rape blossoming in a wash of yellow. This, too, is a post-Permian novelty. Before then the forests were green and green only, without a dash of sharper colour; you will never find a fossil flower in ancient coal. The flowering plants, or angiosperms, first started to flourish only in the Cretaceous, around the time that the coccolithophores were laying down the sediments that would provide the strong curved back of the Downs.

Like the conifers that evolved before them, the angiosperms make seeds which can be spread far and wide, and which can wait in the soil for an opportune moment to germinate and try to grow. Indeed, they make tougher, more durable seeds than the conifers and their relatives. They have also evolved ways of conscripting other creatures into the production and distribution of those seeds, using colour and scents and sugars. Angiosperms have found a new approach to reconciling the stationary way of life with the need to exchange genes with distant strangers and spread offspring far and wide: they get the animals to do the work.

If we watch one of the bluebells under the oak trees for long enough, we will see a bee come to it. These bees lack the urgency one would expect in a creature so famously busy; they seem to me languorous, if not flat-out lazy. They know there will be flowers, just as the flowers know there will be bees, and they can take their time. The mutual knowledge of the bees and the bluebells is a classic example of what is known as co-evolution; each has evolved in the context of the other. Everywhere you see angiosperms and animals you see co-evolution.

A propensity for co-evolution is not the only characteristic which sets the angiosperms apart from earlier plants. They make stronger, harder wood, a facility which lets them hold up their branches through tension, using the wood at the top of the branch like a cable from which the rest of the limb is suspended, rather than buttressing their branches from below, as a pine does. They have perfected the trick of being deciduous, discarding leaves in the winter so that they no longer have to pay for their upkeep when the sun is low and the skies are clouded. They have adapted to niches that other plants have rarely explored, such as that of the creeper. Beneath the thickening cover of new leaves, many of the trees at the foot of the Downs have trunks wrapped in the darker green of ivy making use of the woody scaffold to harvest oblique winter light that the tree proper can't be bothered to use.

In the past hundred million years the inventive angiosperms have taken over the plant world pretty thoroughly. There are still conifers and ferns and mosses, and they dominate some ecosystems, but the angiosperms now make up the vast bulk of the earth's biomass, and are responsible for most of the photosynthesis that takes place on land. According to one estimate, 235,000 of the earth's 275,000 species of living plants are angiosperms. Here again, co-evolution is responsible; a slight change in the flower, a slight change in the insect that services it, and a new species is created. If we accept the idea that the tempo of evolution is naturally higher for animals, with their panoply of behavioural interdependencies, than it is for other creatures, the blossoming of angiosperm diversity in the past hundred million years shows the plants finally catching up. The angiosperms harnessed the animals for their own purposes, and as a result they now evolve at an animal pace.

Wandering further along the base of the Downs' northern scarp, we come across two paths leading up – 'bostals' carved by sheep over the centuries. One leads through hanging woods, one through open country. On the verges of the open one we might see bee orchids which have shaped themselves into simulacra of their co-evolutionary partners; but I fancy the shadier path through the perched wood. The trees here are predominantly beeches, fine and graceful, their leaves newly budded and still light and small, the understorey beneath them uncluttered. Beeches like well-drained chalky hillsides, and the face of the Downs could have been made for them. Here and there, more common near the woodland's edges, sycamores are trying to cut into the action, little seedlings with floppy five-lobed leaves desperate to soak up as many photons as possible before the pointillist spring canopy coalesces into summer's wash of solid leaf.

The path is hard work. During the ice ages, the water in the chalk that makes up the Downs freezes, and this ice-hardened rock erodes sharply. The ice ages have both steepened the scarp and cut deep bays into it. Towards the top of the hanger, the path leads out on to an open slope studded, at the top, with low hummocks that may mark ancient burials. At the wood's edge other species, lower and hardier, have slotted themselves in among the beeches: hawthorns again, gorse bushes, low oak. The crest of the Downs is swept by a sea wind that sculpts bushes into bent old women. In a hot summer of the sort that seems just round the corner salt on that sea wind will burn the leaves of these bushes and the topmost trees, bring a touch of autumn to them months before it is due. But today the air is calm, and summer, though poised, has yet to start; everything is spring green. Scents no easier described than the shapes of proteins wash the air; birds snap up insects on the wing. Magpies flash into the sky, proud as poets.

We haven't risen far – about 125 metres, according to the map – but the world has changed. In front of us the ridges of the Downs are like a pod of whales breaking the surface of a sleepy sea. Behind and below us the Weald stretches out to dim low hills half-vanished in the distance, a patchwork of wood and field. Eastwards along the line of the scarp is the rolling swell of Windover Hill, where kestrels play on the thermal updrafts above a great chalk figure carved into the side of the Downs, the Long Man of Wilmington. Nearer to hand the wave of Firle Beacon rises over Glynde Gap as if caught on the edge of breaking. Paragliders circle lazily around the ancient hill fort of Mount Caburn. The distance is hazy; in the gaps to the south through which the sea would be seen on a clear day there is just a saturated emptiness.

And all around us is the crowning glory of the Downs – the grass, a solid green echo of the sky.

Grass, like the chalk, dates from around the end of the Cretaceous, a new solution to the challenges of photosynthesis. Of the various problems a plant faces when making its way in the world – or, more accurately, when standing its ground – two are paramount: the problem of being eaten and the problem of being overshadowed. Trees use wood as a response to both; the lignin in wood resists digestion, and it provides the structural support needed for height. But though woodiness is an undoubted evolutionary success, there are alternatives, and the grasses have grasped them, becoming a sort of anti-tree in the process.

Instead of reaching for height, the grasses spread wide, and do so as fast as possible; disturbed bare ground will sprout grass before it sprouts anything else. The grasses grow fast and disperse their seed widely; if the trees move in on their territory the seeds will just wait in the soil until storm or fire clears a new opportunity for them, after which they will be the first and fastest to regrow. In contrast to trees, grasses are less about place and more about process; it's one of the things that makes them as global as a plant can hope to be. Sir John Hooker, the great Victorian botanist, found sheep's fescue, a grass typical of the Downs, at a height of over 5500 metres in the Himalayas. Grasses spring up wherever there is fire or wind or disease or flood to lay low some other vegetation – or whenever people do so. In the modern world, people are the grasses' best friends.

Instead of being inedible, the grasses have evolved to tolerate being eaten. They grow their leaves – blades – from a green heart kept close to the ground, a heart that intruding teeth rarely reach. As a result, getting chomped on is not too much of a problem. And by tolerating being eaten, they also, to some extent, avoid being overshadowed. The animals that eat grass will often also eat seedlings of woody plants that are trying to grow in the same place. In grass that is well grazed most invading seedlings will be eaten before they can grow into trees.

On this part of the Downs, sheep and rabbits are happily taking on the seedling-forestalling duties in return for soft green feed. The fields off towards the village of Falmer are full of bouncing lambs, drunk on some secret sheepish joy to which their parents have grown jaded. Rabbits abound at the grass's edges. If we had come at a different time of day we might have seen deer on the grass, as well – but they are more at home in the woodlands, bounding back to shelter when spotted. There are hares, too, but I have yet to see one. And down below us, halfway back to Lewes, stables at a disused racecourse house the grassland's aristocracy; you can see the prints of their iron-shod hooves in the turf.

Though the first grasses evolved in the Cretaceous, probably about eighty million years ago, fossils tell us that it is just in the past thirty million years or so that they have spread around the world. Their proliferation has been managed through co-evolution with ungulates, the hoofed mammals whose stomachs have been turned into complex processing systems for getting the most out of the fairly meagre nutritional raw material that blades of grass – which are mostly cellulose – can offer, conscripting various bacteria to help in the process. The fossil record shows the thick 'high-crown' teeth needed to grind the most sustenance out of grass evolving over time in a variety of different ungulate families, from antelopes to horses to hippos to elephants. By the onset of the ice ages, three million years ago, grasslands had spread to cover about a quarter of the earth's surface, and the ungulates had spread out with them.

Their influence may well also have spread far out to sea. Grasses are full of tiny little opals – crystals of silica which serve to discourage the more rapacious herbivores by making grass that bit tougher to eat. In order to make these crystals the grasses suck silica from the soil, and as a result the spread of grasses has increased the rate at which dissolved silica flushes into the oceans. This seems to have been a great help to the diatoms, the third of the great phytoplankton families that have risen to prominence since the Permian. Unlike

277

the dinoflagellates, which use cellulose, and the coccolithophores, which use carbonates, the diatoms make their armour from silica. Gregory Retallack, a geologist at the University of Oregon, has suggested that the increased availability of silica that came with the spread of the grasses may explain the way in which the diatoms have flourished since the end of the Cretaceous. The coccolithophores have been losing their place in the sun to them. There is comparatively little chalk being made in the oceans these days; once the waves and ice ages have finally laid low the Downs, nothing quite the same will ever rise up again.

Enough of grass in general; what, specifically, do we have underfoot? Red fescue and sheep's fescue for the most part, I believe from what I've read. I couldn't honestly say I recognize them – but I can see that whatever dominates, there are many other species of grass and herb in the turf, too. Some have been blown here by the wind, some were seeded deliberately when this land was taken out of cultivation and returned to pasture a few decades ago. The thin soil on top of the chalk is not well suited to the most common and rapacious grasses – a relatively high level of bicarbonate ions, leached from the chalk below, suppresses the high levels of root respiration seen in the most vigorous species – and as a result the pasture here is far more species-rich than the tracts of heavily fertilized perennial ryegrass that feed grazing livestock on richer soils. Prettier, too, because here where no fertilizer is used, the grass does not grow so strongly as to squeeze out the space for flowers. Within weeks this pasture will be awash with bright white daisies and golden cinquefoil and patches of blue speedwell, and the grain on its stems will have given the grass a purple tinge. By high summer the turf will have a golden edge to its green. Ungrazed stalks will stand out in the sunlight like sharp strokes of etching.

The Downs' grass has a smell, but not one easily described or distinguished. It has its own sounds: today mostly a still soft hum of insects, tomorrow, perhaps, the whispered rustling of the wind. And it has a feel. The combined work of the roots gives the turf a springiness that belies its thinness, a reflex above the soft rock. This is good land to run on, land that responds to the feet even as it clears the mind.

The futon-thin soils were a blessing for the farmers of the Bronze and Iron Ages. The thick clay soils of the low-lying Weald were beyond the capacity of their crude ploughs, and so they chopped down the forests that had spread over the Downs after the ice age and turned the land over to farms and grasses. By the time the Romans arrived, England's chalk uplands were Britain's most densely populated landscapes. Below us, on the path leading down to Falmer, traces of Roman field systems can still be seen cut into the land; back down the scarp, near Plumpton, you can see the signs of Bronze-Age agriculture half as old again. In this, too, the sheep were crucial, not just as grazers, but as enrichers of fallow fields. Sheep pastured on grassland during the day were packed tight into folds at night. There their sharp hooves would trample their dung into the shallow chalky soil, enriching it for the wheat or barley that would follow.

This, after all, is the final importance of the grasses. They do not just dominate the landscapes of prairie and savanna, nor do they just feed our sheep and rabbits and horses. They feed us, as well. Simple of stomach as we are, we cannot graze on their foliage. But we can feast on their grain. From the dawn of agriculture in the fertile crescent of the Middle East to the 'Green Revolution' that saved millions from starvation in the 1960s and 1970s we have domesticated the grasses more thoroughly and more profitably than any other plants. Today well over half the world's food calories come from the grains of domesticated grasses. Rice, wheat, corn, barley and sorghum are the world's dominant staple foods; rye and millet and oats and sugar cane have great importance, too. Despite the relative poverty of the soil there are still wheat fields throughout the Downs today, though few if any are fertilized by sheep. In a month or so they'll be flushed red with poppies. Not long after they'll be golden for the harvest.

The grasses feed us, and feed our livestock. They have shaped, and been shaped by, the world we live in, and they also shape the world inside our minds, and have done since our ancestors first walked the African savanna. Open grassland shows us a world that is both empty and alive. It lets us walk on, or through, a space as open and abstract as the surface of the sea. Grassland like that of the soft curve of the Downs along which we can walk back to Lewes makes us feel, rightly or wrongly, that we can at the same time be in the world and see beyond it, that we can expand into it, that we can play across it, that we can write on it with our paths and our chalk men and our plantations. It's grassland like this, more than any other habitat, that gives us both homes and horizons.

Forests speak of the past. The history of a forest is everywhere, written in the rings of its trees, the curves of their limbs, the tangle of their roots. Forests offer us shelter and nurture, and grant us the protection of invisibility, but their old shadows and obstructions hide things from us. Grassland, process more than place, lives always in the present, its history hidden but its distances opened to inspection. It reveals us as we are and lets us see what is coming. Forests contain the fancies of our night; grasslands open up the prospects of our days. To see the horizon is to look to the future.

The surety of that future is that what is now present will become a past eventually erased. And grass will do the job as surely as a desert or a sea. In time the unremembering lawns will cover us as surely as the red fescue covers the burial mounds that punctuate the ridge of the Downs. All disturbance holds the seeds of the grass that will remove it from our sight: an aftermath, in its original sense, is the grass that returns after the scythe has passed. John James Ingalls, a nineteenth-century senator from Kansas, caught this gentle dissolution when he celebrated the bluegrass of his state as 'The forgiveness of nature', covering the scars of the Civil War's battlefields as surely as it took away the sign and memory of a cart track no longer used.

More than seven hundred years ago, these gentle fields above Lewes were such a battlefield, albeit a less mechanized one. They were churned to mud by pounding hooves, soaked in the blood of knights and yeomen. Henry III, garrisoned in the town below, rode out to resist the descending army of Simon de Montfort, and failed.

History was made here – and within a year or so there was nothing to see but the forgiving grass that persists today, the only hint of blood the faint redness of the grain on the bent stalks, the only hint of depth their lake-like wind-blown ripples. As the poet Basho wrote:

> Summer grasses:
> All that remains of great soldiers'
> Imperial dreams.

From chalky seas to burning savannas

That the Downs are now covered in grass is the work of man; their underlying shape is the work of a long-term fall in carbon-dioxide levels. The chalk from which the Downs have been carved could only have been laid down in the warm seas of a greenhouse world; the carving itself could only have gone on in a frozen one. The slow fall in carbon dioxide that led from one to the other is the most important biogeochemical change in recent earth history. It set the climatic conditions in which our ancestors evolved. And it brought forth a new form of photosynthesis.

The history of the world tells us that climate can change for any number of reasons; climate change does not require a change in the carbon-dioxide level. But the reverse is not true, as far as we know. You cannot change the carbon-dioxide level without changing the climate. The carbon-dioxide level can be seen as setting the rules of the game. It does not set the climate rigidly, but it defines the range of climates available to the planet. A change in that level changes the rules.

We will come to the effects of such changes in the coming centuries later; for now, we take the longer view. The rise of mountains and the increase in weathering rates which crashed carbon-dioxide levels in the Carboniferous and early Permian cooled the world

281

enough for an ice age. During and after the Permian carbon dioxide climbed back from this all-time low, probably helped by an increase in volcanism that accompanied the break-up of Pangaea. In the Triassic, Jurassic and Cretaceous periods – the 180-million-year-long era of the dinosaurs – the level seems to have been pretty high. Today's carbon-dioxide level is 381 parts per million. Bob Berner's Geocarb model suggests that the carbon-dioxide level was more than three times that for almost all the time the dinosaurs walked the earth. The fossil data seem in broad agreement. The dinosaurs lived in a warm greenhouse world. During almost all, if not all, of their reign the poles were ice-free and often forested.

The high carbon-dioxide levels outlasted the dinosaurs, but not by all that much. The slow and mighty impact of India into the belly of Asia raised up mountains that were, and are, sites of massive erosion; huge amounts of fresh silicate rock were exposed to the wind and rain. The chemical weathering of that rock sucked up a lot of carbon dioxide, and the earth cooled down. By thirty million years ago a new icecap had formed in Antarctica. It was in this cooler and drier world that the grasses, which compete well in such conditions, left the forest undergrowth in which they had evolved and tackled the open spaces, pushing aside the fern prairies which had dominated treeless landscapes for hundreds of millions of years.

Lower levels of carbon dioxide do not necessarily explain all of the cooling, or the details of its timing. The climate can and does change even when carbon-dioxide levels stay the same. The rise of mountains and plateaus can alter the atmosphere's circulation, producing droughts and monsoons in new places; changes in ocean currents can alter the way heat moves from the tropics to the poles. The replacement of one sort of vegetation by another can change a region's albedo or the amount of moisture released into the air through transpiration. All these things can alter the climate independently of the amount of greenhouse warming going on. But they do so within the rules set by carbon dioxide.

At this point, it is worth wondering whatever happened to the geochemical thermostat that chemical weathering is supposed to provide. Should not the decline in the weathering rate brought about by global cooling have allowed the carbon-dioxide levels to restore themselves? In principle, perhaps; but oddities of circumstance can hold such general principles at bay, at least for a while. In this case, it seems that the immense capacity for soaking up carbon dioxide brought about by the rise of the Himalayas and their neighbours coincided with a hiccough in the earth's capacity to replenish it. The inputs into the geological carbon cycle mostly come from the edges of tectonic plates: the mid-ocean ridges where new crust is being formed and the 'subduction zones' where one plate sinks beneath another. For tens of millions of years, though, the places where carbonates are laid down in the oceans have been a long way from the subduction zones: most of the earth's subduction currently takes place in the western Pacific, where sea-floor carbonates are scarce. Some carbonates are consumed in the East Pacific, but not enough to make a world's worth of difference.

The fall in carbon-dioxide levels has had effects at the molecular level, as well as the planetary. Low carbon-dioxide levels, like high oxygen levels, increase the level of photorespiration that comes about because of rubisco's unhelpful if understandable tendency to confuse the two substances. By the time the Antarctic icecap was making its presence felt thirty million years ago, photorespiration levels were becoming something of a problem.

It was a problem, though, that seems to have brought with it the seeds of its own solution. One of the effects of photorespiration is to release carbon dioxide. When low carbon-dioxide levels started to make photorespiration a problem, some plants evolved ways to make the best of a bad job by collocating their chloroplasts and photorespiration systems in specialized structures that made the carbon dioxide given off by the photorespiration immediately available to the chloroplasts' rubisco.

Once this step had been taken, the way was open for the evolution of a less costly way of concentrating the carbon dioxide. The first evidence for this new concentration mechanism came from the assiduous research of a woman named Constance Hartt. In 1932 she started working in the laboratories of the Hawaiian Sugar Planters Association Experiment Station, trying to work out how the sucrose in sugar cane was made. By the mid-1940s she had become convinced that a four-carbon molecule called malate was a key part of the process. When carbon-14 techniques came to the Hawaiian laboratory in the late 1940s, colleagues of Hartt's confirmed that, in sugar cane, carbon dioxide was fixed into the four-carbon malate, not the three-carbon phosphoglycerate that Andy Benson and his colleagues had isolated in Berkeley.

The Hawaiian researchers were, quite literally, isolated; their brief accounts of their work were not widely read. Benson took some interest in the issue, but didn't make any progress on it. But in 1965, after the Hawaiians published a rather fuller description of their work, two researchers in the Australian sugar industry decided, over a couple of beers, that maybe they should look into the question and see if they could figure out what was going on. Within a couple of years these two men, Hal Hatch and Roger Slack, had shown that sugar cane and a number of other plants were indeed creating malate in abundance, and that they were doing it as a way of force-feeding carbon dioxide to their rubisco.

In these plants carbon dioxide from the air is stuck to a three-carbon molecule called phosphoenolpyruvate (PEP), producing a short-lived compound that the chloroplasts quickly reduce to four-carbon malate. This malate is normally shipped to cells which are wrapped quite tightly around the leaf's vascular system, the 'bundle-sheath cells'. These are the structures which first evolved as a way of concentrating the carbon dioxide given off in photorespiration. Within them the malate is oxidized to a three-carbon compound called pyruvate, giving off a carbon-dioxide molecule in the process. The pyruvate is sent back to the cells the malate came from, there to be turned back into PEP.

This chemical shuttling can produce a concentration of carbon dioxide within parts of the leaf ten times that of the air outside, recreating the atmosphere of the good old days when the dinosaurs roamed; every bundle-sheath cell its own Jurassic park. Photo-

respiration is reduced, and photosynthesis becomes correspond-
ingly more efficient. There are costs involved – making PEP from
pyruvate uses up ATP. But in the right circumstances the costs are
much less than those involved in salvaging the useless products of
photorespiration.

Because the carbon dioxide is initially used to make a four-carbon
molecule, rather than a three-carbon molecule as it is in the Calvin-
Benson cycle, the plants using this sort of chemical pump to super-
charge their photosynthesis are known as C4 plants. Rowan Sage,
of the University of Toronto, has made a particular study of the C4
plants. He finds their idiosyncrasies a useful way of asking questions
about the bigger picture – about why plants grow where they grow,
and how they respond to their environment. Although it is mostly
found in grasses, according to Sage the C4 approach to photosynthesis has evolved in nineteen different families of plants on more
than forty separate occasions over the past thirty million years. All
of those in which it has been discovered so far are angiosperms, but
they are not particularly closely related. It's a truly striking example
of convergent evolution.*

Its multiple, widespread origins reflect the fact that C4 metabolism doesn't require the evolution of any new enzymes or other
molecular machinery; instead it relies on fine-tuning versions of
enzymes that have evolved for other purposes in a new way. Exactly
which enzymes are used, and how, differs from case to case, as do
the anatomical details of the ways that the leaves are rearranged.
What they all have in common is that they use the same enzyme
to stick carbon dioxide on to PEP. This PEP carboxylase, unlike
rubisco, doesn't get confused by the presence of oxygen, and it works
efficiently even at low levels of carbon dioxide. It plays a crucial role
in the cycle that breaks down sugars in respiration, and probably
originally evolved for that purpose, but it is used in a wide variety

---

* A question exercising various researchers at the moment is whether C4 photosynthesis
could have evolved in the late Devonian as a reaction to the low carbon-dioxide levels then.
If it did, and made photorespiration less of a problem, would that have exacerbated the
oxygen spike?

of other functions, too – including the sensory system that opens and closes stomata. Its advantages over rubisco as a way of getting carbon dioxide into a form where it can be used are so great that evolution has repeatedly come up with ways in which to make use of them. But it only serves as a supercharger to the Calvin-Benson cycle; it can't replace it.

C4 photosynthesis evolved in plants that were particularly prone to photorespiration. And a great many of these plants were grasses. Grasses have a taste for marginal living that leads them into areas that are too hot, or dry, or salty for their woody competitors – precisely the conditions where a C4 metabolism that counteracts the effects of photorespiration is most useful. So thirty million years ago, when the grasses were starting to spread through a world which had become cooler and drier, some of those that ended up in stressful conditions seem to have evolved C4 metabolisms. Later, in response to further change, such C4 grasses would take over swathes of the tropics.

The spread of C4 photosynthesis, like so much else in the world, can be traced through the study of carbon isotope ratios. The carbon dioxide that rubisco doesn't take up the first time is kept in the bundle-sheath cells and presented to the rubisco again and again until, like a fussy but chastened child, the ancient enzyme has eaten up everything that's on its plate. As a result, the organic carbon in C4 plants is much less enriched in light carbon-12 than is the case in normal plants – known in this context as C3 plants. And that isotopic signature is passed on to anything that eats them. If, in some arid hothouse future, C4 plants took over from the C3 grasses of the temperate South Downs, they would change the carbon isotope balance, not just of the foliage, but of the soil, the sheep and the rabbits, too – or, more likely, of whatever new grazers the grasses brought with them.

The carbon isotope data suggest that there was a moderate amount of C4 photosynthesis on the Great Plains of North America as long as twenty-three million years ago; something similar may have been the case in East Africa. But until about eight million years ago there is no evidence of a C4-dominated ecosystem anywhere in the world. Then studies of preserved soils, of eggshells, and of the enamel on fossil teeth show a sudden change in carbon isotope values over large parts of the world. C4 grasses spread with astonishing speed through East Africa, northern India and South America. David Beerling, the Sheffield plant physiologist, calls it 'Nature's Green Revolution'. The fauna changed to match the flora; the new open grasslands favoured big grazers, and big grazers favoured big predators. The photogenic drama on the open vistas of today's Serengeti is a product of the C4 revolution, though the animals playing out the different roles do not come from the same species today as they did then.

Among those forced down new evolutionary pathways by the change in the landscape were the African apes; as the woodlands they inhabited shrank and fragmented, becoming an ever more sparse archipelago in a sea of largely inedible grass, the apes faced a series of challenges that would lead to the evolution of the omnivorous hunting species from which modern humans are evolved.

When this bloom of C4 productivity was first identified as such in the 1990s, it seemed to fit naturally and appealingly into a story about carbon-dioxide levels. The spread of C4 plants happened at a time of further global cooling. And the lower the carbon-dioxide level, the greater the advantage that C4 plants have over their C3 competitors. A carbon-dioxide drop about eight million years ago could thus explain both the cooling and the newfound competitive advantage of the C4 grasses, freeing them from their niches to roam as freely as winds and passing animals would allow.

Unfortunately, in the past ten years three different new techniques for estimating past levels of carbon dioxide from the fossil record have failed to show any significant drop at the time of the C4 explosion. They suggest that the carbon-dioxide drop driven by the rise of the Himalayas bottomed out about fifteen million years ago; though the world has cooled since then, the carbon-dioxide level does not seem to have decreased much. At the time of the C4 explosion, it had been scraping along at more or less rock bottom for about seven million years. If low carbon dioxide had been the cause, the C4 revolution would have come a lot earlier.

There is still evidence for climate change around eight million years ago, though, including a possible intensification of the monsoon system. It seems likely that the climate change triggered the spread of C4 plants. But how could a relatively small shift in climate have such dramatic effects?

The answer, according to David Beerling and others, lies in positive feedbacks. In a monsoon climate, the storms that bring the rains are heralded by thunder and lightning. If an intensification of the monsoon made the dry season drier, then woodlands would start to burn more frequently when the lightning hit them. That would have opened things up for opportunistic grasses. And as the grasses spread, the chances of fire would increase further, because grass burns better than lignin-rich wood. Being flammable is an evolutionary advantage in a grass – a way of clearing the ground for the next generation, and of stopping trees from growing back.

Computer models allow some numbers to be put to the strength of this effect. 'Dynamic vegetation models' predict what sort of vegetation will grow where using the basic principles of plant physiology and the physical characteristics of the places in question – water availability, altitude, latitude, temperature and the like. Over the past few decades they've been developed to a level of accuracy that allows them to match what's seen in the real world pretty well, and that has made them useful for answering questions about things that can't be measured. A good example is 'what is the total mass of carbon stored in the plants growing on the planet today?', to which the models mostly provide answers of half a trillion to a trillion tonnes.

These models include subroutines that account for the biomass-reducing and land-clearing effects of fires on the basis of how dry an area is, how frequent lightning is, what sorts of things live there and so on. A team which includes one of Beerling's Sheffield colleagues, Ian Woodward, has recently been exploring what happens when they turn those subroutines off, but leave everything else, including the climate, the same, thus modelling a magical alternative reality – a world without fire. It turns out that this world is also, in large part, a world without C4 grass.

In the burning world, just over a quarter of the land covered by vegetation is covered by forests; in the world without fire, that more than doubles. A large part of that gain comes at the expense of the C4 grasses, which lose more than half of their territory when the fires are turned off. (Fire also seems to be crucial in keeping Mediterranean scrubland unwooded, a process normally put down to a climate.) This global perspective tallies with experiments around the world which show that when fires are suppressed woodlands can and do encroach upon savannas.

In the world with fire, the more-grass-leads-to-more-fire-leads-to-more-grass feedback seems to have allowed the C4 grasses to turn a relatively small climatic shift into an ecological blitzkrieg, sweeping the forests from their ancient homes. Pollen records show trees in the foothills of the towering Himalayas disappearing more or less completely by about seven and a half million years ago as deciduous forests filled with mouse deer and small primates gave way to open savannas roamed by large grazing animals with thick grass-adapted teeth. Samples of ocean-floor sediment from the Pacific show a vast increase in ash particles at the same time. Part of this is probably due to the changes in wind patterns that came with the increase in the monsoon; but part is very likely the result of an increase in the amount of burning. And in this charcoal ash, trained eyes can discern lots of tiny fragments of scorched grass.

Fire wasn't the only thing the C4 grasses had on their side. Being more efficient photosynthesizers means that C4 grasses don't need to get as much carbon dioxide from the atmosphere. As a result they can contract their stomata and use less water. Less water leaving the stomata means less cooling, so the leaves get hotter – but that's OK. The big problem with heat is that it increases photorespiration, and the C4 plants have that covered. The effect on their neighbours, though, is not so benign. The C4 plants can do the opposite of what the forests did in the Devonian and Carboniferous, drying out the air. They don't dry it right down to desert levels, but they bring it down below the comfort level for woody plants and C3 grasses in hot climates. That lets the C4 plants spread, drying things out yet further. The argument is the same as that underlying the idea of a great humidification in the Devonian, but played backwards.

If a change in climate started off a drying trend, the grasses that benefited from that trend could have kept it going and changed the climate even further. Bill Hay, a now-retired geologist who mixes an ability to collate and master vast amounts of detail with a taste for outrageous hypotheses, points out that by drying out the continental interiors C4 plants would also have cooled them down.

Water vapour is a powerful greenhouse gas, but a short-lived one; water molecules fall out of the sky in days or weeks as rain and snow. To harness the power of water vapour, you have to intervene in a process that produces the stuff. The carbon-dioxide greenhouse effect increases water-vapour levels by increasing the rate at which seawater evaporates; the water vapour thus acts as an amplifier for warming that is controlled by the carbon dioxide. C4 aridification would work in the opposite direction; by restricting the supply of water vapour, it would cool things down. And grasses, being lighter in colour than woodland, also increase the regional albedo, cooling the surface by reflecting away more sunlight.

C4 plants may have left their mark not just on their grazers' teeth and the isotopes in the soil and the ashes in the ocean, but also on the global climate – indeed, Hay speculates, some of the change in the climate that is currently used to explain the C4 revolution may in fact be its consequence as much as its cause. A relatively small climatic shift towards dryness and increased seasonality in the Indian Ocean basin could have led to cooling in Asia and Africa as the C4 plants spread. That cooling might in turn have changed global climate patterns enough to help C4 plants spread in the Americas, too. The idea is certainly speculative, but not implausible.

Not long after the C4 revolution got under way, eight million years ago, sea-ice started to appear in the North Atlantic. Even though average carbon-dioxide levels seem to have remained roughly the same for most of the time since, the world has got colder. If C4 grasses did indeed provide some of that cooling, then they were one of the factors that ushered in the ice ages.

The ice ages

For the past three million years the great ice sheets, kilometres thick, have episodically rolled out over the northern hemisphere and then pulled back, switching the world between longer 'glacials' and shorter 'interglacials'. These ice ages are the best-documented example we have of the earth behaving as a complex, interactive system. You cannot begin to understand them without looking at the earth's atmosphere and oceans, at its chemistry and its physics, at its orientation in space, at its carbon cycle and its pattern of photosynthesis. This makes the ice ages at the same time immensely intellectually exciting and peculiarly hard to get to grips with. Every change bumps up against another; no cause is sufficient in itself. Feedbacks pull the climate back and forth in strange new rhythms – and carbon-dioxide levels, rocked back and forth by changes in photosynthesis, have a crucial role in amplifying them. It is as though the rules of the game changed with the run of play.

It is one of the ironies of geophysics that one of the things that appears to have made the ice ages possible is a process that we normally associate with warmth. Until five million years ago, warm water in the Caribbean and the Gulf of Mexico had two ways out; east into the Atlantic, as today, or west into the Pacific through the as yet unclosed gap between Central and South America. When Panama raised itself above the sea, the warm water had only one way to go. As a result the Gulf Stream and its associated currents became much more powerful than they had been before, pushing warm water up to Iceland and beyond. Warmer water in the north meant more evaporation there, and at high latitudes the evaporated water frequently came back down as snow. Thus the Gulf Stream fed the infant icecaps of the north.

On its own, though, the Gulf Stream was not enough. A change in the amount of sunlight was needed, too. This change was brought about a couple of million years later, not by a change in the sun itself, but by a change in the earth's attitude to it.

The earth's journey through space is not as smooth and simple as one might expect; it is beset by various wobblings. The shape of its orbit fluctuates between something more circular and something more elongated. When the orbit is more circular, the distance to the sun stays the same all year; when it is more elongated, the distance stretches and shrinks. The angle between the earth's axis and the plane in which it orbits – an angle called the obliquity – changes, too, nodding up and down by about three degrees. Finally, the direction in which the earth's axis actually points slowly rotates, a change known as precession. Each of these wobbles has a different rhythm. The orbit's elongation expands and contracts every hundred thousand years. The planet's obliquity nods back and forth every 41,000 years. And its precession follows two slightly different rhythms which average out at about 22,000 years.

To see what all this means, think about how these cycles can change what makes summer what it is. Summer is the part of the year when the hemisphere in which you find yourself leans in towards the sun, a leaning that makes the days longer and lets the sun rise higher in the sky. Depending on the precession, the height of summer may come when the earth is closest to the sun, warming things up further, or when it is further away, which will subtly even out the seasons. If the obliquity is high – if the earth's axis has nodded down towards the plane of its orbit – then the effects of summer will reach further towards the pole than they would otherwise, since the summer sun will rise higher in the sky. At high latitudes the amount of solar energy received at midsummer can vary by as much as twenty percent depending on whether the wobbles all reinforce each other or tend to cancel each other out.

Just under three million years ago the cycles came together in a way that provided the coolest possible summers in the northern hemisphere, and they stayed in synch for thousands of years. The winter snows fed by the relatively warm waters of the North Atlantic were able to last through these cool summers without melting. And so, year after year, the snow and ice grew thicker. The ice-albedo feedback kicked in: as the ice spread, its ability to reflect sunlight back out into space grew more pronounced, cooling things down further. The first modern ice age was under way. Ever since, the ice has waxed and waned in accordance with the orbital cycles.

Vast northern icecaps are bad news for things living in Greenland, Canada and Scandinavia. But why do they cool the rest of the world too? Part of the answer is down to the ice-albedo feedback; the spread of ice cools the world by reflecting more sunlight straight back into space. Another partial answer has to do with water vapour. As more and more water is frozen into the spreading icecap, the amount left in the sea is reduced; the sea level starts to drop all around the world and as a result the surface area of the ocean shrinks. When glaciation is at its peak, the sea's surface is reduced by an area roughly equal to that of Africa. Smaller, cooler oceans and seas mean less evaporation, and so the ice ages are also dry ages. Losing the greenhouse effect of the water vapour adds to their coldness.

And then there is the carbon-dioxide level. Carbon-dioxide levels during ice ages vary with the extent of the ice, amplifying all the other changes. One of the reasons for this seems to be a fertilization of the open oceans brought about by the dryness and the receding seas.

Various places in the ocean offer abundant nitrate and phosphate but no phytoplankton. In the 1930s, a Norwegian oceanographer named Haaken Hasberg Gran suggested that the phytoplankton were absent because there wasn't enough iron to support them. Iron crops up all through the biochemistry of photosynthesis. The cytochromes need it, and so do various parts of the electron transfer chain that leads from photosystem I to the Calvin-Benson cycle. So, for that matter, does photosystem I itself. But unfortunately, because the levels of iron involved are indeed low, and ocean research ships are made of iron, measuring iron levels with enough precision to prove Gran's hypothesis was hard.

In the 1980s the problem was rendered more graphic by brilliantly processed satellite images which used the spectral measurements that picked up the wavelengths associated with chlorophyll and extremely careful modelling of the behaviour of light as it entered and left the oceans to produce pictures which showed where in the oceans there was the most chlorophyll, and thus where the photosynthesis was going on. Combined with maps of nitrate and phosphate, these remarkable pictures made the 'High Nutrient Low Chlorophyll' areas graphically apparent. And at the same time, an ebullient American oceanographer named John Martin made use of 'ultra clean' techniques to get accurate measurements of iron levels in the dead zones. Iron deficiency was indeed a factor – and Martin went on to suggest that it might explain ice-age changes in ocean productivity.

The key to his insight was that the ice ages were also dry. The major source of iron to the mid-oceans is dust from the continents – the tropical North Atlantic is more productive than the southern part of the same ocean because of dust from the Sahara. Martin suggested that the increased amount of dust blown from the drier continents in the ice ages would have made various parts of the ocean more productive. The effect would be particularly marked, he thought, in the southern oceans, where the level of unused nutrients is currently quite high, and where the dust supply might have been particularly abundant. South America takes on quite a different shape in the ice ages. The coastal shelf to the east becomes an extension of Patagonia; had there been any ice-age Argentinians, they could have walked to the Falkland Islands. Iron-bearing dust from these new plains would enrich the sea all around Antarctica. The rate at which the phytoplankton photosynthesized would increase, and that increase in photosynthetic activity would draw down carbon dioxide from the atmosphere. Thus a change in sea level produced by the growth of icecaps in Canada and Scandinavia would lead to a change in the carbon-dioxide level all around the world. And the dust only had to contain a very small amount of iron to work its magic – a hundred thousand tonnes or so. Give me a couple of tankers full of iron filings, Martin used to say, and I'll give you an ice age.

Andrew Watson has since taken part in various experiments designed to test the iron fertilization hypothesis. These experiments – which involve setting to sea in a research vessel, dumping carefully prepared iron overboard and measuring what happens next in as many ways as possible – have proved Martin at least partly right, though sadly he died before the results were in. In the most thorough of them, in 1999, the careful application of a few tonnes of iron to the ocean south of New Zealand produced a bloom of phytoplankton nicely visible from space, a great curling comma of chlorophyll that went on to grace the cover of the journal *Nature*. Meanwhile a less controlled and less well documented, but rather more dramatic, experiment on the same effect has been going on in the North Pacific. In parts of China millions of tonnes of topsoil are being dried out and lost to the wind every year, a natural phenomenon exacerbated by over-grazing and the diversion of water to farmlands. Quite a lot of that topsoil ends up in the ocean. The iron supplied by increasing flows of dust over the past decades has been making vast stretches of the ocean north of Hawaii measurably more productive.

That iron fertilization happens in the modern world doesn't prove that it happened in the glacial world. But there is other, more direct evidence. Cores drilled into the Antarctic icecap, in which the ice gets steadily older as the core gets deeper, provide startlingly well-preserved records of the world's climate history. The most famous of these is the core drilled by Russian and French researchers at Vostok, a Russian base which is more remote and inhospitable than any other inhabited spot in Antarctica. The Vostok core offers a record of oxygen isotopes that reveals the total amount of glaciation (because rain, and snow, tend to be made of water containing oxygen-16, the lighter isotope, the growth of icecaps tends to leave more oxygen-18 in what is left of the oceans). There's also a record of hydrogen isotopes which, interpreted in a similar way, gives a record of past air temperatures. And there's a record of the atmosphere itself. The ice contains air bubbles, and thanks to hard-won advances in laboratory technique, the gas in those bubbles can be analysed.

The record of the climate that has been extracted from Vostok's 3.5 kilometres of ice is one of the most dramatic inscriptions of modern science. It reveals four ice-age cycles, each a hundred thousand years long. In each of them the ice builds up in a number of spurts, with slight relapses in between; sometimes there are three such spurts, sometimes four. After the ice finally reaches its maximum extent, it collapses with an almost unseemly haste – icecaps built up in fits and starts over 80,000 years recede in less than a tenth of that time. An interglacial follows, its length varying according to the details of the orbital dynamics. Then the ice begins to build again.*

And the carbon-dioxide level in the air bubbles follows on in lockstep. It does not trigger the growth or demise of the ice sheets, but it amplifies their effects, warming and cooling the world as a whole. Between the ice ages it sits at about 280 parts per million; at the peak of the ice ages, when the world is, on average, 5°C colder, the level is 180 parts per million.

The striking thing about the Vostok record is its regularity – the way the same pattern repeats, with variations, from cycle to cycle, a pattern as clear in the carbon-dioxide level as the oxygen isotopes that track the icecaps, each cycle slow to build and quick to fall. If you want proof that the earth does indeed work as an integrated system then here it is. It is not a stable system, as the original Gaia

* The nature of the 100,000-year rhythm is unclear. In the earlier history of the ice ages, recorded in ocean sediments but not in the Vostok ice, the dominant rhythm is the 41,000-year rhythm of the obliquity cycle; that cycle has a far greater effect on warming in the North Atlantic than the 100,000-year elongation cycle. Various researchers believe that there is some other process now playing a role that supplements the weak 100,000-year orbital signal: suggestions range from cosmic dust to changes in the dynamics of the ice sheets themselves.

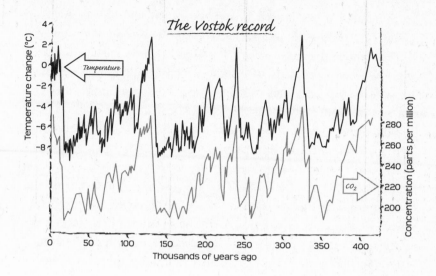

hypothesis suggested – but its behaviour is far from being random. Lovelock has argued that what the Vostok core shows is a system oscillating between two different states. It never manages to stabilize itself in either one of them, but never overshoots them either: the temperature and carbon-dioxide level at the beginning and end of each cycle are spookily similar. Like some primitive prog-rock synthesizer, the system amplifies the smoothly oscillating input of the earth's orbit into a sawtooth shriek with just the same frequency.

The Vostok data have now been confirmed and in some ways superseded by a newer core, Epica, which reaches almost twice as far back; there's talk of an even more ambitious core that will bring back ice more than a million years old. But the iconic power of those four Vostok cycles has yet to be beaten. Particle physicists talk of a promised 'theory of everything' that could be printed on a T-shirt, but they don't yet know what icon to print. The science of the ice ages does – the geophysiological equivalent of a heartbeat pulsing across the oscilloscope of an EKG.

The Vostok ice doesn't just hold the gases of past atmospheres. It holds their dust, too, and reveals that changes in dustiness and

297

carbon dioxide are indeed linked, as the Martin hypothesis suggested. According to models developed by Andrew Watson and his colleagues that have been calibrated to fit this data, increased photosynthetic productivity in the oceans could have accounted for a forty-five-parts-per-million drop in the carbon-dioxide level (others think the effect may be less). At most, that's about half of the difference between the carbon-dioxide level in the interglacials and the level when the icecaps are at their fullest extent.

Watson and most of his oceanographic colleagues believe that the rest of the difference in carbon-dioxide levels is due to changes in the way that carbon dioxide is stored in the depths of the ocean. While we think about carbon dioxide as an atmospheric gas, at any given time most of the carbon dioxide on the earth is in fact dissolved in the oceans. They contain fifty times more of it than the atmosphere. Carbon dioxide in the atmosphere is endlessly being absorbed by the waters at the oceans' surface. Some of it comes out again more or less straightaway. Some of it, having been taken down into the depths by the ocean's circulation, doesn't reappear for thousands of years. As long as the rate at which carbon dioxide comes out of the depths matches the rate at which it goes into them, though, everything stays stable.

In the ice ages, it appears, the rate at which carbon dioxide gets out of the depths of the ocean slows down a little, and so the amount that is stored down there increases a bit. Because the amount of carbon dioxide in the deep oceans is so vast, a small increase there requires a large decrease in the amount left in the atmosphere. Increasing the vast stock of carbon dioxide in the ocean by one percent reduces the relatively paltry amount in the atmosphere by half.

There's not yet any agreement on how carbon dioxide is kept down in the depths, but the position of the roadblock keeping it there is known. There are few places where water from the depths rises all the way to the surface, and the main one is in the sea around Antarctica. Relatively small physical changes in the most southerly parts of the ocean can thus make themselves felt on a global scale:

if such changes restrict the amount of deep water that can get to the surface, they can limit the carbon-dioxide exchange with the atmosphere. Various details of the ice-age climate can be invoked as explanations for the roadblock. Changes in winds might perturb the relevant currents; shelves of sea-ice might isolate the upwelling ocean from the atmosphere; the surface waters might take on a peculiarly stable layered configuration that it is hard for the deep waters to break through. Watson favours the last of these, but is open to the other two. Whichever turns out to be correct, the effect seems likely to be large enough to keep just a bit more carbon dioxide stuck in the ocean depths, and thus reduce the amount in the atmosphere a lot.

Good supporting evidence for the idea that the seas around Antarctica are the key to the ice-age climate comes from the way that ice ages end. Something – as yet, no one knows what – starts warming up the southern oceans. Once that warming has been under way for a couple of centuries, the carbon-dioxide level starts to rise, presumably because the exit route from the deep oceans is reopened. Rising carbon dioxide warms the planet further. The glaciers retreat; the icecaps start to melt; new-risen waves cut fresh cliffs into the seaward edges of the Downs. The great spring begins.

The fertile millennia

The oceans and the atmosphere are not the only carbon reservoirs that play a role in all this. The soil and the biomass of plants on land matter too. And in the depths of an ice age this reservoir gets very low. Photosynthesis doesn't work well when things are dry, or for that matter when they are cold – the membranes in which the photosystems sit stiffen at low temperatures, slowing the workings of the electron transfer chain. And being starved of carbon dioxide obviously compounds the problem. Where now there are temperate forests – or where there would be, had we not replaced them with fields and pasture – the ice ages have tundra swept by newly savage winds. Even in the tropics, the forests contract. Grasses, and later some trees, spread on to the continental shelves revealed as sea level drops, but that growth is not enough to counteract the pauperization of the plants elsewhere.

Today, the plants on the earth's surface contain somewhere between 500 billion and a trillion tonnes of carbon. The soil from which they grow contains more than the same again for a total comfortably over two trillion tonnes, all put there by rubisco and sunlight, and much of it amenable to being turned back to carbon dioxide. In the ice ages, dry and stripped of trees, a significant part of this store of carbon is lost. At least 300 billion tonnes of carbon and perhaps as much as a trillion tonnes is given up by the plants and soils and sucked into the vast cold store at the bottom of the ocean.*

When the most recent ice age came to its end, between 15,000 and 12,000 years ago, all the carbon lost to the oceans was restored to the atmosphere, and some of it, through the atmosphere, to the land. The earth started to grow green again. In the north, pine forests started to reassert themselves, and in doing so warmed the planet further through a biological inversion of the ice-albedo effect. Snow that falls on tundra presents a white reflective surface to the sun; snow that falls on firs ends up on the ground beneath a dark canopy that sucks up sunlight and warmth. The spread of the pines is an important part of the earth-warming feedback that kicks the planet into its interglacial summer; in the words of John Harte, a Berkeley ecologist, the forests don't follow the snowline north – they chase it.

With their tissues warmed by the thickened air, their roots watered by increased rain, their rubisco revived by air richer in

---

* It is possible that because atmospheric carbon-dioxide levels never get below 180 parts per million this figure may reflect the limits of this process; it may mark the point below which terrestrial ecosystems simply can't take up carbon any more, bringing the biological cycle close to a full stop. This would be the sort of feedback system discussed in the previous chapter, on pages 257–8.

carbon dioxide, tropical woodlands previously constrained to small and isolated tracts of land break their bounds and spread, knitting themselves together into the great rainforests we see today. In the temperate zones, woodlands spread from horizon to horizon, blanketing the land. And in some warmer regions, grasses grew unaccustomedly heavy with seed, a change which may have had profound historical effects.

The point at which our ancestors became modern humans, with cultural abilities and complex language skills equivalent to ours today, is much debated by anthropologists. Some favour an archaic origin 150,000 years or more ago; others favour something more modern, perhaps 50,000 years ago. All, though, agree that the change had come about well before the end of the ice age. And yet the first aspect of our cultural capabilities to mark us out from other animals on a planetary scale did not come about until much later. It is only about 12,000 years ago that there is clear evidence of what we would now recognize as agriculture.

It is widely assumed that the delay is due to the ice-age climate – that the world was simply too cold and unpleasant for farming before the great spring. But Rowan Sage, the C4 expert, points out that this is not so. Throughout the ice ages there were areas warm and damp enough for agriculture, including some in Africa, the continent on which modern humans originated. The reason that we had no fields or ploughs then, he argues, is not because we were not able to imagine such tools, nor because the climate would not permit it. It was because the carbon-dioxide levels would not. Before the deep oceans gave forth their hoarded carbon at the end of the ice age, the plants our ancestors would eventually domesticate were simply too spindly and malnourished to be able to feed us in any numbers. It was not the retreat of the ice which permitted the development of sedentary gathering, and thus of agriculture, and thus of civilization: it was the rise in carbon dioxide.

Between 15,000 years ago and 12,000 years ago the carbon-dioxide level rose from around 200 parts per million to about 270 parts per million – a dramatic rise by natural standards, though

incredibly long-drawn-out compared to that which we have engineered ourselves in the past century. Sage has run various experiments
in which crops and their wild ancestors grown at ice-age carbon-
dioxide levels of about 200 parts per million are compared to those
growing at post-glacial levels, which suggest that in a couple of
millennia productivity in C3 plants grew by between thirty and fifty
percent.

The plants grown at higher carbon-dioxide levels made better use
of water, because they didn't need to keep their stomata wide open;
this would have made them more resistant to drought. They also
made better use of the available light, because they could grow more
leaves more quickly at the start of the season, thus minimizing the
number of photons that fell fruitlessly on bare ground. More energy
makes the plants better at getting nutrients, such as nitrate, out of
the soil, and thus allows them to put more protein in their seeds.
And it also makes C3 plants better able to face competition by C4
plants. This is important because C4 grasses are not particularly
amenable to human harvesting: they have small and brittle seeds.
With the exception of millet, the early crops domesticated in the
Near East, China, Africa and the Americas were all C3 plants, not
C4 plants. Maize, now the world's biggest C4 crop, was domesticated
a few thousand years later, and sugar cane later still.

In the ice ages, warm grasslands were dominated by C4 grasses
even more than they are today, because their capacity to concentrate
carbon dioxide gave them even more of an advantage. Experiments
at a carbon-dioxide level of 150 parts per million – which is, admittedly, slightly lower than even the lowest ice-age level – show C4
grasses out-producing C3 grasses by more than a factor of twenty.
Not by twenty percent: by twenty times. However, at 270 parts per
million the advantage is less than a factor of three. C3 plants are
still not great at dealing with C4 competitors, as anyone who has
tried getting crab grass out of their lawn can bear witness. But
they're better than they were in the ice ages.

After the ice age, even before domestication, wild grains would
have become far more fruitful, encouraging previously nomadic

people to settle and multiply. An American agronomist named Jack Harlan has shown that under today's conditions a family would have been able to harvest a year's worth of edible grain from wild einkorn, an ancestor of wheat, in just three weeks. Settled populations and highly productive cereals provided a setting for the development of agriculture that had not existed at any previous time in human prehistory.

In his book *Guns, Germs and Steel*, Jared Diamond argued brilliantly that there have been special places in the history of the world – places which, by luck, had biological resources that other places lacked. Most favoured were the shores of the Mediterranean, the native habitat of thirty-three of the world's fifty-six large-seeded grasses, some of which became the cereals of the fertile crescent. Even before domestication, some of these grains could, in the right conditions, have supported large populations. Africa south of the equator, on the other hand, had not a single such large-seeded grass, and thus nothing from which a cereal could be bred. The places where there were plants and animals suitable for domestication, Diamond argued, thus got a head start in the development of urban civilizations.

What Sage's work adds to this picture is the idea that, for all these lucky places, there was a single lucky time. The fertile crescent is defined in space by biogeography; the fertile millennia are defined in history by plant physiology. After at least 30,000 agriculture-free years, humans developed agriculture in four separate and largely isolated regions within a few thousand years of each other. Soon after the fertile crescent saw its first farming, rice was being domesticated in China, sorghum and pearl millet in the Sahel, and lima beans in the Andes. What the people doing this had in common was shared with all their ancestors: hungry bellies and ready minds. What set them apart from their forebears, but allied them to their contemporaries around the world, was the sudden increase in the carbon-dioxide level that gave them plants which could grow more quickly, more strongly, more resiliently and more fruitfully than they had for the past 100,000 years.

The end of plants

The interglacials are oases of carbon-dioxide richness in the great deserts of the ice ages. Yet even in the interglacials, the environment is not one that photosynthesizers would choose as optimal. The trees on the slopes of the Downs may look vibrant in the spring sunlight, but they are not all that far from the end of their biochemical tethers. Persistently high levels of oxygen, while helpful to animals, mean that in temperate climates C3 plants are constantly losing at least ten percent of their productivity because of photorespiration, while in warmer climates they lose more. And there is not enough carbon dioxide. Though interglacial levels of carbon dioxide are better for plants than glacial levels, higher levels would be better still. Using the sort of vegetation models with which he explored the world without fire, Ian Woodward and his colleague David Beerling have calculated that in the high-carbon-dioxide world of the Cretaceous plant productivity may have been twice what it is today. Since then plants have had tens of millions of years in which to adapt to lower levels of carbon dioxide, but there is only so much that can be done. Carbon dioxide is still their stuff of life. In the final analysis, however efficient they are at using it, plants want more, rather than less. This desire will play a key part in the current carbon/climate crisis; it is also going to be important on far longer timescales.

When the idea that the chemical weathering of rocks provides the earth with a thermostat was first put forward in the early 1980s, Jim Lovelock quickly grasped two implications. The first, as we have seen, was that life would be able to change the setting of this thermostat by influencing the weathering rate. This was part of what made Lovelock start taking a greater interest in the biosphere's past. The other, grimmer, implication bore on the biosphere's future – or lack of it.

The great appeal of the thermostat idea was that it explained why the earth's climate had stayed more or less steady over the planet's history. As the sun had grown hotter the levels of carbon dioxide had fallen, because the hotter the sun, the higher the rate of weathering. The greenhouse effect thus weakened as the sun strengthened, with the resultant climate staying more or less stable.

But as Lovelock quickly realized, the carbon-dioxide level has now become so low that it will soon have nowhere left to fall to. In the Archaean carbon dioxide may have made up more than a hundredth of the whole atmosphere. Now it makes up a lot less than a thousandth, and the plants live in permanent hunger.

In 1982, in a paper in *Nature* entitled 'Life span of the biosphere', Lovelock and his friend Michael Whitfield argued that if the thermostat kept driving carbon-dioxide levels lower, then pretty soon plants would be unable to grow at all. Using a very simple mathematical model for the strength of the greenhouse effect, they suggested that within about 100 million years the thermostat-driven carbon-dioxide decline would starve all the earth's plants to death. With the death of the plants the terrestrial food chain would collapse, and complex life would be wiped from the land.

The idea that the world must come to an end has been proclaimed by science for more than a century. The enunciation of the Second Law brought with it the concept of the 'heat-death of the universe', the idea that there must eventually come a point at which everything has reached the same temperature, and no ordered change – no life – is possible any more. Happily this form of cosmological death is now seen as absurdly distant (and some cosmologists feel it may not be inevitable). The twentieth century, though, brought a closer doom; as the internal workings of stars were deduced, they revealed that the sun would, in time, swell to the size of a red giant, and in so doing either sterilize the earth with its heat or, just possibly, engulf it whole.

This distant prospect led to what is probably the most overused joke in astronomy, in which an eminent astronomer is questioned by a member of the public:

'Excuse me, Sir Martin [or Dr Sagan, or Professor Hoyle, depending when the joke is being told], when was it you said the world was going to end?'

'In five billion years.'

'Oh thank goodness for that – I thought you said five million.'

Aware of the potentially humorous disconnection between the timetables of science and the horizons of humanity, Lovelock and Whitfield were quick to point out that, in human terms, the crisis they were describing was 'infinitely distant'. But in geological terms the coming crisis was pretty close – far closer than the sun's eventual expansion. If they were right it would mean that eighty percent of the future that the plants of the Rhynie Chert had looked forward to had already passed by. It would mean that only half the span allotted to the angiosperms that arose in the Cretaceous remained. It would mean that, for almost all of the five billion years that remained before the sun finally did its big red thing, the earth would be unfit for complex life.

This original paper was, it turned out, too pessimistic. While the strange rhythms of the ice ages drove the carbon-dioxide level lower than it had ever been in the previous history of the planet, there is no reason to think that it cannot mount something of a recovery, over and above the bump that humanity is currently providing. For a while, if left to themselves, the ice ages will continue. But as the continents continue their long slow dance, the large platforms of land in the north on which the icecaps grow will move south, making each new ice age harder to start. As part of the same realignment, ocean currents will change again, warming the poles.

And in the medium term a new subduction zone will almost surely be created along the eastern seaboard of the Americas, while another is likely in the southern Indian Ocean. Both will deliver sea-floor carbonates that have been accumulating for tens of millions of years to the pressure cooker below, and replenish the flow of carbon dioxide to the atmosphere. Much of this may be used up in the weathering of new mountains, such as those which will rise when Australia hits Borneo, or as Africa and Europe continue their merger and squeeze the rocks of the Mediterranean high into the sky. But some may be left over, for a while. The thermostat will be back in business, for a bit; but the sun will still be getting steadily hotter, and so it will be ever easier for weathering to win out.

Another reason for postponing Doomsday was offered by Jim Kasting and his colleague Ken Caldeira when they revisited the lifespan of the biosphere a decade after the Lovelock and Whitfield paper had been published. That original paper had neglected to take into account the fact that C4 plants can continue to grow at quite remarkably low levels of carbon dioxide (ten parts per million), and that some phytoplankton can probably survive down to about the same limit, too. Combining this with a subtler representation of the greenhouse effect, Kasting and Caldeira suggested that plant life might have as much as 900 million years to run. But that run was downhill. From here on in, their model said, the planet's productivity would drop ever lower.

It would take a billion years for carbon dioxide, and the life that relies on it, finally to disappear, Kasting and Caldeira argued, a billion years in which the average global temperature would rise by just 10°C. In the next half-billion years, though, things would really heat up. In the absence of the negative feedback embodied in the thermostat, the surface temperature starts to track the rising output of the sun. But soon enough after that another, positive, feedback takes over, as water-vapour levels rise in the lower atmosphere. Once enough water vapour gets into the atmosphere, its lower reaches become so opaque to infrared radiation that the temperature in the upper atmosphere becomes uncoupled from that at the planet's surface, with the result that a fundamental gap opens up between the rate at which the planet emits long-wavelength radiation from the cool upper atmosphere and the rate at which it absorbs the heat of short-wavelength sunlight. This persistent shortfall in the planet's ability to lose heat means the world is doomed; the earth will boil dry like a saucepan left on the stove, its oceans steamed off into space hydrogen atom by hydrogen atom. It will be transformed into a simulacrum of its sister planet, Venus, where the same process got under way billions of years ago – where the only remaining water sits in clouds of concentrated sulphuric acid, and the metals that boil away in the unholy heat of the lowlands condense as glittering frost on the peaks of the burning mountains.

If life were still around, maybe it could do something about this – perhaps by increasing the planetary albedo. But life will be all but gone, unable to grow by fixing carbon after the carbon dioxide is finally stripped from the atmosphere. Tim Lenton has argued that Kasting and Caldeira underestimate life's tenacity, failing to account for the fact that, as life ebbs from the planet, the chemical weathering rate will decrease. But though Lenton's figures give life a slightly longer lease, its extended stay is at the cost of ever-reduced powers. And there will still come a point when Gaia's capacity to use the endless energy of the sun will be undercut by its reliance on the resources of a finite world. It will be no more a failure than any death after a full life is a failure. But it will be an end.

It may be that a thin strand of bacterial life will cling on after the photosynthetic link to the sun's power is severed – there is, after all, a faint possibility that it has found a way to hang on in the caverns of Mars, and there are some who argue that hardy bacteria might even now have a niche in Venus's battery-acid clouds. But life's ability to shape the planet it lives on will be over. When photo-synthesis shuts down for lack of carbon dioxide to fix, life will lose its relevance to the geochemistry of the planet-wide desert; it will offer the great heat engine no new ways of making entropy. Looking at the planet through a bioscope will no longer bring new insights. The life of the planet will be over. The unwatered rocks will be her tombstone, and there will be no grass to cover the grave.

The long walk

As a boy, with restless energy, Jim Lovelock used to cycle out from south London through all the byways of Kent and Sussex. Now in his mid-eighties, he takes great delight in walking long-distance footpaths with his beloved wife Sandy. As I've been writing this chapter, they've been tackling the Peddars Way through East Anglia. They have walked the southwest coast path – 650km of trail, with almost 30,000 metres of up and down – twice since his eightieth birthday, and perhaps will walk it again. One day I hope to follow them on that. Coasts and ridgeways are the finest walks, with their sense of a world cut in two about you, of a near and a far and a bearing.

Like me, I think, Lovelock feels closer to nature when walking through it than when stopping in it. It is by walking that we discover what there is to see, and where it can be seen from. Walking makes us close to the world through process, not place: through treading on it and moving through it and watching its vistas change. We walked from place to place long before the newly-grain-heavy grasses called on us to stop and stay. We walked the ice-age bareness; we walked the C4 savannas. We didn't walk because we liked the views; but I think we came to like the views because we walked.

Jim knows he has not all that many walks left to him; I don't think that discourages him, nor necessarily does it heighten the pleasure. It is the world and his wife that do that. But he does take some solace from knowing that, on a vastly different scale, Gaia too is now old. Born more than three billion years ago, she seems to have only about a billion left to go. Still full of wonders and dignity, she is also getting tired. She is on a downhill path.

Poised more comfortably in the middle of a life, Gaia's mortality has less resonance for me. But it still has meaning. Having an end makes life on earth something like a story, and something like a walk; to be able to see the distant ending makes sense of things, as the sea makes sense of the Downs. It gives them shape.

It has, however, no practical consequence. None of what we have seen in this glance into the future has any more practical meaning for us than the question of whether the sun swells up in five million years or five billion. It matters not to us, nor to any who are recognizably our descendants; it matters not to our species.

Some species survive all-but-eternities. Some of today's cyanobacteria may be near-identical to some of those that first thrived long before the Great Oxidation Event, and those same species may yet survive until the world runs out of carbon dioxide. Algae almost indistinguishable from billion-year-old specimens Nick Butterfield has found fossilized in the Canadian north are still to be found on rocky foreshores today. The maidenhair trees that in England we sometimes see in arboreta and old, bold gardens are close relatives of those found in fossils from the Permian. But animals in general, and mammals in particular, go forward at a different tempo. The average duration of a mammalian species in the fossil record is just a million years.

And we are not an average mammalian species. As a mantra for finding one's individual place in nature, Peter Horton's Igwajas – In Gaia we are just another species – has much to recommend it. But as an objective statement about our collective future it's just not true. We are not just another species. We are something with more impact than a giant asteroid strike. We are something potentially as profound and disruptive as the advent of animals, or even perhaps as the creation of photosystem II. Our technologies provide a capacity for change fundamentally different to that provided by natural selection. Not just change to the biosphere – though we have achieved, or inflicted, plenty of that, and will do more – but change in ourselves.

The still-nascent ability to reprogram genes, redesign proteins and reshape developmental pathways, combined with the ability to integrate the ways in which computers process information into the ways in which we do so ourselves, seem sure to change at least some of us in ways as fundamental as the development of agriculture or even language. At least some of our descendants will be unrecognizable as human long before our mammalian million years are up. I would be surprised if humans as we understand them last as long as the current extended interglacial. It is possible that we will cease to be; that our potential for change is as profound as any that evolution has wrought does not mean we will be enduring.

Or, I think more likely, we will be swept up in something new. Science-fiction writers, playing with such ideas today as Wells and Verne played with the laughable notion of spaceflight in the nineteenth century, have various names for the thing doing the sweeping – the accelerando, the hard rapture, the singularity. I will just call it the Change.

You may think there will be less Change than I am suggesting, and you may be right, but it is hard for me to see why. It is possible that the sorts of understanding and the skills in creation needed to fundamentally reshape the human being are simply not available to our intellects, or to the ways we act upon knowledge. But look at the progress of the past hundred years – the progress that weighs isotopes sifted by stratospheric chemistry in the Archaean, and reads the genomes of chloroplasts, that lets us build pictures of the atomic machines that pull apart water molecules and weigh the chlorophyll content of whole oceans by staring down from space – and ask yourself if understanding and reshaping the mechanisms of the mind is really likely to be entirely beyond us.

There are some who argue that God, or human nature, will not allow such a Change; but as a disbeliever in one, and a believer sometimes inspired, sometimes perturbed by the sheer diversity of the other, I doubt it. Today we have children whose embryos were chosen according to their genes, we have computers taking stumbling steps towards the mimetic simulation of the tissue of our brains, we have bacteria being designed for programmability, we have implants that allow people to move metal limbs with their imaginations and see through electronic eyes. Do you really think it is going to stop? Our paths point to a posthuman future, and it is a landscape we must address.

John Varley, a science-fiction writer unfortunately not widely known beyond the bounds of the genre, has provided a metaphor for this landscape of the future that I find particularly compelling: the steel beach. The steel beach is an environment we have made, but it is also an alien world we must adapt to like lungfish in the mud. The technological future we inhabit will be ours to adapt to as much as it will be ours to shape, a steel beach no more intrinsically hospitable for all that it is of our own making.

Of the many boundaries that could be dissolved in such a world, one which the imagination of science-fiction writers has explored from time to time is the boundary between plant and animal. In the science-fiction of Olaf Stapledon there are large walking humanoids bedecked with vast masses of foliage, plant-men that my friend Henry Gee shrewdly suspects may have influenced the treelike nature of the ents in Tolkien's *Lord of the Rings*. In the warped and wondrous world of Geoff Ryman's *The Child Garden* humans who learn through viral infection spread their purple photosynthetic skins to the sun on street corners. In his novella 'The Green Leopard Plague' Walter Jon Williams sees photosynthetic skin as a safety net against starvation in a world where death becomes increasingly optional. I'm not entirely convinced by the bioenergetics here, but Williams's purpose is not to provide convincing bioenergetics. It is to play speculation and narrative off against each other in a way that opens the mind, in a way that makes us ask what need always be the same, and what might change. In practical terms, yes, it makes more sense to engineer better plants; but the questions here go beyond the practical to explore the way technology can change both our independence and our interdependence.

The loveliest image in the admittedly scarce literature of human photosynthesis is one of Varley's own – a space-going symbiosis between plant and person. The idea that space is a perfect place for solar power was around long before American engineers realized that the new technology of the solar cell offered a peculiarly suitable power source for satellites. Konstantin Tsiolkovsky, the Russian engineer who implanted the idea of space travel deep in the collective mind of the Soviet period, was fascinated by solar energy and the power it could provide. Tsiolkovsky understood the sun's 'cosmic energy' as the force that powered the biosphere – he was an acquaintance of Vernadsky, a product of a similar intellectual milieu, and cut from the same visionary cloth – and saw life in space as a way of using that ultimate energy source to the fullest. He imagined a moon populated by creatures that walked like animals and photosynthesized like plants, and saw something similar as mankind's cosmic destiny.

Varley took a similar dream and personalized it; he took the great icon of the seventies, the living earth seen over the limb of the lifeless moon, and shrank it right down to the human scale. Something of a hippy himself, he imagined a world in which those jaded and wearied by the pace of future life drop out into green, living space-suits a bit like a Savile Row bacterial mat. These symbionts take carbon dioxide, piss and shit, and recycle them using sunlight. They hang in space like vast thin sails, absorbing the sun, nourishing the person within. The human-symbiont dyads are planets in microcosm, marriages in metabolism. The symbiont completes the human, the human completes the symbiont, giving it carbon, giving it reason, giving it thoughts to think. Alone together they drift among the rings of Saturn, lost in beauty.

I don't necessarily expect this vision to come to pass. But I would not be that surprised if, somewhere in our posthuman future, in adaptations meant for some other planet or for none, photosynthesis came to play a part. Beyond the pull of gravity the difficulty inherent in producing a brain's worth of glucose from a body's worth of sun could be more easily worked around than it is on earth.

In the long run, if human and posthuman intelligence persists and spreads, I expect wonders as strange as Varley's symbionts, or stranger. I don't expect anything that profound in the current century; but I think I already see the buds on the trees, and I expect them to take leaf and blossom within this millennium. Which is to say, I expect them within the lifespan of a reasonably long-lived tree.

Walk east towards Pook's Hill and the Pevensey Levels from Lewes and you will come to Wilmington, where in the churchyard overlooked by the Long Man sits a yew that may be two thousand years old. Men who fought at the Battle of Lewes could have cut their bows from it; now it is celebrated in the Norman church beside it by a wonderful new stained-glass window based on micrographs of yew wood, the sunshine streaming through its representations of stomata like light from the wounds of the living Christ. This yew and its ancient brethren scattered across the landscape could, I increasingly suspect, outlive the human race – or at least see the first flowerings of its successors.

This is not something I had thought of when I chose to construct this book according to the lifespans of men, planets and trees. For all my abstract faith in the future I have to admit it is not a thought I find altogether comfortable. But it is something which I think likely to be true.

The path to Gaia's end, while it doubtless has interesting turns we cannot yet see, has a shape that we can already map out; we can see the sea cliffs in the distance, and the dead world beyond their whiteness. The path of humanity's future, on the other hand, cannot be discerned in any landscape yet glimpsed. We cannot see more than a few steps down it, or get any firm sense of its direction or length. We are in the woods.

But for my purposes here, though, what has been said so far is enough. The rest of this book is not concerned with futures far away in their significance, if perhaps not so far away in time. It is about the future here at hand, the future course of the carbon/climate crisis into which we have already slipped. Before we change humanity, we have the unfinished business of the earth to attend to.

PART THREE

In the span of a tree's life

A green thought in a green shade.

ANDREW MARVELL,

'The Garden'

It would, of course, be nice and noble to say that I pursued research into photosynthesis because it addresses the important questions of an alternate energy source and food supply. But it wouldn't be true. I have no pretensions to be a do-gooder; I simply enjoy research and it fulfils an inner need. That it sometimes addresses a question of practical importance, and engenders support is fortuitous and lucky.

GEORGE FEHER

CHAPTER EIGHT

Humanity

Bastille Day
The discovery of photosynthesis
The phlogiston cycle
Changing leaves, changing worlds
The power of humanity
The carbon/climate crisis

If rain, which is so necessary to vegetable productions, and if fire, which is so necessary to animal life and the internal constitution of this earth, have such an intimate connection with the atmosphere of the globe, it is surely an object for the philosopher, to trace the chymical process of nature by which so many operations are performed.

JAMES HUTTON

Bastille Day

On Bastille Day, the day the jail walls came down and the journey to and past terror began, I sit in the shade of trees, looking out over beautifully tended grass to a serpentine lake. The summer sun is tempered by a soft breeze; there's a gentle hum of lawn-mowing in the distance. Nothing is quite still; nothing is going anywhere.

The Marquess of Lansdowne's family has lived here at Bowood House, in the county of Wiltshire, for two and a half centuries. In 1757 the first of them to live here, the first Earl of Shelburne, asked the great landscape gardener Capability Brown to have a look at the estate. Brown assured the earl he knew of no finer place in England, and though one suspects that was what he said to all the earls – and to the other members of the nobility and gentry, about 140 in all, whose grounds he was commissioned to landscape – the flattery may also have held a subjective truth. Brown saw all landscapes as good, and at the same time saw all landscapes as improvable; there was nothing finer than one which might soon be worked on, one which might soon have its potential – its capabilities, as he would put it – brought out.

The first earl died before actually giving Brown a commission to work on the estate. His son, who succeeded him, would become the first Marquess of Lansdowne and also, briefly, prime minister; he was the man who brought the American Revolutionary War to its formal conclusion. The first marquess greatly enlarged Bowood, adding a 'Big House' to the existing 'Little House' and decorating them sumptuously, and commissioned Brown to refashion the estate seen from their windows. He also brought to Bowood the two men who discovered the role of plants in the renewal of the air and the role of sunshine in the life of plants: the English minister of religion Joseph Priestley and the Dutch physician Jan Ingenhousz. Both men spent summers here. Both may have sat where I sit today, enjoying the view that Brown created with typical if somewhat disturbing industry, uprooting old avenues and planting new trees by the thousand, drowning the village of Mannings Hill in the process of creating the lake.

Sitting here in the shade I am nestled between two geometries, geometries that inspire thoughts of the past out of which our future is growing. There's the open geometry of the lawn-dressed land sweeping down to the lake, the land that Brown sculpted into soft curves and subtle saddles with crafty precision. And there's the stranger, denser geometry of the trees above me, the solid shade of which belies their irregularity, even deformity. On one tree, a Spanish chestnut, the foliage has been stripped from all the upper branches. Those leaves that are left to it are bunched around its waist like a great green nut threaded on to the rifling of its deep-fissured bark. The other tree, a cedar of Lebanon, is more dramatic still, a hobbled giant twisted by centuries of intervention. Its south-west side is almost stripped of limbs; the boles that remain have been trained to curve to the east, much against their inclination. The trunk has been thwarted into some sort of pendulous kinkiness high above the ground, passing through a three-dimensional chicane that my eye finds hard to resolve. It is a tree interrupted. The platforms of foliage that cedars are famous for, hanging in the air like plates on the arms of an octopus waiter, are missing: high branches leap out in all directions like feathers from the shoulders of a showgirl. Only at the very top is the classic flat spreading form of the tree apparent – and this crowning cap sits oddly askew.

Off balance and magnificent, the cedar stands more alien than a dinosaur, an ancient form of life as large and as unlike our own as the world can offer. Its function is as sure as its form is strange. Where the trunk first rises from the ground it is smooth and powerful, as wide as I am tall. On the boughs that hang down towards the lawn, the light, bright cones stand oddly pert among the needles, like trulli on an Apulian hillside. Its bark is even, smoothly scored, uninterrupted. Its leaves are clustered in tight, purposeful whorls, dark green with a hint of blue. The water flows up its vascular system, the carbon dioxide diffuses through its stomata: the electrons flow along its membranes, and the hydrogen ions through them, in just the way they should. It works.

The first of the two geometries I am between, the geometry of the landscape at large, reflects the place allowed for nature in the reasoned world Brown worked in. His art was shaped by what he saw as science, not by whimsy. As he wrote to his friend, the Reverend Thomas Dyer, 'Place-making, and a good English garden, depend entirely upon principle and have very little to do with fashion; for [fashion] is a word that in my opinion disgraces Science wherever it is found.' The principles were those of pleasing topography, carefully crafted lakes, and delicately planned plantations, 'so much Beauty depending on the size of the trees and the colour of their leaves to produce the effect of light and shade so very essential to a good plan'.

In Brown's work the original landscape was at once the raw material and the inspiration; his aim was not to create something new, but to let that which was there already be presented in a more beautiful and pleasing way. Improvement came not through the addition of artifice; nature, Brown and his clients felt, was the true guarantor and font of beauty. But it was in the nature of nature that it could be improved by man. Its melodies could be strengthened and enriched by properly chosen harmonies. Contours were not to be followed exactly, shorelines not to be aped by encircling paths; there were 'lines of grace' for contours and 'lines of beauty' for paths, waving and winding in just the right way. There were no calculations to this science of place; the techniques that allow algebraic analysis of such geometries had yet to be invented. But then it was a time when the now-firm link between science and mathematics, the link that makes the predictive numerical model such a fundamental part of the scientist's trade, was still unforged. The sciences of chemistry, or of natural history, were as innocent of mathematics as the science of landscape. Brown did not make models of nature – he surveyed it and then changed it to his designs. He made nature into a model of itself, a representation both simplified and heightened.

Brown's geometry is what we want from the world; the second geometry, the geometry of the cedar and the chestnut, is what we get. It is a geometry laid out in time as much as in space – and time is something which cannot be marshalled within the boundaries of estates, or governed by its owners' will. This geometry is shaped by men with saws, and concerns about what boughs should overhang the kitchen-garden wall; but it is also shaped by drought and wind and disease, by circumstance and accident. On Bastille Day of all days, we know that history is not something that can be laid out according to the aesthetics of the ruling class.

In the late 1760s and the 1770s the cedars of Lebanon in the Royal Botanic Gardens at Kew, which were among the first grown in the country, had just begun to produce seed, and a great many cedars were planted at Bowood. True to his sense of himself as someone bringing out the best in a nature already present, Brown usually restricted himself to British trees for large plantings – lime, elm, Scots pine, oak, chestnut, beech. But he would allow himself foreign stock, too, especially as a way of producing specific highlights and features, and he became particularly fond of the cedar of Lebanon. The one I sit below, I feel sure, is one of his.*

As such, it is marked out as a survivor. Cedars from the Levant

---

* I'd like to think the bare-topped green-belted chestnut is also one of Brown's, but while stout in the trunk it is nothing like as impressive as another specimen of the same species a hundred yards or so along the path, a great big drum of a thing sitting on the grass as solid as a rock at sea, its dark glossy green leaves enlivened by starbursts of fading yellow catkin, its shade laced with a distinctively sexual smell. That tree was planted as a seedling by the third marchioness, and if a tree of 1825 outdoes so comfortably the one under which I am sitting, can mine really be older? That said, the third marchioness's tree is a seedling from the great Tortworth Chestnut, a behemoth which may be more than a thousand years old, a tree already written up as remarkable in the seventeenth century. Perhaps its Bowood offspring shares in its vigour, and has thus outgrown a longer-established neighbour.

thrive and grow tall in the rain and rich soil of England. The cedars that Gilgamesh the King hacked down in their thousands and those whose timber built the temple of Solomon would have been small compared to those that grace the parks of England. But if the English soils and rains are to the trees' liking, the Atlantic storms are less so, and a larger tree is an easier pushover. Rough winds have not been kind to the cedars of Bowood. Originally there were hundreds. Twenty-five years ago, there were still dozens. After the storms of 1984 and 1990, there are only a handful.*

New trees replace the old. A little way off is a hornbeam just twenty years old that could in time outstrip the cedar and the chestnut. It is about the same age today as the twisted cedar was when the Bastille fell in 1789, more than a sapling, less than a tree. And it serves as a reminder that parts of the cedar are still that age today. We all know that trees grow in rings, the outer ones younger than the inner; the rings record the years and their weather with surprising fidelity, and have revealed much about the history of climate. But focusing on the rings in two-dimensional cross-sections hides the three-dimensional truth – trees preserve their past forms in near entirety.

That truth is beautifully brought out in the work of the Italian artist Giuseppe Penone. Where Brown went out of his way to hide nature's utility (fenced fields are not to be seen in his landscapes) Penone celebrates that utility's reversibility. He works with big brute beams of wood, the sort of thing that you might expect a warehouse to be made of, choosing a particular ring in the cut end of the beam and then chiselling his way up that ring. What in cross-section is just a curve becomes a complex three-dimensional surface. The hard rectangular geometry of the beam is lost, and a graceful, fragile sapling is revealed; what were knots when the material was commoditized wood become the stumps of young branches as the wood is restored to the form it took in early life. Penone undoes the work

---

* The great storm of 1987, far better remembered by most of the English, passed the Bowood cedars by; history is strange.

of the sawmill, undoes the work of the sun, and releases something both young and old, an imprisoned sapling that had lived but little and seemed long gone. In wood, Penone makes historical sense of the cliché of a sculpture locked away within its brutish material.

As Faulkner wrote, the past isn't dead. It isn't even past. The sapling that stood here when Europe was rocked by revolutions political, industrial and scientific, is still there today. If I could reach through solid wood I could grasp it in my hand. But the world that sapling lived in, a time of ideas and ideals more vivid than those we are accustomed to today, has gone, swept away in part by the new sciences and technologies to which it gave birth.

It is not just that politics has changed, not just that economics has changed, not just that society has changed – though of course they have, to such an extent that finding a park still owned, seven generations on, by the descendants of the aristocrat who first bought it from the crown feels a remarkable thing. The biogeochemistry has changed, too. The air the cedar's needle leaves take in today is much richer in carbon dioxide than that on which it was reared in the days of Priestley and Ingenhousz. And the climate that rocks its branches and has bested its peers is changing, too, as that carbon dioxide traps more of the sun's heat in the earth's atmosphere.

This strange, uneven tree has lived through some of the most rapid environmental change the world has ever seen, through change that has made the atmosphere itself as artificial as a Capability Brown landscape, though sadly without the benefit of aesthetics or design. And if the cedar lives to be twice its current age – no one yet knows how long a cedar can live in England – it will see much more. We do not know how much change there is in its future; we do not know the extent to which that change will be modified by purposeful action, tamed and civilized. This magnificent tree is destined for a history as confusing, and as man-made, as the pruning that has shaped its limbs and trunk.

If, that is, it does not fall in the storms to come.

The discovery of photosynthesis

Joseph Priestley came to Bowood ten years after its capabilities were realized, and lived in Brown's landscape for most of a decade. He'd have known the cedar by the garden wall in the youthful form that's now buried in its core; the cedar, for its part, fixed carbon dioxide from his lungs into its wood, where the molecules thus made remain today, just a couple of hundred tree rings in from the carbon the cedar took from me on Bastille Day. And yet so great is the number of molecules we all exhale in a life, and so great is the number that a tree makes use of, that that is no particularly special link. If some scraps of Priestley's breath are fixed in this tree, then the same is true of all the trees of England. Over his life, he breathed out countless trillions of times more molecules of carbon dioxide than there are trees in the world, molecules which, within a matter of years, were spread evenly over all the airs of the earth. There is something of Priestley everywhere; there are traces of his breath in every plant in the world. Your breath and your great-grandmother's and Joseph Priestley's and Uncle Tom Cobbleigh's are intermingled throughout the biosphere. Trees are made from the breath of the world, not just the breath of those who seek their shade.

To Priestley, this talk of molecules and carbon dioxide would have been gibberish. The vision of radical equality in the face of creation, though, might well have struck a chord. A non-conformist minister and teacher born in a village outside Leeds, Priestley was devoted to liberty, both religious and political; to happiness, especially the happiness of his family, friends and flock; and, though not really accepting that the three could be separated, he was devoted to learning. He had a prodigious capacity for learning, and a striking love of it. His faith in its power to improve the world was equal to, and indistinguishable from, his faith in God.

Educated at a dissenting college, because his faith put him outside the bounds of the Established Church, he was not able to attend an

English university. After starting a small school he found that he liked teaching, as well as learning, and went on to teach the sons of the merchants and businessmen of Manchester and Liverpool at Warrington Academy, the oldest and most respected of England's non-conformist seats of learning.

Priestley was a teacher of language and grammar, but he had an interest in science, as well, and at Warrington in the 1760s that interest began to blossom. He became a friend of Josiah Wedgwood, the brilliantly innovative potter, and a member of what would become known as the 'Lunar Society', a circle of provincial friends interested in both understanding and changing the world. They were men of scientific curiosity, technological ambition and, often, entrepreneurial flair. To their Tory denigrators, such men were 'projectors', throwing absurd and fanciful schemes into the future as a lantern throws patterns of light out into the night; today we can see them as the men whose factories, canals and ideas were creating the industrial revolution.

Along with Wedgwood the Lunar men included Matthew Boulton, the princely industrialist who ringingly informed a visitor to his factory that 'I sell here, Sir, what all the world desires to have – Power'; James Watt, the melancholic Scot whose steam engines allowed Boulton to make that boast; William Small, a sweet-natured, dry-witted and immensely learned physician; James Keir, an honourable and likeable chemist, army officer and glassmaker; and Erasmus Darwin, an enormously corpulent physician, poet and polymath, who translated Linnaeus into English and wrote vast epics on themes from natural history.* They were Lunar because they met on the Monday closest to the full moon, the time of the month chosen so the moonlight might illuminate their journeys home. Priestley was in utter sympathy with their view of the world, and came to take a leading role in their industrious enlightenment.

In the mid-1760s Priestley set himself the task of writing a history of investigation into electricity. He went south to London to uncover

---

* Erasmus was Charles Darwin's paternal grandfather, Josiah Wedgwood his maternal one.

the latest research, becoming a friend of Benjamin Franklin in the process, and undertook to replicate and take further all the experiments which he could – an undertaking that included flying kites in thunderstorms. His *History and Present State of Electricity* was published in 1767, the same year that he moved from Warrington to become the minister to a congregation in Leeds. This appointment allowed him to return to ministry and religion, which he considered his principal calling, but did not stop him from further research.

His new home was next to a brewery, and Priestley's imagination was caught by the strange air which stood over the tuns of fermenting beer, a heavy air which, when smoke was added to it, could be observed to cascade down the sides of the open vats and pool on the floor. Priestley recognized the characteristics of this heavy air, in which a candle would not burn, to be those of the 'fixed air' that the Scottish chemist Joseph Black had produced by heating up 'magnesia alba'. We would call that magnesia a carbonate, and the fixed air as air rich in carbon-dioxide gas, but again the terms would have been meaningless to Priestley, since his science held no place for carbon, for oxygen, or for the notion of a gas, whatever that might be, comprised of two parts of the latter to one part of the former.

Priestley performed all manner of experiments on this fixed air, at least once contaminating a whole batch of beer in the process (even so the brewers were happy to help). He was particularly absorbed by the lethal effects the fixed air had on mice, and how they could be protected from the harm it did them.* Mice, he noted,

---

* Robin Hill, who in his later years took a great interest in Priestley's life, suspected that the concern with asphyxiation was spurred by reports of the 'Black Hole of Calcutta', an infamous episode of incarceration in which 123 of 146 people in hot, airless confinement had died. Perhaps so; then again, perhaps Hill was projecting back into the eighteenth century his own uncomfortable experiences with the deadly gases of the Great War. It is hard to ignore the degree to which the great twentieth-century scientist identified with his predecessor. Hill, too, was from a non-conformist family, and reasonably radical in his politics. Hill laid particular stress on Priestley's friendship with Benjamin Franklin, and in one passage from a letter of Franklin's to Priestley it is easy to imagine Hill hearing echoes of his pacifist friend Emerson. 'The hint you gave me jocularly, that you did not quite despair of the philosopher's stone, draws from me a request, that when you have found it, you will take care to lose it again; for I believe, in my conscience, that mankind are wicked enough to keep slaughtering one another as long as they can find money to pay the butchers.' If Hill's research brought to light the

like candles, needed fresh air (indeed candles needed a prodigious amount – 'a gallon a minute'). A mouse left in a glass jar would die well before starvation set in, and if a candle had been allowed to burn out in the jar before the mouse was introduced, the mouse would die much more quickly. In both cases the jar would end up containing Black's fixed air.

In 1771, Priestley decided to see whether the effect plants had on the air was the same as that of mice and candles, somehow using up what was good about it. After all, the Anglican churchman Stephen Hales, one of the great English natural philosophers of the first part of the century, had shown that plants, like animals, needed to take up air in order to breathe, and had developed a specialized piece of equipment, the 'pneumatic trough', with which it was possible to study this process. Hales saw the plant's leaves as analogous to the animal's lungs, a way for them to draw 'some part of their nourishment from the air'. To Priestley's surprise, though, a sprig of mint kept in water under an upturned jar did not die for want of aerial nourishment. Nor did the air in which it was living become unwholesomely 'vitiated'. Unlike the 'fixed air' exhaled by a mouse, the air from the mint jar would invigorate a flame, not douse it.

Priestley found he could burn out a candle in a jar, put in a sprig of mint, and after a while find the air so well renewed that he could burn a fresh candle in it. Indeed mint actually grew better when provided with tainted air in which animals died and candles guttered. Describing his experiments in the *Philosophical Transactions of the Royal Society*, Priestley reported that 'In no other circumstances have I ever seen vegetation so vigorous as in this kind of air, which is immediately fatal to animal life'.

As the winter of 1771 came on, Priestley found that these results became harder to replicate. The next summer, though, he resumed the work, and became more convinced than ever that the plants

---

unfortunate occasion on which Priestley's loving daughter sought to help her father in his absence by cleaning out all the bottles in his laboratory, regardless of what was in them, Hill must surely have shivered in sympathy at the thought.

restored the air 'injured by respiration', not just in little laboratory jars, but in the world at large. As he wrote to the Royal Society:

The injury which is continually done to the atmosphere by the respiration of such a number of animals, and the putrefaction of such masses of both animal and vegetable matter, is, in part at least, repaired by the vegetable creation. And, notwithstanding the prodigious mass of air that is corrupted daily by the abovementioned causes; yet if we consider the immense profusion of vegetables upon the face of the earth growing in places suited to their nature, and consequently at full liberty to exert their powers, both inhaling and exhaling, it can hardly be thought, but that it may be a sufficient counterbalance to it, and that the remedy is adequate to the evil.

Priestley's work on airs of various sorts caused quite a sensation: the Lunar men loved it. On a tour through the Midlands the following year, Franklin blamed an attack of fever and gout on the expeditions to ponds and dank ditches on which Darwin and Boulton dragged him in the name of the new aerial research. There were also practical applications to the work: Priestley had found ways of getting fixed air into water and making it fizzy. His apparatus for making such soda water, commercialized by other hands, became all the rage. And there were broad, philosophical implications: his discovery of the role of plants had brought the way of the world into the realm of reason in a new way. As Franklin put it, 'That the vegetable creation should restore the air which is spoiled by the animal part of it looks like a rational system.' The restoration of the air provided a providential purpose for all plants. Sir John Pringle, the President of the Royal Society, drew particular attention to the idea when awarding Priestley the Copley Medal, one of the society's highest honours:* 'From the oak of the forest to the grass

---

* Pringle himself had received it twenty-three years before for his contributions to military medicine (he was the man who invented the terms 'septic' and 'antiseptic', and who first suggested that field hospitals should be given some special exemption from attack by artillery). The medal was awarded to Robin Hill 214 years later.

of the field, every individual plant is serviceable to mankind; if not always distinguished by some private virtue, yet making a part of the whole which cleanses and purifies our atmosphere. In this the fragrant rose and deadly nightshade cooperate.'

It was at around this time that, at the invitation of the second Earl of Shelburne, Priestley moved from Leeds to Bowood. Shelburne, a radical Whig politician from a family with a tradition of scientific interests, did not object to the fact that Priestley was by now moderately notorious in some circles for the 'rational dissent' of his Unitarian views. He employed Priestley to tend his library, educate his sons and help with research on parliamentary questions, providing him with a handsome salary and with lodgings in both Bowood and London. In was in this new employ that Priestley published his *Experiments and Observations on Different Kinds of Air* in book form, and made what some histories judge his greatest discovery. Heating a 'red calx' of mercury with a powerful magnifying glass, he produced a new air in which candles burned bright and mice thrived, and which, when inhaled, left his chest feeling 'light and easy for some time afterwards. Who can tell but that, in time, this pure air may become a fashionable article in luxury. Hitherto only two mice and myself have had the privelege [sic] of breathing it.'

But this new air was more than a potential successor to soda water as a chemical tonic. It was to become the instantiation of a crucial part of Priestley's understanding of the world. The other airs – inflammable air (hydrogen), nitrous air (nitric oxide), phlogisticated air (nitrogen), alkaline air (ammonia) – were all seen, to some extent, as diminished or polluted versions of normal air. This new air was an improvement on it. The improvement, Priestley realized, was due to a lack of phlogiston, a substance, or essence, that was at the centre of chemical discourse at the time. Phlogiston is not something easily explained in our modern worldview; nor, for that matter, did it have a fully agreed definition in Priestley's world. It was either allied to, or identical with, the principle of fire, depending on which chemist you listened to. Burning things gave off phlogiston, not as a side product, but because that was the essence of what it was to burn; they ceased to burn when they ran out of phlogiston, or when the air around them could absorb no more of the stuff (this was why it was impossible to burn things in jars from which the air had been pumped out; there was nowhere for the phlogiston to go). Heated by the focused sun, the red calx of mercury was drawing phlogiston out of the air above it, leaving it 'dephlogisticated'. And this, Priestley realized, was also what plants must be doing. Candles, mice, and all other fires and animal spirits released phlogiston into the air; plants removed it.

The man put down by history as Priestley's rival, the French chemist Antoine Lavoisier, came to a different understanding of much the same experiment. In Lavoisier's eyes, the heating didn't cause the calx to absorb something from the air; it caused it to emit something. Lavoisier called that something *oxygène*, a name that reflected what he saw as the substance's proclivity for creating acids. He saw the fact that, after heating, the mercury weighed less than it had before as crucial evidence. Lavoisier's view of chemistry, in which substances such as oxygen, nitrogen and hydrogen combined with each other and with other elements in ways that had predictable effects on their weights, would fairly quickly evict phlogiston from its place at the heart of the subject; when Priestley died in the early nineteenth century he was almost the only believer left.

In Britain and America Priestley is sometimes seen as the discoverer of oxygen – indeed there is a plaque to that effect at the entrance to the small room at Bowood where Priestley worked. But while it is true that if we examined the substance Priestley made there today we would call it by that name, the name is not all there is to it. Priestley didn't believe that there was anything coming out of the calx and into the air, and he really did believe in phlogiston. To see Priestley as the discoverer of oxygen would not be seen as a posthumous honour by the man himself – more as a mark of disrespect, allowing him precedence only in a world framed in his rival's terms.

Nor did Priestley discover photosynthesis, though again it is sometimes claimed that he did. While he did discover that plants

331

revived the air after animals vitiated it, he did not, at first, realize that they needed sunlight in order to do so. In retrospect, this seems slightly odd. After all, Priestley was well acquainted with the earlier work of Stephen Hales: he used experimental arrangements based on Hales's to capture and purify gases; his measurements of phlogistication used properties of nitrous air first described by Hales; and he had taken to heart Hales's notion that the leaves were the lungs of the plant. But he seems to have ignored Hales's speculations about the role of light, in which he asked: 'May not light also, by freely entring the expanded surfaces of leaves and flowers, contribute much to enobling the principles of vegetables; for as Sir *Isaac Newton* puts it in a very probable query, "Are not gross bodies and light convertible to one another?"'

Hales's idea was current enough in the middle of the eighteenth century for Jonathan Swift to single it out for lampooning in the satirical portrait of the Royal Society found in *Gulliver's Travels*; in Lagado, at the Academy, Gulliver comes across a 'projector' who claims he will soon be able to extract sunbeams from the cucumbers into which they have fallen and transfer them to glass jars from which they could later be released to cheer up summers that fell below par.

For all this Priestley seems not to have appreciated the role of light in the activity of plants. Instead he became sidetracked by the 'green matter' he found coating the glass of some of his jars. When there was 'green matter' present, dephlogisticated air was given off even in the absence of plants. The green matter was a film of algae, but Priestley, though he inspected it through a microscope, did not realize this and convinced himself that it was neither animal nor vegetable. Glassware coated with green algae, which came to be the hallmark of photosynthesis research in laboratories such as Warburg's and Calvin's, was in Priestley's a red herring, distracting him from the properties of plants he had sought to study.

The person who put the parts of the puzzle together correctly was Jan Ingenhousz. Like Priestley, Ingenhousz began his scientific research when already established in a neighbouring field; as a medical man he had become an expert in inoculation, using preparations made from the scabs of the infected to protect his patients from smallpox. His skill in the procedure brought him to the attention of the crowned heads of Europe, and the Austrian Empress Marie Theresa awarded him a lifelong pension. (She had hoped he might prove an amusing courtier, as well as her children's saviour, but in this the nice but apparently rather dull doctor disappointed her.) As had been the case with Priestley, Ingenhousz's scientific interests began with the study of electricity, and again as had been the case with Priestley this interest led him to become a friend of Franklin. The two of them toured the English Peak District in 1771 and visited Priestley in Leeds just before he undertook his first experiments with plants.

Ingenhousz was impressed by the ideas about the providence of plants and the balance of nature that Sir John Pringle expounded when awarding Priestley the Copley Medal. Returning to England in the summer of 1779, after five years in Vienna, he carried out a series of experiments on the properties of plants. Unlike Priestley, who rather delighted in trying whatever came to mind and making the most of accidents – James Watt called it 'his usual way of groping about' – Ingenhousz's experiments appear to have been thoroughly programmatic. The book that grew out of them, *Experiments Upon Vegetables*, which was dedicated to Pringle, argued that plants could indeed improve the air, but that they would do so only in daylight. He distinguished the need for light from a need for heat by putting leaves in water that was warm but shaded, and in water still cold from the well but brilliantly illuminated. He recorded that young leaves were not as powerful as old ones, and that only green tissues (leaves and shoots) were any good at all. He observed through his microscope that the green matter which grew in glassware left in the sun was formed of cells that looked like plant cells, and thus asserted that it was in fact vegetable. He put forward the idea that plants formed dephlogisticated air through some sort of transmutation of water. He also noted that plants respire as well as photosynthesizing – that in the darkness, they produce fixed air.

Later he was to observe that soil in which no plants are growing will do the same thing too.

These published insights seem to most people sufficient cause to say that Ingenhousz discovered photosynthesis. Priestley himself, writing to a friend about Ingenhousz's work, mentioned the considerable difference between day and night as something 'he hit upon and I missed'. Later, though, Priestley became rather less generous, claiming that he had thought much the same as Ingenhousz had, at the same time, though he realized he had not published as fully or promptly. But while at some point in that summer Priestley may have thought much the same as Ingenhousz did, at other points he thought differently. Ingenhousz followed his line of enquiry unswervingly.

If Priestley did not discover oxygen or photosynthesis, what did he discover? You might argue nothing – and that a stress on discovery, like a stress on Nobel prizes, is one of those unfortunate enthusiasms that hides the real nature of science and its history from us. You could say Priestley was a victim of his success; that he helped start a scientific cavalcade of free thought which went on to change the way the world was seen so thoroughly that Priestley's own concepts no longer had meaning. Or you could say that he discovered something so basic that we don't have a single word for it – the fundamental complementarity of plants and animals, the relationship that goes beyond just being eaten, the relationship at the heart of photosynthesis and the carbon cycle.

In the 1780s Ingenhousz returned to Vienna; at the end of the decade, though, when a revised version of his book was translated into French, Ingenhousz headed off to Paris to stay as Lavoisier's guest and promulgate his ideas, which were now expressed in terms of oxygen and carbonic acid (carbon dioxide), rather than phlogiston. He arrived on the day that the Bastille fell. Fairly soon the man who had been Marie Antoinette's childhood doctor decided that France was no longer for him, and left for the Netherlands before fetching up again in England. In 1798 he became ill, and the Earl of Shelburne invited him to move to Bowood, where he could be cared for. He remained there, amid the landscape laid out by Brown, using the library built up by Priestley, until his death the following year.

The French revolution was also a turning point for Priestley and Lavoisier. Priestley had left Bowood in 1780 to take up a ministry in Birmingham. From that point on he did almost no work on plants; much of his energy went into his ministry, and the reforms in education and politics that he saw as going along with it as part of the programme of rational dissent. For Priestley, the fall of the Bastille was a great event. He did not care for the way that the French did their chemistry – as well as disliking Lavoisier's disbelief in phlogiston, he feared that obscure technical language and expensive equipment would make chemical learning less accessible to the common man. But as a progressive and a radical, he liked French politics, and their effect on English politics. 'Have you seen [Tom Paine's *The Rights of Man*]?' he wrote to Wedgwood. 'It is most excellent, and the boldest publication that I have ever seen.'

On 14 July 1791, on the second anniversary of the fall of the Bastille, a 'Church-and-King' mob burned the radical preacher's meeting houses and then ransacked and set fire to his home, destroying his library, uprooting his garden, smashing his instruments and laboratory glass. He was soon forced to emigrate to America, living out his last ten years in Northampton, Pennsylvania, where he enjoyed the friendship of Thomas Jefferson, former pupil of his Lunar colleague William Small.* (Jefferson also owned a copy of Ingenhousz's book, inscribed to him by the author during the tumultuous Paris summer of 1789, that now sits in the Library of Congress.)

Lavoisier fared less well. His eminence as a man of science could not, in the end, outweigh the fact that he had also become very rich as a tax farmer – someone licensed to collect revenue and cream off

---

* It was in the aftermath of a meeting at that Northampton house in 1874, held to honour the centenary of Priestley's *Experiments and Observations on Different Kinds of Air*, that the American Chemical Society was founded.

a commission. It was on the money he made from this occupation that he built much of his career; it was said that he had revolutionized chemistry 'with the balance and the bank balance'. As the revolution progressed a background in tax farming became an increasing liability; being one of the men thought to have blocked Jean-Paul Marat's membership of the Academy of Sciences did not help. Lavoisier was sentenced to death in May 1794, the judge delivering the sentence with the famous opinion that 'The Republic has no need of philosophers'.

Within ten years of Lavoisier's death – in the year that Priestley died in America – the first, heroic era of photosynthetic discovery was brought to a close. In 1804 a Swiss scientist, Nicolas-Théodore de Saussure, building on the work of his teacher, Jean Senebier, provided a reasonably definitive account of what plants did with sunlight, one which was to stand, in its basic form, for more than a century. Saussure had learned his science in the post-Lavoisier world; he was happy with carbon and oxygen and nitrogen, and had no time for phlogiston. He was also keen on precise quantifications; as a young man, he had helped his father climb Mont Blanc in order to make measurements of atmospheric phenomena, and went on to climb other Alps in similar pursuits.

Saussure argued that, in the light, plants took up water and carbonic acid – which, though it was not known as carbon dioxide, was known to be composed of carbon and oxygen. At the same time they released oxygen, which Saussure assumed came from the carbonic acid. In the dark, they absorbed oxygen, combined it with carbon stored within them, and gave off carbon dioxide. Other nutrients that plants need – nitrogen, phosphates, the salts of some metals – they got from the soil, and these made up a few percent of the plant's mass. But the vast bulk of the plants came from the water and the air.

Although the 'humus theory' that the carbon in plants comes from the soil persisted in some circles for decades afterwards – a tribute to the enduring illusion that soil makes plants, rather than the other way round – Saussure's synthesis was soon widespread

336

and taken as authoritative. No one added a great deal to it until the second half of the nineteenth century, when starch and the chlorophyll-stuffed chloroplasts entered the story. It was not seriously amended until the 1930s, when van Niel, Hill and Kamen showed that the oxygen came from water.

The phlogiston cycle

Saussure's synthesis capped a wonderful period of discovery; the principles of chemistry had conquered the working of the plant. But in the process, something was lost. Our picture of the world was rendered much more accurate, but in its scope it was impoverished. Carbon, oxygen and the other entities in Lavoisier's laboratory were simply substances, things reducible to weight and proclivity. The phlogistic principle had been part of something grander. It had had providential significance and civic implications: it governed the virtue of the air, which was a major concern in an age haunted by 'mephitic' vapours and the threat of consumption. It was a principle in constant flow, from animal to air to plant to animal, and such freedom of flow had great virtue in the mind-set of the Lunar men and the radical eighteenth century; it was important for blood to flow freely, for trade to flow freely, for ideas to flow freely, for the lines of beauty in pottery and the lines of grace in a landscape to flow freely. For a devout materialist like Priestley, who disbelieved in the spiritual (part of his non-conformity was a refusal to credit the Holy Spirit or the divinity of Christ), the flow of phlogiston was the working of divine providence in the world.

The person with perhaps the grandest conception of the flow of phlogiston was a Scot, James Hutton. Hutton was a man with an early love of chemistry, a degree in medicine and a background in farming who found fame in geology. He was a key figure in the magnificent intellectual flowering of the Edinburgh Enlightenment

– a friend of the economist Adam Smith, almost without doubt an acquaintance of the philosopher David Hume, and an intimate of Joseph Black, the chemist who had first defined the qualities of the 'fixed air' given off in combustion. Hutton was also a friend of James Watt, and through Watt he knew the Lunar men, visiting them on his trips to England, sharing their enthusiasm for the improvement of nature and the transformation of the land; like Wedgwood, Darwin and Boulton, he was an investor in canals, the great land-changing projects of the day. It may well have been through his interest in the Forth and Clyde canal that he met Watt, who worked on canal projects as a surveyor, though it is also possible they met through their mutual friend Black.

Hutton's interest in the workings of the earth was in part the fruit of practical experience. He had inherited two small farms in southern Scotland, and wanted to make money from them. Scottish agriculture was relatively primitive at the time, and like many other rational 'improvers' Hutton looked elsewhere for insights into how things could be done better. He travelled to East Anglia and to Holland to learn the latest farming techniques, returning with light ploughs that he thought would serve better than the heavier Scottish machinery and with plans for new crop cycles that would enrich the soil. Impressed by the work of Jethro Tull he introduced turnips as a crop for cattle to eat, rotating their cultivation with wheat, barley and clover pasture in the Norfolk manner. He introduced cultivated grasses, and recommended the practice of ploughing crops back into the soil when it needed enrichment.

The soil was Hutton's main concern. When, towards the end of his life, he brought together his writings on farming in the long and unwieldy *Elements of Agriculture*, he devoted more pages to crop rotation and its effects on the soil than to anything else, and gave almost as much attention to the origin and fertility of soils. He was particularly interested in, and concerned by, erosion, investing great effort in levelling out his fields to diminish its impact. The result was a farm of surprising trimness and efficiency: according to one commentator at the time, 'Dressing the land, drilling and hoeing turnips, rolling, and all the operations of husbandry were done to a degree of neatness and garden-like culture, which in farming had not been seen before. Persons of every description came from every quarter to gratify their intellectual curiosity, as well as to get information. The profits of the undertaking was said to have amounted to 600 per cent.'

Improving the land improved Hutton's finances; it also inspired him philosophically. It was clear to him that soil was not just removed by erosion, but also formed by it. Soil, he was sure, must come from solid rock. But if this were the case, then there was a problem: while erosion could take soil away indefinitely, its capacity to make new soil was constrained by the finite supply of rocks on higher ground to whittle away. Once erosion had laid low the high places of the earth there would be nothing from which new soil could be made. Given that the earth was the creation of a provident God, this was not an acceptable outcome. And so Hutton imagined that there must be a great cycle in the history of the earth in which soil eroded away into the sea was buried, turned to rock, and at a later date lifted up to form new mountains.*

This idea that earth must recycle itself led to what is normally seen as Hutton's greatest contribution to geology – the notion of 'deep time'. In a biblical culture, it was natural to measure the history of the earth on the same scale as the history of its human inhabitants. The earth was thus thought to have lasted only a few thousand years. Hutton's ideas about the recycling of the earth's crust opened up far greater vistas. The earth had cycles to its history just as the planets had orbits in the sky, cycles whose turning, in Hutton's most famous phrase, showed 'no vestige of a beginning, – no prospect of an end'.

---

* Though Hutton's argument depended on God, in a modern form it would not necessarily have to do so. Following Andrew Watson's application of the anthropic principle to Gaia, one could simply argue that, because we are here to inhabit it, the earth must be habitable over long periods of time, and that there must thus be a recycling of sediments into rock and rock into soil. It is not a coincidence that Watson has prints of illustrations from Hutton decorating the walls of his office.

Hutton's idea of the endlessly self-renewing earth stretching back into untellable depths of time was informed by his medical training and physiological interests. The earth, like the human body, was a physiological mechanism; it had a purpose – the maintenance of life – and a mechanism by which to effect that purpose. For the body, it was the circulation of the blood; for the earth it was the recycling of the necessities of life – of water as rain, of soil as rock. And in both cases, the best analogy with which to explain these things was one that came from a workshop. From its philosophy to its pornography, the eighteenth century was the age of the body as machine. Ingenious automata captured the habits of life in mechanical emulation; theories of association showed that thought itself had mechanisms that could be predicted; in the persona of Fanny Hill, John Cleland sang new hymns to the power of the piston. In the body, the circulation was driven by the heart, a pump, and powered by respiration, a combustion intensified by the bellows of the lungs. The great cycling mechanisms of the earth, Hutton believed, were analogous to the greatest machines of his age – the steam engines being developed by his friend Watt. And they were powered in the same way: by coal, which was a peculiarly powerful fuel because it was imbued with large amounts of phlogiston.

Hutton did not reject Lavoisier; he merely saw his ideas as incomplete, unable to capture the role of light and heat. For that, Hutton thought, you still needed phlogiston. He saw it as a sort of fixed sunlight – 'the solar substance' – which Ingenhousz's 'accurate investigation of the vegetable oeconomy' had shown to be created and stored in plants. The plants were cooled by the process, and impelled to reduce carbonic acid to its constituent carbon and oxygen. Animals used up phlogiston, becoming warmed by the process and combining carbon and oxygen into carbonic acid. This was how life on the earth's surface was maintained, and not until the days of Mayer and Boltzmann and thermodynamics, a lifetime later, would there be a scientific description of the way that the world endlessly replenishes itself remotely as compelling. While it is profoundly ahistorical to do so, it is tempting to see phlogiston as some sort of marker reserving the place that entropy takes in our modern view of the carbon cycle – a hard-to-grasp side-effect of fire and heat and the work they do, mysterious and yet at the same time a real part of the world, a driving principle. Without the extra dimension added by such an idea, the chemistry of plants and animals was just a matter of recipes and balances.

And as well as providing the phlogiston needed by animals, plants exported phlogiston to the depths of the earth in the form of coal. In the depths, 'The consumption of phlogistic substances is a great and necessary operation in the oeconomy of the world. There is constant fire in the mineral regions; – fire which must consume the greatest quantities of fuel; the consolidation of loose materials, stratified at the bottom of the sea, depends on the heat of that fire.'

The earth, in a sense, was a huge animal, the physiological heat of its deep interior derived from the complementary coolness of the plants on its surface.

The ancient plants in the coal fields being opened up to exploitation at the time Hutton was writing – one of his neighbours borrowed money from him to develop a seam near their farms – were thus evidence both of the earth's constant recycling and of the existence of a subterranean fuel supply adequate to power that recycling. The rate at which phlogistic coal was buried in sediments would set the speed of the geological cycling. The cycles would turn as fast as the heat from the coal could drive them. And the cycling would go on for ever, or at least for as long as there was a hot sun in the sky and cool trees in the woods. Sunlight and mineral fire were the means by which the earth was kept eternally habitable, and phlogiston was the bridge between them. Volcanoes were the smokestacks of the great engines of the world.

As with much of the period's scientific theorizing and projection, there is something both magnificent and quaint about the idea of coal-powered geological cycles. They have the same strange charm as Erasmus Darwin's plans to change the course of the winds over Britain by means of a vast range of windmills, or to abolish war by setting the navies of the world to the task of towing icebergs to the tropics as a way of ameliorating the climate there and thus furthering the course of civilization. They have, of course, the same weakness, too – the problem of being pre-thermodynamic. Without a science of energy you cannot say which grand schemes are possible, and which are not. Without defined concepts of energy and entropy you cannot measure the difficulty or ease of changing the world. There is no way to know that if plate tectonics really were powered by coal, ten trillion tonnes of coal reserves would be used up in just a couple of centuries, widening the Atlantic less than Capability Brown widened the river at Bowood.

The lack of a grasp of thermodynamics makes it impossible to know how you will change the world. But it doesn't make it impossible to change it. The Lunar men – 'The friends who made the future', in the historian Jenny Uglow's excellent phrase – showed us that. Wedgwood resolved they should 'astound the world with wonders', and they did. Their canals mobilized serious sums of capital, hard-won political will and huge workforces, rebuilding the country's landscape on a new scale, changing the fundamentals of its trade, linking the provinces where coal and iron were mined to the manufactories and markets that needed them. Farming was never again to dominate the country's economy in the way it had before.

Watt's great steam engines provided the power to keep the mines dry, to drive the bellows of the foundries' blast furnaces, to crank the spinning looms. Ever more engines of ever-greater power choked the skies with their smoke. The effort to improve them contributed hugely to the eventual development of that missing science of work and power and energy, and once the conceptual tools of thermodynamics were at hand, further improvements could be planned and executed with precision. Within fifty years, the country-shaping canals began ceding their place to the self-motivated steam engines of the railways. Fifty years on from that, the steam engines started to give way to steam turbines, far better at generating electrical power. Fifty years on from that, the locomotives had started to lose out to the internal combustion engine of the car and the lorry – and the aircraft.

The fossil-fuel-powered future which the Lunar men anticipated and helped to set in motion changed the way we live. It also changed the world we live in. Hutton was the first man to look at coal seams and see them as stores of a 'solar substance' that had been warehoused in the unfathomable depths of time; he was the first to see a direct link between the fuel powering the economy of industrializing Britain and the fuel powering the 'wise oeconomy of nature'. He was wrong about the nature of the link; nevertheless, he might well have suspected that vastly increasing the rate at which that fuel was used would affect both those economies. He would certainly have been fascinated by the change in the atmosphere that would be prompted by releasing stored sunlight so profligately – by building single power stations which can burn coal faster than the carboniferous swamps of the entire planet could create it, and releasing the buried carbon of a hundred million years in just a few centuries. As he wrote in one of his essays on phlogiston:

What an object for natural philosophy to understand the intention of nature, in giving an atmosphere to this earth! If rain, which is so necessary to vegetable productions, and if fire, which is so necessary to animal life and the internal constitution of this earth, have such an intimate connection with the atmosphere of the globe, it is surely an object for the philosopher, to trace the chymical process of nature by which so many operations are performed.

And if Hutton would have relished the idea that mining phlogiston could change the atmosphere, the Bowood cedar planted when he was still a farmer went one better. As it grew from the young sapling a hundred yards from Priestley's experiments to today's twisted adulthood, it recorded the changes in the atmosphere as they happened, tree ring by tree ring, needle by needle.

OLIVER MORTON

Changing leaves, changing worlds

Ian Woodward, I can't help thinking, would have got on rather well with Joseph Priestley. Woodward is a Yorkshireman brought up in the Sussex Weald who has now returned north to Sheffield as a professor. He has a friendly manner, a ready laugh and an enthusiasm for plant physiology that has blossomed into an interest in the global effects of vegetation. We met him in the previous chapter, turning the world's wildfires on and off in his computerized vegetation models like an affable Loki. He is fascinated by the range of scales such models bring together, uniting the chemical reactions within the leaf and the genetic processes within the nucleus with the texture of the whole landscape and the atmosphere of the planet.

It is easy to imagine men so happy with models as being distanced from the real world, and indeed they sometimes are. Woodward, though, is not. The physiological processes that power the world in his models are processes that he has studied in the lab and in the field. And one of them is one that he discovered. Woodward is one of those men who has enjoyed the brief experience of looking at a sample in a laboratory and realizing it has unequivocally revealed something about the world at large that no one else has ever known. Many people enjoy rich and fulfilling scientific careers without ever experiencing such a moment – but I doubt if there's a one of them who doesn't envy those who have. It is in such moments that the magic of doing science, the marriage of the subjective and the objective, is at its most intense. In such moments a universal truth is at the same time a thought as yet unvoiced; something that's not yours at all, in the long run – something that will end up utterly independent of you – but that is, for that moment, no one else's.

Any such discovery is a thrill, even if the truth it reveals is a subtle one, uninteresting to all but experts. Woodward's was of broader significance. One afternoon in Cambridge he became the

344

first person to understand one of the subtler ways in which plants tell us about the past and react to the present.

Woodward started his research career trying to understand the differences between highland and lowland plants. What was it that plants on plains didn't do that those on hilltops did? Why wouldn't the mountain plants do well in the valleys? While studying bilberries in the mid-1980s – fair-sized bushes in low-lying areas, small intense shrubs on the Scottish hillsides where he did his fieldwork – he noticed that the mountain varieties weren't just smaller, greener and far more inclined to large root systems. Their leaves were also studded with stomata in unusual profusion.

In his laboratory in Cambridge Woodward started to grow bilberries in various different environmental conditions, methodically working through a checklist of ways life above might differ from life below. It wasn't temperature, it wasn't moisture, and it wasn't light. Then he tried carbon dioxide. The results were remarkably clear-cut. The plants grown in low levels of carbon dioxide had many more stomata, with the exact number depending on the degree of carbon-dioxide deprivation. Twenty years on, Woodward lights up as he remembers the thrill. It was 'one of those days when your hair feels prickly at the base of your neck – one of those "yes!" days'.

The reason the bilberries differed was that the thinner air of the mountain tops, while containing the same proportion of carbon dioxide as the air of the valleys, contained less in absolute terms. The increased number of stomata was a way of offsetting the gas's scarcity by making it easier for what there was of it to get into the leaf. Plants added more stomata to their leaves at altitude for the same reason that people add more red cells to their blood at altitude – so as to provide the tissues that need a constant supply of the gas with as much as they want.

Woodward talked the results over with Nick Shackleton, a Cambridge palaeoclimatologist with whom he shared an interest in carbon isotope studies. Shackleton pointed out that atmospheric carbon dioxide didn't just vary with altitude – it also varied over time. Before the industrial revolution there was less of it: old leaves might be able to illustrate, or even quantify, the difference. In the botany department's herbarium, which has samples of over a quarter of a million leaves from all over England carefully preserved between sheets of paper, Woodward put together a selection of leaves from oaks, beeches, poplars, hornbeams and lindens that had been collected over the past two hundred years. With great care he spread dilute nail varnish over their surfaces to make a sort of death mask; peeling the vinyl off, he carefully counted the impressions that the pores had left. As he had suspected, the newer the leaves were, the fewer stomata they had. By comparing the density of the stomata on the old leaves with the densities he found when he grew similar species under low-carbon-dioxide conditions in the lab, he was able to estimate the carbon-dioxide level in the atmosphere in which those leaves had grown up. On the basis of his portfolio of leaves from the English Midlands he was able to argue that, between the early days of the industrial revolution and the 1970s, the carbon-dioxide level had risen from 280 parts per million to 340 parts per million.

David Keeling had been tracking carbon dioxide's ups and downs in the air over Hawaii for almost thirty years by this stage, so a rise in carbon dioxide was not news. And there were far longer records from ice cores, records that stretched back to the ice ages. Even though it was hard to read the carbon-dioxide levels in the topmost ice, where the snow is not as thoroughly compacted as it is down below, the ice cores seemed to show a rise in carbon dioxide over the past two centuries. Woodward's meticulous work – built on leaves collected by generations of botanists with a similar love for their subject – provided an independent confirmation of that rise, agreeing quite closely with the ice-based estimates. And the rise it was confirming was spectacular.

In the 200 years that divided Woodward from his earliest samples the world's carbon-dioxide level had risen by sixty parts per million. As far as can be seen from the ice record, that is three times larger than the largest changes in the past 11,000 years. It is about three quarters of the difference between the level at the height of the ice ages and the level in 1750. And since then, things have moved on. Today's plants are growing leaves with fewer stomata than those they grew back when Woodward published his work twenty years ago. Today, carbon dioxide stands at a level of about 381 parts per million. In terms of carbon-dioxide levels, the difference between our world and that of James Hutton and Capability Brown is greater than the difference which separated them from the last glacial maximum, when the landscapes of Britain were shaped not by ingenuity but by a kilometre or so of ice.

Woodward's techniques didn't just confirm changes in the recent past; they opened up new possibilities for examinations of the deeper past, ways of looking at atmospheres that pre-date the oldest ice cores. Over the past twenty years stomata-counting has taken its place as one of the various approaches used to estimate carbon-dioxide levels in the geological past – to try and tie them to the spread of grasses eight million years ago, say, or to the rise of the Himalayas thirty million years before that.

The method is not entirely straightforward. Other factors influence the number of stomata, though they have less effect, and there's also the problem of where the leaves come from. Beneath the canopy of a forest, where the air is still and the soil is respiring, the carbon-dioxide level will be higher than it is in the open air above and the stomata will reflect this. In his herbarium work, Woodward used only leaves from branches that also had flowers on, since flowers are found in sunlight, not tucked away beneath the canopy; that way he could be reasonably sure he was sampling leaves that had responded to the open air. Those working with fossil leaves don't have that luxury. And they also don't have the advantage of being able to compare their samples with leaves from close relatives grown in the lab. The further back in time you go, the less sure you can be about how the plants responded to carbon dioxide, because the plants whose fossil leaves you're looking at become ever less closely related to the ones you've studied in the lab.

All these differences lead Woodward to caution against putting too much weight on counts of stomata as measurements of carbon dioxide over deep time. But if interpreted carefully, and with all the caveats observed, they are a useful tool. David Beerling, whose office is just round the corner from Woodward's, and his colleagues use them quite a lot. According to their studies carbon-dioxide levels are higher today than they have been for twenty million years. The changes made in just a few centuries of fossil-fuel burning are the sort of changes normally associated with geological timescales. We have entered a new geological era of our own making. After the Pleistocene – the period of the ice ages – and the Holocene – the ten thousand years of interglacial warmth that followed – we have entered the Anthropocene: the climate of man.

And we have a lot more fossil fuel left to burn, should we choose to. If we really wanted to, we could get the carbon dioxide up to the level it was at when the Himalayas rose up. If we really wanted to, we could get back to the levels that the dinosaurs enjoyed in the Cretaceous. And in doing so we could change the way the world works even further than we have already, leaving our mark on every leaf.

The power of humanity

The current build-up in carbon dioxide is just one of a suite of changes humanity has brought about in the workings of the planet over the past 250 years – through industry, through agriculture, through our own deliberate fault and through unknowing accident. A useful way to put the whole suite into context is to think in terms of power. Since the thermodynamic revolution of the mid-nineteenth century power – that which Matthew Boulton sold and all men desired – has been defined as the rate at which energy is used over time, and today it is measured in units named after Boulton's colleague Watt. A human metabolism typically runs at about eighty watts.* A leaf in the sunshine stores energy at a rate of less than a milliwatt per square centimetre.

When we talk about planets, we need to talk about watts by the trillion – terawatts. All told, our industrial civilization runs at a power of about thirteen terawatts (13TW); that is the rate at which we use energy in our cars and factories and washing machines and combine harvesters and central-heating systems and hairdryers and personal stereos. This is a truly enormous amount of energy. Imagine Jim and Sandy Lovelock, instead of stopping at the end of the 650km southwest coast path that snakes around Dorset, Devon and Cornwall, choosing to continue walking all the way around the coast of mainland Britain, and in doing so passing a major nuclear power station on every single kilometre of their path: that's 13TW.

How does this output compare with what nature can do? In a crunch, quite poorly. A major volcanic eruption such as Mount St Helens has a power of about 50TW, while a large earthquake such as the one which caused the Indian Ocean tsunami of 2004 releases energy at a rate of over 2000TW. But these geological events are short-lived (seconds for the earthquake, hours for the eruption) and rare (for events of this magnitude, one every few decades). Our industrial 13TW is 24/365. It should be compared not to the earth's savage spurts but to the average rate at which heat flows from the hot interior to the surface. That flow of heat, the primary driver of all the planet's internal processes, is about 40TW.

So our industrial civilization already uses about a quarter of the power needed to drive the great engines of plate tectonics. And as the underdeveloped countries get over their 'under-', the energy used by their citizens is sure to increase. If everyone on the planet used energy at the level that the average European does today, we would be matching the earth's internally generated 40TW already (and the Americans among us would have had to decrease their current energy use by a third). If a world of nine billion people – a

---

* Up to 500 watts if the human involved is a racing Lance Armstrong, the nearest thing to a machine made of flesh outside eighteenth-century erotica.

bit less than one and a half times as many as there are today – used energy at the rate that Americans now do the engines of civilization would consume almost 100TW.

In the coming century, there is every chance that the powers of civilization will outstrip the powers driving plate tectonics in terms of pure energy. In terms of actually bringing about change we have already surpassed them. Take the issue that motivated James Hutton – erosion. Over the past 500 million years, the rate of erosion on the continents – which ultimately depends on the rate at which tectonic processes lift up new mountains – has been equivalent to the loss of a layer twenty-four metres thick every million years, plus or minus about ten metres. Current rates of loss due to human activity are hard to judge but it seems sure that agricultural land is losing soil at many times that rate,* and that the overall rate of sediment loss from the continents is something like three times the long-term average due to geology alone. It has been calculated that the rate at which the crust is reshaped purely for construction purposes is now greater than the average rate of natural erosion over geological time.

Our industry, though, is not the most significant of human activities in terms of sheer power. The rate at which sunlight is turned into biomass by a plant is called the 'net primary productivity', 'net' because it doesn't include the energy the plants fix into sugars but then use up in respiration, since that leads to no net gain in biomass and no net loss of carbon dioxide from the atmosphere. Net primary productivity, being a flow of energy over time, can be expressed in terms of power. Estimates of the net primary productivity of all the earth's ecosystems run at about 130TW, 70TW on land and 60TW in the oceans. If you consider just the ecosystems under human control – all the croplands, pastures and forestry plantations, and a proportion of natural forests calculated on the basis of how much

---

* A fair estimate for the United States might be a current rate of loss equivalent to a few hundred metres per million years; in developing countries, loss rates equivalent to 1000 metres per million years – that is, a millimetre per year – may be widespread.

humans exploit them – you find that between 15TW and 30TW of the world's productive land-based photosynthesis now takes place on human terms, thanks largely to the immense productivity of our fertilized farms. Our use of the oceans may add on another 5TW. The flow of solar energy into human-dominated ecosystems is thus considerably larger than the total amount of power used by all our machines.

This does not mean that all that energy ends up as food. The energy in the earth's two-billion-tonne crop of cereals, by far the biggest factor in agricultural production, is stored away at a rate of about 1TW, less than ten percent of the total.* After adding on the energy stored in the rest of our agricultural products, and in the wood we build with and burn and pulp into paper, and in the fibres from which we make our clothes, you still find that most of the net primary productivity in human-dominated ecosystems does not end up with humans. But that doesn't mean that it is wasted. The primary productivity we don't extract from our croplands and pastures as harvest goes to build the roots and strengthen the stalks that hold the grain, to feed the bacteria that keep adding nitrogen to the soil and the worms that keep aerating it; it allows fungi to break old organic matter down into nutrients for new growth and songbirds to sing, insects to buzz and field mice to rustle in the grass. To get the benefits of biology, you have to pay the costs of keeping things alive. The primary productivity that stays in agricultural ecosystems is part of that cost.

Unfortunately, what we leave behind does not – cannot – cover the whole bill. In terms of the chemicals within it, as opposed to

---

* 1TW means about 160 watts per person for six billion people, which given that we have metabolisms in the 80-watt to 100-watt range is not bad. The energy we get in the form of non-cereal crops is on top of this – but a significant part of the energy in cereal crops ends up being fed to animals, not humans, so that we can eat meat and drink milk and scramble eggs, and the conversion from vegetable to animal has a cost in terms of energy. And then many of us in rich countries eat more than we need. So while it is satisfying to know that the flow of energy through the world's farms is adequate to supply the flow of energy through the population's metabolisms, it's also necessary to bear in mind that the current dispensation does not spread the energy out as evenly as we might wish.

the energy that drives it, a natural ecosystem is pretty close to a zero-sum game. It can't lose any of the chemicals it needs faster than it can replenish them by fixing them from the air or by breaking down soil for them, by receiving them dissolved in water or blown in as dust on the wind. If you want truly sustainable agriculture you have to mimic this economy; everything that comes out of the ground has to go back in. On small mixed farms it's possible to get pretty close to this ideal. On large industrial ones it's not, and the shortfall has to be made up with imports from elsewhere. It is because we take a great deal of protein from our farms, the nitrogen from which, in most of the world, we then flush down the toilet, that we have to have a vast nitrogen fertilizer industry.

Humanity's involvement in the nitrogen cycle has, over the past two centuries, become one of the fundamental facts of life on earth. It has grown up in tandem with our involvement in the carbon cycle, and both need to be grasped for an understanding of the extent of our influence on the biosphere. In the eighteenth century, enlightened farmers like James Hutton started to use the 'Norfolk rotation' – wheat, turnips, barley, clover. Its advantage was that symbiotic bacteria in the roots of the clover fixed nitrogen and thus enriched the soil. By the middle of the nineteenth century the Norfolk rotation and increased use of manure meant that farmers in England were getting wheat yields three times higher than those their forebears had enjoyed. Without this growth in productivity, it would not have been possible to feed the growing population of the cities. Ever-better farms made the ever more crowded cities of the industrial revolution possible.

The *Organikers* of the nineteenth century could see that artificially fixing nitrogen would be a great boon, and many of the best of them worked on it. Le Châtelier, Vernadsky's teacher, might have cracked it metaphorically had it not cracked his apparatus literally, setting off a rather dramatic explosion in the lab. In 1909 Fritz Haber, a friend of Einstein and Willstätter, devised a way of turning hydrogen and nitrogen into ammonia using a constant flow of the gases over a hot catalyst at high pressure. Carl Bosch at BASF, then the largest chemical company in the world, re-engineered Haber's quite demanding laboratory process – it operated at a hundred times atmospheric pressure, and at about 500°C – into something that made industrial sense. Within four years Bosch had a plant producing twenty tonnes of fixed nitrogen a day. During the Great War Germany's nitrate production – needed for armaments as well as to boost the blockaded country's agriculture – increased more than tenfold, and at its end the Treaty of Versailles required BASF to license the Haber-Bosch process to the victors. By the 1930s, the chemical industry was fixing almost a million tonnes of nitrogen into fertilizers every year. Today, less than a century after Haber's experiments, it fixes about a hundred million tonnes a year. Roughly two percent of the energy used in our 13TW civilization is spent on fixing nitrogen.

In the nineteenth century agricultural improvements allowed industrial intensification in the cities. In the twentieth century industrial improvements allowed agricultural intensification on the land. Nitrogen fertilization is largely responsible for increases in average cereal yield from 750kg per hectare in 1900 to 2.7 tonnes per hectare today. According to calculations by Vaclav Smil of the University of Manitoba, who has made a fascinating study of the topic, forty percent of all the nitrogen in protein eaten by humans has been through an industrial nitrogen fixation process; other things being equal, today's farmland would be hard put to provide protein to half today's population without the fixed nitrogen provided by the chemical industry.

This massive distortion of the nitrogen cycle has its costs. Roughly half the nitrogen added to fields doesn't make its way into crops. Instead it either runs off into rivers as nitrates, or is given up to the atmosphere. One of the nitrogen-bearing gases soil can give off, nitrous oxide, is a greenhouse gas 300 times more powerful, molecule for molecule, than carbon dioxide. Nitrates washed off farmland into rivers disturb all sorts of ecosystems, often overstimulating algal growth to such an extent that water gets depleted of its oxygen; the nitrates from America's farmlands that drain down the Mississippi create a vast 'dead zone' in the Gulf of Mexico as a result. As important, if not more so, the use of nitrate fertilizers breaks the link between raising livestock and growing crops. When manure was a major source of fertilizer, mixed farming made sense – think of the tightly penned sheep on the Downs trampling their dung back into the grass. Cheap nitrate fertilizers have allowed cereal crops to be farmed in monocultures on a scale that would never previously have been possible, and this has had vast effects on the economics and ecology of farming.

Without today's input-heavy agriculture – inputs not just of nitrogen but also of phosphates and other fertilizers, of calcium carbonate to regulate acidity, of pesticides that allow the efficiency of monocultures, of fuel for mechanical ploughing and cultivating and harvesting – the world's population could not be fed on today's acreage of farmland. That doesn't mean that today's agriculture has been intensified in the wisest ways possible, or that it is, in the long run, sustainable. It might well be replaceable with something equally intense but more environmentally acceptable. But using today's methods, abandoning inputs would mean increasing our croplands even further than we already have.

Which is a lot. About eighty percent of the land now bearing crops around the world has been converted into farmland since the beginning of the eighteenth century; the area devoted to pasture has risen too, though not quite as much. This change has had huge effects on various aspects of the earth's energy balance. It has changed the albedo, as croplands are normally lighter than woodlands, and thus reflect more sunlight. And it has also changed the water cycle, which has its own effects on the climate.

The atmosphere, being transparent, picks up relatively little heat from sunshine; it is warmed by the earth below, not the sun above. Most of this warming is in the form of infrared radiation which the earth gives off and the atmosphere absorbs. But a significant amount – about twenty percent – is transferred in the form of 'latent heat'. When a liquid evaporates, it takes energy from its environment, thus cooling it; 'sensible' heat, which can be felt, is converted into latent heat, which can't (the distinction was originally made by James Hutton's dear friend Joseph Black). When water that has evaporated from the earth's surface condenses to form droplets at high altitudes, the latent heat is given back to the air, increasing the temperature aloft.

Crops, in general, do not pump water into the atmosphere as prolifically as woodland does; their leaf area is smaller and their roots are shallower. Replacing woodlands with crops thus changes the rate at which latent heat is transferred from the surface to the atmosphere. And this change, unlooked for and little commented on, is huge. Deforestation has reduced the amount of water that plants put into the atmosphere by about 3000 cubic kilometres a year. That represents a reduction in the flow of latent heat of about 240TW. In terms of energy, the drop in heat flow produced by clearing land for agriculture is larger than the earth's total net primary productivity. The magnitude of the change in energy flow that can be brought about by changes in vegetation like this is what makes ideas about climate change driven by the spread of C4 plants seem plausible.

The effects of agricultural change in the flow of heat into the atmosphere are on a par with those of the climate's most famous hiccoughs – El Niño events. El Niños are felt around the world; they cause droughts in Brazil and southern Africa, wildfires in the American southwest and good harvests in the midwest. Perhaps most importantly they interact with the rhythms of the Asian monsoon, sometimes catastrophically.* All these effects can be traced back to changes in the transfer of latent heat from the warm waters of the central Pacific to the air above them similar in magnitude to the change in latent heat transfer that has come about as a result of deforestation and the spread of farming in the tropics. That change in land-use may well have had climatic effects on a similar scale; the difference is that the land-use change is permanent, rather than

* The link with the monsoon, it is worth noting, seems to have been weakening recently, and there is some indication that this is due to global warming.

intermittent, and so its effects are harder to discern than those of the on-again-off-again El Niños.

And this is still not the largest of our accidental effects. The sun illuminates the earth at an astonishing rate of 170,000TW. About thirty percent of that is reflected straight back out. About nineteen percent is absorbed by the atmosphere, mostly by water vapour. Just over half is absorbed by the surface. All the world's plants and algae working together manage to make use of about a third of one percent of that energy. The rest just serves to heat things up; it warms the water of the oceans and the surface of the land. There it drives the winds and ocean currents, which flow with far greater power than plate tectonics can muster. The rate at which heat moves up the North Atlantic is 700TW, almost three times all the world's gross photosynthetic energy storage, let alone its net productivity.

Heat from the surface is lost to the atmosphere, mostly through infrared radiation and latent heat. The atmosphere, specifically the greenhouse gases in the atmosphere, absorb this heat and radiate it back out, warming the surface. The warmth we feel from above comes from all the sky, not just the direct sunlight. The warmth of the nights is warmth the earth has lent to the sky and is getting returned.

But the sky does not return all the heat to the surface. Some of the energy is radiated outwards, rather than inwards. As we have seen, it is eventually lost to space at almost exactly the same rate that incoming sunlight is absorbed. Almost, but not quite – and a slight imbalance in a torrential flow can matter a lot. Overall, the sun provides the earth with about 340 watts per square metre, about 240 watts of which are absorbed by the surface and the atmosphere. Adding greenhouse gases to the atmosphere allows it to absorb more of the outgoing heat from the earth's surface, and then to radiate more of it back down to us. As a result the earth is now absorbing more energy than it is emitting. This effect is known, in climate-science circles, as a 'forcing', and it, too, can be expressed in watts per square metre. The forcing due to the increase in carbon dioxide that has taken place since the Bowood cedar was planted is currently put at 1.66 watts per square metre.

That sounds fairly innocuous: 1.66 watts per square metre is just a bit more than the energy it takes to light my living-room with an energy-saving light bulb. But that 1.66 watts applies to every square metre of the earth. That's the equivalent of 4000 living-rooms for every man, woman and child on the planet, with a light bulb in each of them. All told, the increase in the greenhouse effect due to the build-up of carbon dioxide between the eighteenth century and today adds up to almost 850TW.

Add in the other human greenhouse gases, the enhanced methane from rice paddies and cattle and landfills, the nitrous oxide from fertilized farmland, and you get up over 1000TW. By interfering in a fairly marginal way with the flow of Vernadsky's cosmic energy, the by-products of our industrial civilization trap almost a hundred times more energy than that civilization actually uses.

The carbon/climate crisis

What to call this extraordinary state of affairs, in which we add almost a hundred watts of warming power to the planet for every watt we make deliberate use of? It is most often discussed under the rubric of 'global warming', but this lacks both accuracy and a sense of danger. The change in the climate brought about by our enhancement of the greenhouse effect is not going to be a monotonic warming, and its effects will differ markedly in different parts of the world (some bits, though very few, might conceivably cool). Warming also sounds both gentle and broadly pleasant; what we are doing is unlikely, for most people, to be either. Lovelock, deeply alarmed on the subject, prefers to speak of 'global heating'.

'Climate change' avoids a stress on temperature, which since altered patterns of rainfall are likely to be the main disruptors of health and happiness is probably a good thing. But climate change is somewhat vague, and open to the critique that climate is, as we have seen, always changing, though rarely if ever at the pace it is today. What is more, there are many people for whom the overtones of the word 'change' are primarily positive. 'Climate crisis' or its agitprop cousin 'climate chaos' come closer to the bill, though chaos, while a stronger word than change and with fewer happy connotations, places an emphasis on unpredictability that may not be helpful. While we cannot predict in detail what is to come, we can use our knowledge of the earth system to sketch its broad outlines with at least some confidence.

My preference is for crisis over chaos – I like the word's medical associations with the breaking point in a fever, and the fact that it stems, in its Greek roots, from the concepts of choice and decision. And I like to stress the role of carbon. This is a crisis not just of the climate, but of the carbon cycle and of the other cycles – notably the hydrological cycle – with which the teeth of the carbon cycle's gears are engaged. Its tendrils reach through the stomata of every plant and into the Calvin-Benson cycle.

To get a feel for the complexities of the carbon/climate crisis, drive a little way outside Urbana, Illinois, the college town where Robert Emerson and Eugene Rabinowitch came together to study photosynthesis after the Second World War, to a rather special field of soy beans. Look at it from the road and you might not notice anything unusual; but look at it from above and you'll see small pylons in among the plants, pylons arranged in a set of fairy rings that cast a spell on the crops inside. The crops in the rings grow a little quicker than their fellows on the outside, but for much of the season the differences are minimal, at least to the inexpert eye. Unless, that is, the inexpert sees in the infrared. Look at the field in the infrared and the fairy rings jump out in bright colours, a couple of degrees warmer than the rest of the crop.

The researcher in charge of this field is Stephen Long, an English plant physiologist who is now the Robert Emerson Professor at the University of Illinois. Long started his career studying the effects of peculiarly stressful conditions on photosynthesis; then, in the 1980s, he started to look at the effects encouragement might have, specifically in the form of increased carbon dioxide. It had long been known that, in the lab, elevated carbon-dioxide levels lead to better growth: commercial greenhouses pay good money for carbon dioxide to make use of this fact. Give plants more carbon dioxide and, all other things being equal, they will probably increase their net primary productivity. But experiments in controlled laboratory conditions don't necessarily pick up all the factors that might be expected in the real world. For that you need controlled out-of-the-laboratory conditions, which are the purview of a technique called Free-Air Carbon-dioxide Enrichment, or FACE.

A FACE experiment is a little bit of the near future, or for that matter the distant past, brought into the present. A circular patch of growing plants is ringed with sensors that measure how much carbon dioxide is in the air and with tubes that can squirt out a little more and so top up the levels as needed; computers controlling the squirting counteract the tendency of the wind to blow the stuff away. In the past ten years there have been about a dozen FACE experiments around the world, looking at natural woodland (loblolly pines and aspens), tree plantations (poplars and sweetgums), grasslands, pastures, cereal and other crops (such as Long's soy beans), deserts and Italian vineyards (a fine piece of prioritizing). It is in some ways an impressive list, in others a rather dispiriting one. Almost no work has been done in the tropics, either for crops or for natural ecosystems. Given that in fifty years at least eight billion people will be relying on crops growing in a heightened level of carbon dioxide, you might think there would be rather more research into how those crops were likely to grow outside the temperate zone. Compared to the total amount of money spent on crop research, a few plots around the world grown under FACE conditions seems a far too modest use of the technology that can best predict the impacts of the future. As Long puts it, 'No agricultural company would use the techniques we use to judge future world food security to predict the performance of a new product.'

FACE experiments don't predict future temperatures. That is the business of the world's most contentious computer models, the general circulation models (GCMs) that are central to climate research. Basic physics is enough to tell you that, all other things being equal, trapping outbound infrared radiation in the atmosphere will cause the world to warm. In the nineteenth century the great Swedish chemist Svante Arrhenius argued that changes in carbon-dioxide level might have been responsible for the changing climates of the ice ages, and went on to point out that, by adding carbon dioxide to the atmosphere, industrial civilization was probably warming the planet, a change he expected to be slow – at nineteenth-century rates of production, he thought, it would take thousands of years to double the carbon-dioxide level – and beneficial, making the world warmer and wetter.* Working out how this would actually happen, though, was hard. The earth's climate is not a simple thing like a poker in a fire or a kettle on a stove. Arrhenius felt obliged to take into account just one of its complexities – the fact that warming by carbon dioxide would lead to increased evaporation, and thus increased warming by water vapour – and that on its own turned a fairly simple calculation into months of thankless work with pencil and paper.

The problem is that adding more energy to the climate system doesn't just warm that system up in a rising-tide-floats-all-boats sort of way. It creates a whole new system. Sensible heat turns into latent heat and back again; heat gets transferred from places where it can't be radiated back out into space to places where it can. The system is riddled with feedbacks that either amplify or counteract the general warming trend. The system's sensitivity – the degree of warming it provides for a given change in the greenhouse effect – simply can't be calculated up-front. It can only be extracted after the workings of a whole new climate system have been observed or laboriously simulated, which is what GCMs do.

* Others agreed with him; the German physical chemist Walther Nernst, who as well as framing the Third Law of thermodynamics also patented an early light bulb and pioneered the use of nitrous oxide as a high-performance fuel additive, talked enthusiastically to his young colleague James Franck about the possibility of setting fire to coal seams in order to get their warming carbon dioxide into the atmosphere as quickly as possible.

GCMs work on some of the world's most powerful computers, but despite that they are in some ways woefully limited. They cannot simulate the world on scales fine enough to capture all the processes on which climate depends – the specific ways in which clouds form, say, or the all-important subtleties of the ways warm surface water is mixed into cooler underlying water in parts of the ocean. As a result they have to rely on rules of thumb to give the gist of what the processes achieve, and these rules of thumb can lead them astray. Despite this, though, there is a growing consensus from the models about how sensitive the climate is to carbon dioxide: in a world with twice as much carbon dioxide as the pre-industrial world, the models predict an average global temperature 3°C higher than that of the eighteenth and nineteenth centuries, though there is a margin of error. Malte Meinshausen, a researcher at the Swiss Federal Institute of Technology, has compared a number of GCMs in order to study the implications of a future in which, by the middle of the twenty-first century, the atmosphere contains a steady carbon-dioxide level of about 550 parts per million, roughly twice the pre-industrial level. The chances of the temperature rising above 2°C in such circumstances were between sixty-eight and ninety-nine percent. The chance of overshooting 3°C was between twenty-one and sixty-nine percent. A 3°C warming would in all likelihood be enough to put large parts of the Greenland icecap into near-irreversible meltdown, and though that would not put anyone in imminent danger, as the melting would probably take many centuries, it would in the long run raise sea levels around the world by as much as seven metres.

It is not average temperatures, though, that matter most. Nor, for most of the world, is it sea level. The population displacement that would be necessitated by a severe rise in sea levels over a matter of centuries would be vast – but people have moved around in vast numbers for the past two centuries, too. The biggest danger is to food supply. If areas that once were warm and temperate become hot and arid, farmers, especially poor ones, may be hard put to cope. Movements in rain belts, monsoons and drought susceptibility, all put into the context of trying to feed more people from often degraded land, are the biggest risks of the greenhouse world.

At this point, a deficiency of the GCMs becomes obvious. They deal, for the most part, just with physics and chemistry – much of the earth's biology is just left out. If you put the biology back in, things look even more complicated, and more alarming.

The most obvious biological effect on the system is that higher carbon-dioxide levels will, all other things being equal, be good for plants. This has led to a belief in some quarters – often but not exclusively fostered with the help of money from oil companies – that the biosphere will ameliorate the carbon/climate crisis by soaking up excess carbon dioxide, and agriculture will boom as a result. Atmospheric observations show there is indeed some truth to this. The rate at which atmospheric carbon-dioxide levels are rising is significantly lower than the rate at which our power stations and deforestations are pumping the stuff out. Some of the difference is down to carbon dioxide dissolving in the ocean. The rest is being taken care of by plants, algae and cyanobacteria, which are storing the carbon away somewhere where, for the time being, the world's respirers can't get at it.

According to the latest report by the Intergovernmental Panel on Climate Change (IPCC), industrial processes (mostly fossil-fuel use, but with a small contribution from the manufacture of cement) are now adding seven billion tonnes of carbon to the atmosphere every year. In the late 1990s, the most recent period for which it has yet been possible to derive the figures, land-use changes such as deforestation were adding between 0.5 billion and 2.8 billion tonnes of carbon a year on top of that. However, today's atmospheric levels of carbon dioxide are rising at a rate of only 4.1 billion tonnes a year. Half the carbon we emit isn't staying in the atmosphere. Some is being absorbed by the waters of the ocean (which are becoming acidified as a result); some is being fixed into organic carbon by phytoplankton; and somewhere in the range of one billion to four billion tonnes is being stored away in the soils and biomass of the continents.

There is a lot of uncertainty in these numbers. But in the past 150 years the IPCC says that at least a hundred billion tonnes of carbon put into the atmosphere by people has been taken out by plants, about a quarter of the total, and a bit more than that has been absorbed by the sea. As the world warms up, the sea will become a less efficient sink – gases dissolve better when it's cooler. But the land sink could increase, because higher levels of carbon dioxide in the atmosphere will encourage the plants to fix more of it. The magnitude of this fertilization effect is one of the things that FACE models try to predict, but there are other ways of tackling the problem too – model-based methods such as those used to study the 'world without fire'.

To make such predictions, you need a climate model which will predict the climate for a given carbon-dioxide level, and you need a model of the biosphere and its carbon cycle that will tell you how much photosynthesis goes on in a world with that climate and that level of carbon dioxide. You then drive these lashed-together models into the future by feeding them a scenario for human carbon emissions. As the carbon-dioxide levels rise the climate changes: as the climate changes the carbon cycle changes: as the carbon cycle changes the carbon-dioxide levels change. The key question is whether that feedback loop is negative or positive. Does the biosphere act as a brake on change by sucking up more carbon, or does it exacerbate things by giving the stuff off?

If photosynthesis were the only factor playing a role, the feedback would be negative: more carbon dioxide would mean more carbon fixation. But the carbon cycle is subtler than that. For one thing, the amount of photosynthesis depends on the plants getting watered, too, so if rainfall changes it can go down even as carbon dioxide goes up. More importantly, as the world warms up the rate at which the microbes in the soil respire goes up, too, producing more carbon dioxide. While increasing carbon dioxide on its own, without warming the world, would lead to increased amounts of carbon-dioxide storage on land and in the oceans, increasing the temperature while keeping carbon dioxide stable would lead to decreased storage.

So what happens when you vary carbon dioxide and climate together? In 2006, a consortium of climate-modellers looked at how eleven different climate models, each coupled with a model of the carbon cycle, responded when they were all fed the same scenario for carbon-dioxide emissions through the twenty-first century. Some of the carbon-cycle models were digital vegetation models like those from the world without fire, others used different approaches that accentuate different parts of the cycle. But they all agreed, at least qualitatively. By the end of the century the net effect of climate change was to drive carbon dioxide out of the biosphere and into the atmosphere. The feedback was positive: the living world was accentuating the changes man was inflicting on it.

The models did not agree on the size of the feedback: in the one where things changed the least, the effect was to increase carbon-dioxide levels in 2100 over what they would have been if the biosphere and oceans had kept on working as they do today by just twenty parts per million; in the most dramatic the increase was 200 parts per million, twice the rise in atmospheric levels from the industrial revolution to today. The models also differed as to the mechanism responsible. In some of them decreased photosynthetic productivity was the dominant factor, in others increased soil respiration. Some saw increases in storage in the ocean but not on land, some saw that pattern reversed. Given the range of results it is impossible for all the models to be right, and a racing certainty that none of them is capturing the whole picture accurately. And it's worth noting that the scenario for industrial carbon emissions over the century used when making the comparisons was a fairly extreme one, with the emission rate rising by a factor of four and a final atmospheric level, before taking in carbon-cycle changes, of three times the pre-industrial level and more than twice the level today. But for all this, the fact that a range of different models working in different ways all pointed in the same direction is undoubtedly sobering.

What about the parts of the biosphere we depend on most – our crops? The main message of the FACE studies carried out to date is that for cereal crops, life in the open under heightened carbon-dioxide conditions is not as much of a blessing as it had seemed during experiments in closed chambers. Indoor work predicted cereal yield increases of about twenty percent under doubled carbon dioxide. FACE work has seen increases of seven percent for rice and eight percent for wheat. As the carbon-dioxide level is pushed up further, the returns start diminishing. Predictions that were once made of crop yields rising by a third under late-twenty-first-century carbon-dioxide levels now look very unlikely.

According to Long, it seems that the plants, rather than treating the added photosynthetic efficiency offered by extra carbon dioxide as an invitation to work harder, are instead treating it as an opportunity to put a bit less effort into getting the same results. One of the most striking features of the FACE crops is that they contain a good bit less rubisco than normal crops do. Because rubisco works better at higher carbon-dioxide levels, the FACE plants make their lives easier by not making so much of it. This may suit the plants well, but by making the plants' leaves less protein-rich the rubisco-reduction raises a problem for those who eat them. Cattle grazing on carbon-dioxide-fertilized pasture will probably need to eat more grass for the same protein intake.

There is also some evidence that the nitrogen changes are not just down to getting by with less rubisco; in various crops there seems to be less protein in the non-photosynthesizing seeds, as well. Arnold Bloom, a professor at the University of California's most agricultural campus, Davis, has developed a theory that carbon-dioxide fertilization makes it harder for plants to take up nitrates. Nitrate assimilation, he thinks, depends more than has previously been realized on chemical changes brought about by photorespiration. If photorespiration is reduced when carbon-dioxide levels are raised, then nitrate assimilation might be reduced as well. To counteract such an effect fertilizer might need to be applied at even higher levels, or in the form of ammonia; FACE experiments suggest that ecosystems where soil nitrogen is found in the form of ammonia, such as pine forests, respond to carbon-dioxide fertilization quite enthusiastically.

There may be other factors counteracting carbon-dioxide fertiliz-ation, too. Long's fairy rings are able to squirt out ozone as well as carbon dioxide. In the lower atmosphere ozone is produced by the action of sunlight on other pollutants, and the rate at which this happens is expected to rise in many regions of the world over the next fifty years. Long's SoyFace results suggest that a twenty percent increase in ozone, which some consider possible by 2050, could cut soy yields by twenty percent, enough to wipe out any carbon-dioxide-fertilization benefit and drive yield down below today's levels.* Ozone tests on other crops have not yet been carried out.

And then there is water. Earlier I said that the striking thing about Long's soy plots, at least to those of us with infrared eyes, was their temperature – a few degrees warmer than the surrounding crops. This extra warmth is not due to the greenhouse effect of their little fairy ring of carbon dioxide – it takes a whole atmosphere's worth of greenhouse gas to warm things up. The warmth comes instead from the carbon dioxide's physiological effects on the plants. The higher carbon-dioxide level in the FACE rings means that the soy growing there can get away with fewer stomata, or with keeping those that it had more tightly shut. Lessened 'stomatal conductance' means less transpiration and thus less cooling; the sensible heat in the environment stays sensible, instead of turning latent. And so the FACE rings end up a couple of degrees hotter than the soy outside.

There are clearly advantages to lowered stomatal conductance: it makes plants more efficient in their use of water, and other things being equal it makes it harder to damage them in droughts. There are doubtless times and places in the decades to come where this effect will be valuable. But in others it could damage plants badly. Hotter leaves photosynthesize less well, and plants don't seem to have evolved any direct way of dealing with this problem. The systems that control the stomata are beautifully evolved to trade off

* Plant-breeders have tended to assume that they have been managing to produce ever more ozone-tolerant varieties as the ozone level has risen, thus negating its effects, but Long's work has shown this is not true. Plots planted with the latest types of soy and those planted with seeds released thirty years ago show equal decreases in yield under elevated ozone levels.

photosynthetic productivity against water loss, keeping a carefully controlled optimum. They don't seem to pay any attention to overall temperature. Indeed, at high temperatures the stomata actually close down a bit, heating the leaf more; higher temperature means less photosynthesis, thanks to rubisco's heat sensitivity, and since less photosynthesis means less need for carbon dioxide the stomata throttle themselves back. While such effects could be moderately bad news for all sorts of plants, they could be particularly damaging for rice, which cannot flower above a certain temperature.

As we saw in the discussions of the spread of C4 grasses and of agriculture, a change in stomatal conductance doesn't just change the situation for the plants involved; it also changes the climate of the region they grow in. A heavily agricultural region such as the American midwest could be warmed up beyond the level models now suggest by a carbon-dioxide-induced drop in stomatal conductance, especially as less water vapour in the atmosphere means fewer clouds and thus a lower albedo. The hydrology would change as well – water that is no longer being sent back to the sky by plants will be drained off the land in other ways. The flow in rivers would increase.

This is not a theoretical possibility – it is apparently already happening. In a study published in 2006, researchers from the UK's Hadley Centre for Climate Change and colleagues elsewhere used models to try to explain a measured increase in the amount of fresh water flowing off the continents over the twentieth century. The record of the century's warming climate alone did not explain what they saw; nor did adding in the effects of changes in land-use. But when they added in the effect of the reduced stomatal conductance that would have come about because of the century's hike in carbon-dioxide levels the fit got much better. In statistical terms, the runoff changes were most reasonably attributed to changes in the stomatal conductance.

Changes in stomatal conductance, like changes in land-use, will probably be more important at a regional level than a global level. But some regional changes can have global repercussions, as shown by another set of studies at the Hadley Centre. These teamed up a GCM and a digital vegetation model – the rather wonderfully named 'Top-down Representation of Interactive Foliage and Flora Including Dynamics', or TRIFFID. The most spectacular result was that the forests of Amazonia – the world's largest set of tropical forests – disappeared almost entirely. A substantial part of the warming in the model, as is the case in many GCMs, took the form of El-Niño-like disturbances in the temperature of the Pacific Ocean; a quasi-permanent pool of warm water formed where there usually isn't one today, and brought with it changes in the atmosphere's overall circulation like those seen in El Niño events. During El Niños northeastern Brazil is hit by drought, and so it was in this simulation. But in the greenhouse world the effects were more profound and widespread. In parts of the region the closing-down of stomata staved off the worst effects of the drying by keeping water in the soil; but in others the stomata exacerbated things by reducing the amount of cloud cover and rain yet further. As the forest died and was replaced by grassland, stomatal conductance and rainfall dropped lower, since grass draws up less water from the soil. No rain, no rainforest. By 2100, the amount of Amazonia that the model treated as broad-leaf forest had dropped from more than eighty percent to less than ten.

This was a peculiarly dramatic regional effect – and one with global consequences. According to the model, 160 billion tonnes of carbon was lost to the atmosphere as the forest turned into grassland and the soil beneath it was impoverished. That is a loss equivalent to more than two decades of fossil carbon emissions at today's rates.

The TRIFFID results for the twenty-first century are far from a sure thing. The climate model they were built into, like all climate models, doesn't capture all the details of today's world let alone tomorrow's; it appears to have a tendency to make northeastern Brazil rather too dry even under today's conditions. The model may also exaggerate the rate at which soil respiration increases with temperature. And it may miss abundant extra carbon storage in other parts of the biosphere. Pairing the Hadley Centre GCM with

TRIFFID produced the most dramatic results in the comparison of eleven such pairings that looked at the overall strength and sign of biosphere and ocean feedbacks which one could take to imply that that particular pairing is oversensitive. But it highlights the basic truth of life's role in the coming centuries of change. The biosphere will not remove the problem by sucking up all our carbon dioxide. It will, however, add new dimensions to the issue as plants grow more luxuriant but less nutritious, and as they start to dry out the air above them.

And it will have effects in the oceans that could complicate the picture yet further. One of the more remarkable discoveries to have come out of Gaian thinking is the role that the photosynthetic plankton play in cooling the planet. In the 1970s, working on first principles and the promptings a sense of smell sharpened by decades in chemistry labs gave him when walking along the seashore, Jim Lovelock suggested that various elements washed out of the terrestrial biosphere in river runoff might be returned to it in the form of volatile organic compounds made by plankton. This turned out to be true; much of the iodine and sulphur that plants use on the continents comes to them from organic compounds excreted in the oceans. Life in the ocean increases the habitability of the land. And the sulphur compounds that are shipped from ocean to continent in this way play a crucial climatic role in the process.

For clouds to form, the water vapour in the atmosphere needs little specks of something to cling to; without these 'cloud condensation nuclei' it's hard to produce the droplets from which clouds, and eventually raindrops, will form. Over the continents there's all sorts of gunk in the air for the water to cling to. Over the oceans, there's rather less. But when plankton pump out sulphur compounds, that sulphur will oxidize in the atmosphere and clump together to form little aerosol particles which are just what the water vapour needs in order to form clouds. The plankton in the oceans, most notably the coccolithophores, put out about twenty-four million tonnes of sulphur in this form every year – more than twice the amount of sulphur emitted by all the world's volcanoes. A significant amount of the cloud cover over the oceans owes its existence to the sulphate aerosols produced by the phytoplankton, and those clouds tend to cool the surface below them.

The plankton that produce this sulphur tend to prefer things somewhat cooler, and Lovelock believes that part of the reason that they produce the stuff in the first place is to encourage the formation of sun-shading clouds. And at the moment, the coccolithophores are living in a peculiarly warm ocean compared to the glacial conditions that have been the norm for most of the past three million years. It is possible that in a warmer world, these plankton will suffer, and that as their numbers decrease so will their sulphurous emissions, and thus so will the cooling clouds they engender.

Alternatively, the carbon dioxide itself could get them. If you were focused solely on climate change, rather than the whole carbon/climate crisis, the fact that the oceans soak up a fair whack of the carbon dioxide emitted could be seen as a good thing. But it is having a significant effect on the oceans themselves: it is making them more acidic. Normally – even in the sudden changes at the end of an ice age – atmospheric carbon dioxide never rises fast enough for this to happen. The increasing acidity it might lead to is counterbalanced by reactions with the carbonates in the oceans' sediments. But with the current meteoric rise there is simply not enough time for those reactions to take place, and the surface waters are getting significantly more acidic. Even if the increase in carbon dioxide had no climatic effects at all, ocean acidification would still be a significant global problem, as no sea creatures have evolved for more acidic conditions. There could thus be a lot of ecological damage. And the carbonate-clad, sulphur-pumping coccolithophores look like the sort of creatures that could be expected to suffer the lion's share of that damage. If they do, then their cooling clouds will go with them.

The phytoplankton are not the only photosynthesizers that add aerosols to the atmosphere. The terrestrial biosphere produces fragments of wax and leaf, pollens and spores not to mention free-flying microbes – all told perhaps as much as a billion tonnes a year. Again, this bio-dander may have climatic effects, though the air over the continents does not have the purity that makes planktonic sulphur the only game in town for clouds forming over parts of the ocean. More climatic complexity may come from the various organic chemicals produced by plants – most intriguingly methane. The apparent discovery in 2005 that plants produce methane in a previously unknown way has been something of a biogeochemical bombshell, and what if anything it means for the future course of the climate is unclear. But it may provide a solution to a question that has been troubling Bob Spicer for decades. When climate models are applied to the Cretaceous, they predict very severe winters in continental interiors where the fossil record shows that there were thriving forests. Some regional warming process or other was at work that the models are not currently capturing. That the forests would be making methane to warm themselves sounds unlikely – but the links between the climate and the biosphere are sufficiently complex and subtle to make ruling out such a link foolhardy. After all, thirty years ago no one thought plankton blooms could make clouds.

The carbon/climate crisis is almost unbearably complex, and far from fully understood. In general, there seem to be more ways for the biosphere to exacerbate the rise in temperatures than to ameliorate it; the IPCC now suggests that, for a business-as-usual type of scenario, biological effects such as the release of soil carbon could increase the temperature by an additional degree or so by the second half of the century. To think that running ever more carbon through an ever-warmer biosphere won't trigger at least one major earth-system surprise in the course of the next hundred years is to take an incredibly complacent attitude to life's complexities.

Like the cedar that shaded me at Bowood, the biosphere has been warped at the hands of men. A change rooted in the eighteenth century is shaping its growth into something which, while still magnificent, is strangely constrained and oddly off balance. Happily, we know how to support it. We can use guy ropes to stabilize it, and insert great props to hold up its heavier boughs. And we can find ways to stop pumping carbon into the climate.

CHAPTER NINE

Energy

A world of wedges
Sun among the poplars
Postscript: Incorrigibly plural

EDMUND: Percy . . . What you have discovered, if it has a name,
is some green.
PERCY: Oh, Edmund, can it be true? That I hold here, in my
mortal hand, a nugget of purest green?
EDMUND: Indeed you do, Percy, except, of course, it's not really
a nugget, more of a splat.
PERCY: Yes, well, a splat today, but tomorrow, who knows? Or
dares to dream!

Blackadder II
('Money', by RICHARD CURTIS and BEN ELTON)

A world of wedges

My father's father, whom I never knew, was a clerical worker at a coalmine in South Wales. My late father spent some of his youth guarding convoys that carried fuel, among other things, from America to Britain, and from Britain to the Soviet Union. He went on to work with my godfather, Sidney Colwell, on the establishment of a system of emergency oil reserves in Britain. At the beginning of the North Sea oil boom he was for a couple of years an independent economic consultant to the oil business at large. My first introduction to the world of geology came from the maps that he put up on the wall of a room that I thought my playroom and he thought his office. The importance of these strange dark patterns strung out between the coasts of Scotland and enticingly frilly Norway fascinated me; my father put up with my chuntering about the possible relationships between the bigger and smaller blobs with more than a modicum of patience.

Shortly afterwards he took a job in the Oil and Natural Gas Directorate of the European Commission, which provided him with an office of his own. My first memories of Brussels are of Sunday mornings on which only cars with even-numbered licence plates (one week) or odd ones (the next) were allowed to drive, a reaction to the 1973 oil crisis. My boyhood imagination became more convinced than ever that oil was an important thing. One of the first stories I can remember writing at school had both a refinery and an offshore rig in it. It is almost needless to say that the refinery exploded in a way that a ten-year-old imagination found quite satisfying, though the bad guys were thwarted in their attempts to set fire to a whole oilfield.*

Two generations of a family clothed and fed by the fossil-fuel business is not exactly common – but it's far from a twentieth-century rarity. To take another family not entirely at random, one of my wife's grandfathers had a gas station in central Minnesota, a Sinclair station where the choice of a big green dinosaur as the company logo unambiguously announced the fossil nature of the fuel for sale. The dinosaur's effigy still clings magnetically to our fridge. Her father, though, ended up working in another line of the energy business; before retiring, he was a public servant responsible for asking the people of Minnesota where they might like their windmills.

A huge amount of the life and strife of the twentieth century revolved around the getting of fossil fuel – of coal and, more dramatically, of oil. So important have such fuels been that they have become almost synonymous with energy. But this elision obscures the truth. Fuels are not a source of energy – they are just a way of storing it. And the finite store of sunlight that fossil fuels represent is inevitably going to run out. If the people of the developing world are to use energy at roughly the rate that people in most developed countries use it today, we need to quadruple our current power generation capacity. That energy needs to come from somewhere other than the deep past. It needs to come from something available today.

Exactly how soon we will come up against the finitude of fossil fuels is a matter of fierce debate. The distinguished and inspiring Princeton professor Kenneth Deffeyes, who taught me to use a magnetometer on my one and only field trip as a geology student, and who was immortalized in John McPhee's geological classic *Basin and Range*, says that the peak level of world oil production will have been passed by the time this book is in your hands. Others, including economists with whom I think my father would have sided on the matter, believe the peak is much further off. But I don't think anyone

* Like James Hutton, I was unconvinced by the notion that oxygen is necessary for the burning of fossil-fuel reserves, and sure they could manage without it.

doubts it will be passed while this book remains in copyright. It is near-impossible to imagine oil production at the end of the century being more than a fraction of what it is today.

A decline in oil does not mean the end of fossil fuels; there is methane and there is a great deal of coal – about 700 billion tonnes in measured, recoverable seams at reasonable depths, conceivably ten times as much if we dig deeper or mine more subtly. But while 700 billion tonnes of coal would last a couple of centuries at today's rate of extraction, if you used it to run a post-oil, post-gas 40TW economy it would last you for just a couple of decades.

And if you used the coal the way we use it today, you'd put 600 billion tonnes of carbon into the atmosphere in the process. In carbon emissions, it would be the equivalent of putting every tree and shrub and meadow on the planet to the torch.

We have relied increasingly on fossil fuels in the two centuries since Priestley and Hutton; we cannot conceivably rely on them for the two centuries to come if we want to let everyone on the planet share the benefits that the wealthy nations enjoy today. And the mere act of trying would incur an absurd environmental price. No one knows how damaging a given degree of climate change will be; we do not know how the biosphere will respond, and we do not know how easy it may be for farmers around the world to adapt to the change in rainfall patterns. It may be that doubled carbon-dioxide levels are not in fact all that bad, if the people of the world have the resources needed to adapt to them. But if the rapid economic growth that developing countries want is provided by a fossil-fuel economy, doubling is not what we will get. By the end of the century we will be well on the way to tripling.

Projections of emissions, and of the climates they produce, get almost indecipherably complex; they are far more detailed than they can conceivably be accurate. The IPCC has developed forty different scenarios for future patterns of fossil-fuel emissions, economic growth and land-use, and graphs of these tracings of the future are known in the trade as 'spaghetti plots' for good reason. Happily, though, the gist of what needs to be known, and what might conceivably be done, can be expressed in a diagram that can be drawn on a paper napkin. Indeed at first it was.

The napkin in question was in a cafeteria at Princeton University, on the day that President Bush's science advisor was coming to an afternoon briefing at the university's Center for Energy and Environment Strategies. Robert Socolow, head of what has now become the Carbon Mitigation Initiative, wanted a straightforward frame for the discussions they were going to have. And as he talked the problem through he realized how simply the issues could be expressed.

Taking a napkin he sketched out a couple of axes, a timeline for the next fifty years running from left to right, the annual level of fossil-fuel carbon emissions running up the side. Then he added a horizontal line to represent a steady carbon emission rate of seven billion tonnes a year, which is to say, roughly today's rate. This was a rough approximation of what various models say about the emissions that can be permitted if the atmospheric level is to be capped at 500 parts per million.* The emissions levels in those models actually rise a bit in the early part of the century before starting to drop, and in the second half of the century they would have to drop yet further; but over the next fifty years a straight line is a fair summary of the gist.

Then Socolow added another line, this one sloping up from left to right; a business-as-usual emissions scenario, one in which emissions grew steadily from about seven billion tonnes of carbon today to just over fourteen billion tonnes in fifty years' time. The triangle contained between those two lines represented the carbon emissions that the world needed to forgo in order to stand a chance of eventually stabilizing the atmosphere at 500 parts per million of carbon dioxide. It was a triangle fifty years long and seven billion tonnes of carbon a year high; it represented 175 billion tonnes of carbon.†

* Add on the effects of methane, nitrous oxide and other greenhouse gases and that is roughly equivalent to a doubling of pre-industrial carbon dioxide.
† The area of a triangle is half the base times the height: in this case, ½ × 'fifty years' × 'seven billion tonnes per year'. The years and 'per years' cancel out, leaving 175 billion tonnes. The way that units can be cancelled out like this is one of the lasting joys of back-of-the-envelope science.

As a final flourish, Socolow cut the triangle up with six more lines, so that it looked like a fan in the act of being snapped open. Each of the seven segments was a wedge fifty years long and a billion tonnes a year high: twenty-five billion tonnes of carbon.

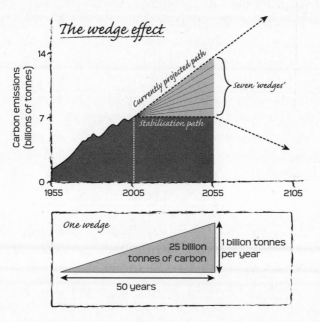

This is a rhetorical diagram, not one to analyse for details. It is a simple graphical way of visualizing the problem, and of representing its solutions as inherently, if not incorrigibly, plural. There is a tendency among proponents of all alternative-energy technologies to claim that theirs is the one true path, the right solution. By breaking the problem down into a set of quantified goals, Socolow's wedges are a corrective to this tendency, encouraging people to ask not 'is this technology better than that one?' but 'how many wedges, if any, of carbon-avoidance can this technology provide?'

Here are some of the fifteen possible answers Socolow and his colleague Stephen Pacala came up with in the paper that introduced the notion of wedges to the world. To keep twenty-five billion tonnes of carbon out of the atmosphere over the next fifty years you could:

379

double the fuel efficiency in two billion cars from nine litres per hundred kilometres to 4.5 litres per hundred kilometres (thirty miles per gallon to sixty miles per gallon), or alternatively halve the distance people actually travel in those two billion cars; double the number of plants burning coal while at the same time almost doubling their efficiency, from thirty-two percent to sixty (which means using different sorts of generator); build four times as many methane-powered generating stations as we have today, instead of building efficient coal plants with the same generating capacity; triple the world's installed nuclear-generating capacity, again at the expense of coal; build four million one-megawatt windmills, which would cover an area of 300,000 square kilometres, in order to make hydrogen to fuel cars; increase the world's installed base of solar cells the best part of a thousandfold; stop all tropical deforestation and plant new forests over an area equivalent to a tenth of Africa to soak up ten billion tonnes of carbon.

Each of those heroic objectives is just a single wedge. To start the work of stabilizing the climate at the equivalent of twice the pre-industrial carbon-dioxide level – not forestalling all climate change, but keeping a lid on it somewhere in the 2°C to 4°C range – you need to do them all, or to do other things on an equally grand scale. And if you choose the first three options – less oil in cars, better coal-fired power plants, more gas-fired plants – you're still looking at running out of fossil fuels. That's not a reason for eschewing those options. In some ways they are the low-hanging fruit, in that building gas-fired power stations is easy, and efficient cars aren't that hard either. But in the second half of the century, the point at which we would have to go beyond wedges and start actually cutting emissions, rather than just holding them steady, we will have an absolute need to harness non-fossil options.

It may be possible to get two or three wedges, even more, out of improved efficiency – we are remarkably profligate with energy, and could use it far better. But though efficiency is under-pursued as an energy policy (especially in America) that neglect does not hide a panacea. Real changes in generating are undoubtedly needed as well

– as are measures to live with the climate change to which we are already committed, and which cannot be avoided.

There are three main non-fossil energy options – or two and a half, depending on how broad a sense of the term fossil you have: solar, nuclear fusion, or nuclear fission. Nuclear fission, the process named by Bill Arnold and used in today's nuclear reactors, is the one you might see as a half. The energy in uranium is obviously not fossilized in the way the energy in coal is – it is not stored sunlight. Instead it is stored starlight. Uranium is made in supernovae; its ready-to-burst nuclei carry their energies over vast depths of time and wide swathes of space. There was uranium in the cloud of dust and gas from which the solar system formed, and the earth's share of that uranium has been heating the planet's depths with the fossilized death-throes of alien stars for four and a half billion years.

The earth's uranium has been steadily used up and never replenished. This is not, in absolute terms, a great deprivation. There is a vast amount of uranium still left in the earth's crust and mantle; there are four billion tonnes of the stuff dissolved in the oceans. But useful uranium deposits – ores that are at least 500 parts per million uranium – are relatively scarce. One estimate is that there are just seventeen million tonnes of uranium in such ores. That's enough for more than a few wedges, but not enough to run an expanded world economy for centuries once the fossil fuels run out.

There are other ways of building nuclear reactors. They can be designed in such a way that the reactions that liberate energy from uranium-235 – the fissile isotope – turn uranium-238, which is of no use for producing energy, into plutonium, which is pretty good at it. In principle, 'breeding' plutonium from the otherwise rather useless uranium-238 increases the amount of energy you can get from a kilogram of uranium almost a hundredfold. In practice the effects are not quite as impressive, but advanced breeder reactors could eke out seventeen million tonnes of uranium for a fair few centuries, even in a 40TW world.

There are, however, two problems. One is that no one has yet built a commercially viable breeder reactor. This is in large part because of the other problem; plutonium is dangerous stuff. It is much easier to make a nuclear weapon out of a nuclear fuel based on plutonium than one based on uranium. This is why it is illegal for private companies, even regulated utilities, to produce plutonium in the United States. Laws can be changed and that particular one is under review; but a world fuelled by plutonium seems likely to be a significantly more dangerous place.

The nuclear reactors currently generating sixteen percent of the world's electricity are in some ways a success story. Though they have been expensive to build, they can boast an admirable safety record. While Chernobyl was a terrible catastrophe, it has not been the norm, and though it led to thousands of deaths, the burning of coal leads to more, both through mining accidents and through the effects on health of soot. The problems associated with disposing of nuclear waste from today's plants are not yet acute, nor will they necessarily become so. And if the electricity those plants produce were being generated from coal we would be putting 2.3 billion tonnes more carbon into the air every year – a much bigger waste problem. Providing new and better-designed replacements for nuclear stations coming to the end of their useful lives makes sense for those countries already using nuclear power. A wedge's worth of new nuclear power would require three times that investment, with three new nuclear stations for every old one taken off line: a new nuclear power station every month for fifty years at a cost that would comfortably exceed a trillion dollars. As a carbon-reduction measure this is not necessarily a bad idea. Some, including Jim Lovelock, are positively enthusiastic about an expansion that big or bigger. But it is hard to see today's types of nuclear reactor as the primary power source for a 40TW world, and it is hard to get enthusiastic about the as-yet-unproved breeder reactors that might be able to meet that challenge.

Then there is the other sort of nuclear power – fusion. A particular fraternity of physicists has dreamed of harnessing fusion for more than fifty years. Because fusion reactions produce energy by adding together small nuclei like those of hydrogen, rather than breaking down big nuclei like those of uranium, their fuel is virtually unlimited. Wherever there's water, there's fuel for fusion. However, the technical challenges have proved virtually unlimited as well. Stars are able to squeeze the relevant nuclei together using their immense bulk; lacking the wherewithal for such extreme pressurization, earthly physicists have to knock the little things together with greater vigour instead, which means heating them up to millions of degrees – much hotter than the heart of the sun – while keeping them trapped in cunningly contrived magnetic fields. So far it has not proved possible to do this in such a way as to consistently get more energy out of the fusion reactions than you put into the systems that heat and confine the hydrogen.

A five-billion-dollar international project called ITER is aiming to put this right by 2020, but even if it is a success it will be a long way from a commercial prototype. And though in a star almost any form of hydrogen can be used, reactors like ITER and its first few generations of successors will have to have some of their hydrogen in the form of tritium, the heavy isotope first made by Luis Alvarez at the Rad Lab at the same time that Martin Kamen was working on carbon-14. There's no natural source of tritium on earth, so the necessary fuel will have to be made in the reactors through a breeder reaction using lithium. That adds further to the engineering challenges.

To physicists, fusion power's immense challenges are part of the fun. But however much fun they have there's no way that we will see ITER's successors providing a wedge of carbon-dioxide relief in the next fifty years. Even in the second half of the century fusion reactors working in the ways being discussed today will be big, centralized generators that cost a vast amount. They may well look a bit out of place in a world that, for all sorts of reasons, is moving towards smaller and more decentralized ways of providing itself with power.

If there were no alternative, fusion could power a post-fossil-fuel world at 40TW or more, though we might end up putting a lot of carbon into the atmosphere while we waited for the reactors to come on line. But there is an alternative. We do not have to wait fifty years for a fusion reactor; we have one that's just eight minutes away, as long as you move at the speed of light. The sun provides the earth with more energy in an hour than humanity uses in a year.

With two exceptions, all the forms of energy that we think of as renewable are solar. Wind power is solar, because the sun's heat drives the winds; wave power is solar, too, because the winds drive the waves. Solar cells – photovoltaics – are obviously solar, as is the biomass of plants burnt as fuel. Hydroelectric power is solar, too, because it is the sun that drives the hydrological cycle and gets the water up into the hills from which it is then obliged to run back down. The only exceptions are geothermal energy and tidal energy; though the sun does have a role in the tides, it is through its bulk, not its radiance. But while these non-solar renewables have their niches, those niches tend to be geographically constrained. Britain could do well out of tidal power, because its convoluted coastline is surrounded by shallow seas in which there are dramatic tides. Bolivia is not so lucky. Iceland's hot springs are a wonderful source of clean energy; but they are of little help to Belgium, a land whose buried reserves of drama are not expressed in geysers.

In terms of electricity generation, it is the world's dams that currently tap into the greatest share of the sun's energy. Wind is a diffuse and in most places intermittent power source; water gathered into a drainage basin is more concentrated and reliable. A litre of water going through the turbines of a dam can give you hundreds of times more energy than the equivalent mass of air flowing through a windmill. Dams provide twenty percent of the world's electricity supply, and are the dominant source in some developing countries. It is hard, though, to imagine a wedge's worth of new dams, and there is no way hydroelectricity could take over as the world's primary energy producer; the flow in all the rivers on all the continents has been calculated as providing only a dozen terawatts or so.

In electricity terms, dams are the solar champions; in absolute terms, though, it is biomass that provides humanity with the most solar energy. For millennia, biomass provided almost all humanity's energy, either burned for heat or fed to draft animals for work; more recently the wind in the sails of ships and mills, and water in their wheels, provided a modest supplement. According to Vaclav Smil of the University of Manitoba, tireless analyst and enumerator of the world's energy use, it was only at the beginning of the twentieth century that fossil fuels overtook biomass as a source of energy. But even in the fossil-fuel century that followed, biomass use continued to grow. The UN's Food and Agriculture Organization estimates that, at the turn of the twenty-first century, 1.8 billion tonnes of harvested wood were burned as fuel around the world, generating heat at a rate just under a terawatt. The true extent of biomass burning is greater than that, though, because people don't just burn wood from cut trees. They burn all sorts of vegetation, including stuff that they gather from the woods and what's left in the fields after crops have been harvested, not to mention animal dung. Smil suggests that the total level of biomass use may be getting on for twice the figure for harvested wood alone. There are dozens of poor countries, including most of sub-Saharan Africa, in which biomass represents more than half of the total energy supply. Almost half the world's population relies on biomass for its household energy needs.

At the moment, this biomass burning is for the most part inefficient. It also produces a great deal of pollution, not just in the form of carbon dioxide but also in the form of other gases and soot. Indoor air pollution from open fires and poorly designed and poorly fuelled stoves sickens millions of people every year, with young children and their mothers hit worst. As a source of lung disease, it is not far behind tuberculosis and smoking. It also has effects beyond the home: thirty percent of the energy used in India and Indonesia is in the form of biomass, and the way that the soot this puts into the atmosphere absorbs incoming sunlight has recently been recognized as a significant, if poorly understood, factor in the region's climate.

Looking at the pollution, disease and non-remunerative work attendant on gathering and burning biomass in poor countries it is hard not to wish there was much less of it going on. There are enthusiasts, though, who would wish for a lot more, as long as it was done better. They think there are a couple of wedges of carbon to be saved by growing biomass crops both in developing countries and developed ones. Their argument is that biomass plantations are carbon neutral, in that the carbon given off after the plants have been harvested and turned into fuel is the same as the amount of carbon taken up when those plants were growing in the first place. A mixture of new technologies and new global priorities, they go on, could allow plants grown for their energy to replace a significant amount of the world's fossil-fuel usage.

Bioenergy plantations are rarely beautiful, and many people, including many environmentalists, have a natural impulse to prefer just planting straightforward forests instead. But in terms of the carbon cycle, a plantation in which the crop harvested for fuel is quickly replaced by new growth is a much better deal. A forest uses up carbon while growing, but once full-grown it mostly just sits there, its carbon-dioxide inputs and outputs pretty much in balance; a bioenergy plantation, because it is endlessly being used up, endlessly draws down more carbon. So growing a forest offsets a fixed amount of coal or oil, while setting up a bioenergy plantation can offset a coal-fired power station, say, or a set of oil wells. It uses the biosphere not just as a static store of carbon, but as a dynamic defence against its increase.

If carbon fixed by contemporary plants this way is to replace carbon fixed by their ancient ancestors, it needs to come in similarly useful forms, so that it can be used for something more than just fuelling cooking stoves. The most common forms of plant energy discussed at the moment are the straightforward burning of wood in power stations, the extraction of oils from crops to form biodiesel, and the fermentation of biomass to make ethanol or other alcohol fuels.

Ethanol is the most visible of these fuels. On the back of subsidies to the corn industry ethanol production in the United States has risen steeply in the past ten years; more than ten percent of the petrol for sale in the United States is blended with ethanol. Unfortunately, while this makes political sense thanks to the power of the corn lobby and the candidate-selecting voters of Iowa, it makes little environmental sense. Corn needs a lot of fertilizer and pesticides, and making those takes energy. Farm equipment uses fuel. So does fermenting and distilling ethanol. One recent calculation suggests that the energy required to produce a given amount of ethanol in the United States represents ninety-one percent of the fuel's oomph: you get out only ten percent more energy than you put in. More generous estimates give a payback of twenty-five percent, or even a touch higher. To run all the cars in the United States on ethanol made this way would require the country to import far more food, to face losing vast amounts of its soil – and to go on using a vast amount of energy to turn all that corn into fuel.

This does not mean there is no rational future for American biofuels. It is claimed that new technologies are making it possible to derive ethanol from the woody parts of plants, the cellulose and perhaps its relative hemicellulose. This improves the overall energy yield. It could allow ethanol to be made, for example, from the cornstover left over when the corn kernels have been harvested. At present this stover is largely used to protect the land from erosion – to use it for fuel would mean cutting down erosion in other ways, for example by abandoning the plough in favour of 'no-till' farming. Since ploughs produce carbon dioxide by providing aerobes in the soil with access to the oxygen needed to eat its stored organic carbon, a transition to cornstover-ethanol production and no-till farming could be a climate win-win.

Another possibility is to move away from corn, which needs to be farmed intensively, to something like switchgrass. Switchgrass was a common sight on the prairies before they were turned over to agriculture, and the idea of bringing back a native plant to provide green fuel is obviously an attractive one to some conservationists. Unfortunately for this romantic notion, a recent comparison by

Stephen Long and colleagues seems to show that an Asian grass, miscanthus, also known as elephant grass, out-performs the native in pretty much every respect, producing twice the tonnage per hectare. Because it is a sterile hybrid it won't spread like a weed, as many alien species do; its only disadvantage is that it can't be used as forage. Miscanthus grown in such a way as to minimize erosion and the need for inputs such as fertilizers while maximizing the amount of carbon that it puts into the soil might be a quite attractive energy crop. Calculations by David Tilman of the University of Minnesota suggest that better still might be to abandon monocultures and grow mixed prairies for biomass; you get a lot of cellulose that way, and compared to monocultures mixed systems lock away much more carbon in soil. A mixed-prairie system could both produce fuel and lock away carbon.

In 2005 a joint report by the American Departments of Energy and Agriculture suggested that if bioenergy were to take on a significant role – defined as replacing thirty percent of the nation's present petroleum consumption – then there would have to be a billion tonnes a year of biomass to work with. To provide that with miscanthus you would need roughly a quarter of current croplands. It might be possible to expand the American – and European – bioenergy sectors quite a lot with perennial grasses like miscanthus, techniques for making ethanol out of cellulose, and low-impact, low-input farming techniques growing prairie plants on currently unused land. But a billion tonnes would be an incredibly ambitious target, while at the same time offering less than a wedge's worth of global effects.*

The greatest potential for bioenergy lies in the tropics, where the sun is. The Brazilian ethanol industry, unlike its American cousin, produces many times more energy than it consumes, thanks largely

---

* It would have pleased Bill Pirie to see that the mechanical extraction of leaf protein from biomass crops to provide food for livestock is an integral part of some of these plans; feeding the world on mechanically reclaimed rubisco was one of his pet projects, and he tried the products of his various protein mills out on himself and his guests, noting no side-effects other than 'slight faecal greening'.

to the huge productivity of the sugar cane that it relies on. The idea that such success could be replicated throughout the tropics is currently being touted enthusiastically by Peter Read, an unassumingly iconoclastic Englishman now living in New Zealand. Read points out that a square kilometre of Brazilian eucalyptus can store sunlight at about five times the rate of any crop in temperate climes. A million square kilometres of plantation working at half that efficiency could store away almost a terawatt (the usable energy that came out would be rather less, depending on what form it was used in). This is the sort of energy that can start to make a difference at the planetary scale.

But there are drawbacks. The most obvious is that there is not an unlimited supply of arable land for growing such crops. This means that bioenergy crops are both limited in their ultimate potential and, in the near term, in competition with food crops. While bioenergy crops have not been the only factor in the food-price rises of recent years, they have played a role; and while high food prices have the benefit of spurring investment and production, without transitional help they do undeniable damage to the health and welfare of many people. The British environmental writer George Monbiot was one of the first to argue that in Read's world the cars of the rich would be fed at the expense of the bellies of the poor, with cash-crop biofuel plantations taking over croplands and despoiling the earth as they go: "Those who worry about the scale and intensity of today's agriculture should consider what farming will look like when it is run by the oil industry".

That said, there is land that can be worked more intensely, including a fair amount in the sunlight-rich tropics, and there is land that might be used for bioenergy crops that is not suitable for food crops. But substantial capital would be required to create workable plantations and the world has not, to date, shown much enthusiasm for making vast investments in developing countries in ways that benefit all concerned. Big investments in developing countries tend to be linked to the extraction of raw materials, and they tend not to do much good to the citizens of the countries involved. It is easy to imagine bio-energy cash crops as a new form of exploitation.

Read is charmingly unfazed by such criticisms. He argues that if the plantations are seen as a social or economic burden by local people they will always have the option of getting rid of them – fields full of fuel are, by definition, vulnerable to arson. There is also an argument that this resource is different from those which have been stripped out of developing countries in the past. For one thing, the solar source is extensive, not intensive, and its exploitation cannot be tucked away in a few well-protected mines or well-heads by buying off local elites. For another, this resource is renewable, pouring down from the sun without cease – though if it is to be ceaselessly exploited the soil will have to be managed sustainably, and that is a significant challenge. In a developing country where the people had a functioning democratic voice and a just land-tenure system, biofuel plantations might have something to offer. Unfortunately, that currently rules out quite a lot of places.

Read has another argument, though, which is very thought-provoking. Bioenergy use on a global scale could actually bring down carbon-dioxide levels. The coal and oil industries are already quite interested in the technology of 'carbon capture and storage' – taking the carbon dioxide out of power-station chimneys before it hits the atmosphere and pumping it into storage deep below the ground, either in aquifers or in depleted oil or gas reservoirs. When used with fossil-fuel technologies, carbon capture and storage aspires to be a carbon-neutral technology – it gives you the energy while leaving the atmospheric carbon-dioxide level unchanged. Installing such technology on a terawatt of coal-fired generating capacity, and finding the reservoirs to put the carbon dioxide into, would save the world a wedge of carbon.*

But a bioenergy programme using carbon capture and storage goes beyond carbon neutral. At the end of the process the users

---

* More ambitiously, it might be possible to turn surplus carbon dioxide into solid carbonates through reactions with rocks such as serpentine. This would be a near-perfect way of fixing the long-term carbon cycle – taking carbon out as fossil fuel, using it, and returning it as carbonate. Unfortunately, we don't as yet know how to get carbonate production up to anything like the necessary scale.

have more energy than they started out with and the atmosphere has less carbon dioxide. If the carbon/climate crisis were to prove even worse than we now think – if, as we motored past 450 parts per million sometime in the 2020s, we started picking up signs that 550 parts per million would bring with it something really terrible, like a total failure of the summer monsoon, or a runaway dieback of the Amazon forest, or the release of gigatonnes of methane from now-frozen permafrost – then bioenergy with carbon capture and storage would be pretty much the only option available for taking carbon dioxide out of the atmosphere. It would be a way of going back in time and changing the direction of the emissions curve. Read points out that you don't even need empty oil fields or aquifers to store the carbon in. If you smoulder the wood into charcoal, producing some heat and some hydrogen in the process, and then work the charcoal into the soil, you have locked the carbon away for centuries, because charcoal takes a long time to break down. You'll also have improved the soil: adding charcoal, if done correctly, increases the availability of nutrients and the storage of carbon in other forms, too. This could be done on a small, local scale, though it might also be possible to build it into the environmental management programmes for plantations over a large region. If David Tilman is right about the carbon-storing properties of prairies, they too might become carbon-negative fuel sources; and other ways of making carbon in soils less likely to be respired could become parts of standard agricultural practice world-wide.

There's one other technology that lets photosynthesis take carbon dioxide from the atmosphere and bury it away: iron fertilization of the oceans. If you put iron into surface waters in such a way as to encourage phytoplankton blooms, and if an appreciable fraction of the biomass in that bloom sinks down into the depths where it won't be respired back into carbon dioxide, then you can draw carbon dioxide down from the atmosphere. You don't draw out any useful energy in the process, but you do get the carbon dioxide tucked away in the deep ocean for a few centuries. As we saw, though this mechanism isn't a complete explanation for the decline of carbon-dioxide levels in the ice ages, it does seem to be part of the story.

To the sort of minds that like to imagine re-engineering the world – to those who sympathize with Erasmus Darwin's taste for cooling oceans with icebergs and moving wind patterns with windmills – iron fertilization seems rather attractive. And while there is a tendency to shrink from the imaginings of such minds, especially among those with a sensitivity to hubris, they are in fact a useful resource. It is often by thinking through the implications of how things could be different that we begin to understand why they are as they are; such fancies offer the same intellectual openings that modelling counterfactuals like a world without fire do. What's more, we are already rearranging the world; we're reworking the carbon cycle and the nitrogen cycle, we're changing the flows of latent heat, we're shuffling sediments around the continents. We're just doing it in a decentralized, slapdash way. The idea that we might do it better should not be rejected for an unworkable if understandable desire that we not do it at all.

Yes, there are areas in which we should learn to lighten the load we place on the environment. But there are others where strengthening the environment's load-bearing capacity has to be seen as a plausible alternative. We will soon be eight billion people. We will want food, and we will want the products of energy-intensive industry. We will want to move around the world which is our home and our principal source of wonder. We will need a healthy biosphere to keep the biogeochemical cycles of the world turning as we do so. We have to take those desires as givens, and arrange the world to fit them to the best of our capacity. We can oppose current patterns of intensive agriculture as long as we find other ways to feed people. We can constrict fossil-fuel use as long as we don't restrict our fellow humans to poverty as we do so. We can't let a romantic idea that nature should be free to carry on regardless dominate our thinking; nature is everywhere under our influence already. To escape the sight of our works you have to go to Antarctica – almost everywhere else on the continents you'll see at least one aircraft in a clear-skied day. And even on the ice, the air you breathe will have the signature of industry upon it.

We are on the flight deck, and we are alone. We are at the controls, and we have no option but to use them. And we know where we want to go. The fact that we have only a dim idea of how to fly means we must act carefully and thoughtfully, not that we must not act.

Carefully and thoughtfully, though, are crucial words, and in the case of many ocean-fertilization schemes they do not apply. Chucking iron overboard wherever you happen to be doesn't help. You need to do it where the biomass created will sink down into the depths, rather than feeding hungry little animals in the shallows. In practice this seems to restrict the technique's applicability to the southern oceans around Antarctica. Many scientists look on this prospect with horror, fearing that it might seriously perturb all sorts of aspects of the ecosystem. A few, though, are more favourably disposed. Andrew Watson, who has taken part in a few of the fertilization experiments, is one of those who thinks that, with further study and strict monitoring, it might be possible to deploy iron fertilization as a protective strategy. However carefully it is done, though, the total capacity for carbon storage in the southern oceans may be disappointingly small. In the experiments to date the amount of carbon storage for a given amount of iron has been a hundred times less than the first crude models predicted. Iron fertilization, if safe, might help at the margins, but it is not going to save the world.

Bioenergy schemes on the scale that Peter Read imagines, on the other hand, could make an appreciable difference to the carbon/climate crisis. But they, too, pose geophysiological risks, as well as economic and social dangers. The change in the rate of latent-heat transfer into the atmosphere produced by huge plantations in Africa, for example, would be likely to influence the Indian Ocean's summer monsoon. And there would also be a risk of such plantations, if they were economically beneficial, encroaching on virgin forest. Leaving aside the loss of species that that could entail, there's a strong risk of losing overall resilience, too. To get the benefits of biology, you have to pay the costs of staying alive. To get the benefits of agriculture on a large chunk of the world's surface we probably have to pay the cost of leaving much of the rest of it relatively untouched, so that it can keep on doing the various things our simplified farm ecosystems don't. We don't know what proportion of the earth's pre-industrial ecosystems is needed to keep things vaguely on track, but further disturbing them on an epic scale is a risk that needs to be well judged.

This is why, in many ways, the best answer to the problem of renewable energy is also the one which is conceptually the simplest: the direct conversion of solar energy into electricity or fuel. From the 1950s onwards, it has been clear to the designers of spacecraft that, for isolated objects orbiting in space, solar power is the way to get them the surprisingly low number of watts that they need. The challenge now is to take the approach that works for our spacecraft, amplify it by a factor of a hundred billion or so, and apply it to the isolated object we all orbit on – to turn our planet from something with a past to mine to something with an insolation to exploit.

Sun among the poplars

In a park outside Paris, I watch a line of poplars shimmering in the wind, their leaves scintillating like the sound of a shaken tambourine; I listen as Bill Rutherford leads a cosmopolitan table of earnest men in a discussion of the mechanisms underlying photosystem II; and I think of Los Alamos.

There are two models for the direct harvesting of sunlight – the photovoltaic cell and the leaf – and they are vastly different. A photovoltaic cell is a pure, peculiar, and unnatural type of stone, fashioned with skill in impeccably clean industrial foundries. It is contrived, but not complex; two or more pieces of silicon, or some other similar material, with an extremely carefully controlled level of impurities. When a photon excites an electron within this semiconductor, the electron becomes free to travel, and so does the 'hole' that it has left behind. The flow of electrons in one direction and holes in the other will generate useful current for as long as the light keeps shining. There are no mechanisms, no features, no moving parts save the conducting electrons themselves.

A leaf, by way of contrast, is remarkably complex, from the ribs providing structure to the cells that open and close the stomata to the chloroplasts themselves and their magnificent membrane-bound machinery of photosystems and twirling ATP-makers and cytochromes. But it is nothing like so pure, nor so fixed in its properties. It is edible, not mineral. It grows on its own from dirt and air, which is good. But it is not long for this world.

One is made in foundries, one grows from seed. One lasts decades, one a season. One is expensive, one is cheap. One is efficient, one is not. One makes current, one makes sugar. And though the semiconductor definitely has the edge, neither really fits the bill as a way of powering a civilization. Existing photovoltaics are too expensive, and they produce electric power only some of the time. If they are to be broadly useful, they have to be made much cheaper, and some way has to be found to store the energy they produce. Leaves are already cheap, but they are inefficient; run-of-the-mill photovoltaics can turn fifteen percent of the solar energy hitting them into electrical energy for our use, while the best plantations struggle to do a tenth as well. As we have seen, to run even today's world on biofuels, let alone tomorrow's, would take enough space and water to be a hugely disruptive undertaking. And leaves don't make electricity, which much of the world works off: to get them to do so, you must burn them and drive turbines with the heat, losing even more in terms of efficiency, and introducing a new need for capital expenditure. Yet that trick of making themselves from scratch, of putting together all that complex machinery without being told to – that is undeniably neat.

The challenge that faces us is to find new technologies that sit in the space between the photovoltaic cell and the leaf – new hybrids of industry and nature. To make leaf-like things that generate alternative fuels, or even, conceivably, electricity. To make industrial systems that capture the benefits of structural details as small as the fold of a protein – ideally learning some of those neat self-assembly tricks in order to produce such details spontaneously. We need to work on a whole gamut of solar conversion technologies: from those that make oils to those that make hydrogen to those that make electric currents; from those that need a lot spent on raw materials but pay it back through their efficiency to those that can be grown any time and anywhere for minimal cost; from those that power a city to those that power a telephone. We need choices.

The men sitting with Bill Rutherford in that park outside Paris are part of a programme aimed at providing one of those choices. They are experts in the physical properties of the strange metal-bearing structures at the business end of some catalytic enzymes. In particular, they are students of the peculiar cluster of manganese atoms that breaks water into oxygen and hydrogen in photosystem II, and of the arrangement of iron atoms that unites hydrogen ions and electrons to form molecular hydrogen, H_2, in enzymes called hydrogenases. This knowledge puts them at the centre of a pan-European collaboration called Solar-H devoted to finding ways of producing hydrogen from sunlight. This conference centre outside Paris is the setting for Solar-H's first big meeting.

Hydrogen is not a source of energy, which is why it didn't feature in the discussion of wedges. Instead it is a way of storing energy – a fuel. In a world where there are ready-made fuels lying around, a fuel that has to be made before it can be used is not an attractive proposition. In a world in which all fuels will have to be made – the world to come – hydrogen could have much to recommend it. It can be oxidized remarkably cleanly, leaving behind nothing but water. It is also ideal for use in fuel cells, devices in which the energy from redox reactions is turned straight into electrical current. Hydrogen fuel cells, their numerous proponents argue, could be to the twenty-first century what the internal combustion engine was to the twentieth: the principal power source for things too mobile to be plugged into the mains. A 'hydrogen economy' based on fuel cells, the argument goes, would be the logical culmination of a centuries-long dropping off of the carbon content of our fuels. In wood there is more carbon than hydrogen, in coal the two are pretty evenly matched, in oil there is twice as much hydrogen as carbon, in natural gas four times as much. A pure hydrogen economy would be carbon-free.

The raw material for a hydrogen economy is all around us in the hydrogen that is bound to oxygen in water molecules. To turn that hydrogen into fuel, though, you need to apply enough energy to break the water molecules down and release pure hydrogen. Hence the need for the Solar-H project, which is centred on finding ways to mimic the sunlight-driven water-splitting mechanisms found in photosystem II.

The idea of biological hydrogen production dates back to the early 1970s, and the oil shock that left the streets of Brussels stripped of Sunday drivers. In Martin Kamen's lab at UC San Diego – always on the lookout for new things to do with photosynthesis – a team of researchers demonstrated that a mixture of chloroplast preparations made from spinach, ferredoxin (the protein that ferries electrons from photosystem I to the Calvin-Benson cycle) and hydrogenase purified from *Clostridium kluyveri*, a bacterium named after the pioneering Dutch microbiologist who was Cornelis van Niel's teacher and mentor, would produce hydrogen. Daniel Arnon had shown that ferredoxin could feed electrons to a hydrogenase a decade before, but this was the first time that anyone had shown you could produce a system in which both the electrons and the hydrogen ions came directly from photosystem II.

John Benemann, the lead author on that paper, is still in the biohydrogen business thirty years on; he presented some of his more recent results at the Montreal photosynthesis congress in 2004. Biohydrogen is only one of the interests he pursues as an independent consultant – the number of pies in which he has fingers can be gauged by the ceaseless ringing of his beaten-up old cell phone, down which he barks in an accent that still has its traces of Mitteleuropa. Much of his time is taken up with the microbiology of landfills. But he keeps a keen, if jaundiced, eye on the field that he helped to start, and which now gets a walk-on part in many discussions of future energy prospects.

The problem, ever since that first paper, has been that when photosystem II makes hydrogen ions it also makes oxygen, and highly reactive oxygen is not something you want anywhere near the sensitive molecular machinery that brings about biological reductions. Keeping oxygen away from their nitrogen-fixing enzymes is a perennial problem for cyanobacteria. In biohydrogen production the oxygen has the potential to mess up the hydrogenase very badly indeed.

Nevertheless, there are ways round the problem. Tasios Melis, now a professor at Berkeley, became fascinated with photosynthesis as an undergraduate in the late 1960s and has studied it ever since. In the early 1970s, as a young graduate student in Florida, he was in the same department as Hans Gaffron, the man who in the 1930s first conceived of the photosynthetic unit, and who had also discovered that when algae that have been sitting in the dark are first illuminated they make a little hydrogen. The Calvin-Benson cycle takes time to warm up, and until it does so the cell needs an alternative sink for the electrons that the photosystems start pumping out as they are struck by the sun. Hydrogenase provides that outlet. But the effect is transitory and tiny.

It was not until twenty years later that Melis discovered how to make algae produce so much hydrogen that it will bubble up out of their culture medium. The secret was starving them, specifically of sulphur. Without sulphur, the algae were hard put to make proteins, and in particular the spare parts needed for photosystem II, which is endlessly being damaged by oxygen radicals. So photosynthesis levels in the cells dropped down low, and the cells began to get most of their energy through respiration. With the production of oxygen by photosystem II diminished and the use of oxygen in respiration increased, the oxygen levels in the cells dropped, providing an anaerobic environment in which hydrogenase could get to work. And at the same time, the cells shut off the Calvin-Benson cycle; starved of sulphates, they were only interested in persisting, not growing, and so they didn't need new organic molecules. What's more, rubisco that's not being used becomes a plentiful foodstuff for the starving cell. But though the Calvin-Benson cycle gets shut down, there are still electrons being pushed out by the photosystems, and they will only generate sorely-needed ATP if the electron transfer chains have a place to put them. That is where the hydrogenase truly comes into its own, not as an early-morning safety valve, but as a key component in the cell's pared-down life.

Melis has now shown that you can get significant amounts of hydrogen out of a simple photobioreactor – a big bag or flask of algae – by starving the algae into anaerobic hydrogen production, then feeding them to allow them to replenish themselves, then starving them again. The problem is that this system can only work if the photosynthetic rates are low, at least compared to respiration rates. One solution, which Melis himself is pursuing, is to beef up the cells' mitochondria by going into the nucleus and rearranging the genes that apportion importance to the different organelles – effectively renegotiating the endosymbiotic contracts of the late Archaean. The more powerful the mitochondria, the higher the level of respiration; the higher the level of respiration, the more oxygen is used up; the more oxygen is used up, the higher the rate of hydrogen-making photosynthesis can be raised. Increasing the algal respiration rates means not just beefing up their mitochondria but also feeding them something to respire with. Melis favours acetate, a two-carbon compound that is produced by all sorts of fermentations. The ideal is to feed acetate to the algae at a rate that keeps their respiration high enough for lots of hydrogenase-powering photosynthesis to take place under its oxygen-absorbing cover. Some of the members of the Solar-H collaboration are looking at similar approaches: others are interested in finding hydrogenases that are less vulnerable to oxygen, or adopting and improving on the mechanisms by which cyanobacteria keep their nitrogen-fixing mechanisms oxygen-free. An even more ambitious goal would be to forsake the idea of hydrogen altogether, and try and engineer the methane-producing mechanisms from archaea into a photosynthesizer, thus developing a solar-powered system for turning biomass into methane. Methane made directly from atmospheric carbon dioxide would have almost all the advantages of hydrogen – and unlike hydrogen could be fed straight into the existing distribution and storage infrastructure.

It is almost inconceivable that generating hydrogen in any of these ways could be as efficient in energy terms as using photovoltaic cells to power a system in which an electric potential rips water molecules apart with brute force, a process known as electrolysis. But it is possible that it might be done with a lot less capital outlay, if the photobioreactors could be made cheaply enough and, rather surprisingly, if the algae could be convinced to spend a little less of their energetic capital on producing antenna complexes. Individual algae selfishly want large antennae, so that they can keep working even at low light levels. But large antennae mean low overall productivity. Large-antenna'd algae at the surface end up intercepting more sunlight than they can digest, and have to radiate away the balance as heat. With small antennae no cell gets more energy than it needs, and thus much more of the light is used productively.

The problem with trying to restrict the antenna size – a problem noted almost forty years ago by Bessel Kok, one of the pioneers of the study of oxygenic photosynthesis – is that there's always a risk that some of your carefully engineered small-antenna'd algae then will revert to the large-antenna'd form. When they do, those algae will do a bit better than the others, because they'll be able to photosynthesize even when they're near the bottom of the pond. They'll have more offspring with the same advantage, and the mutants will quickly take over. This is one of those cases where the benefits of biology are offset by the costs of staying alive, specifically the cost imposed by evolution, which will try and undo your careful work if individual organisms can benefit that way. Any control on antenna size has to be engineering in a foolproof way, because natural selection will favour any mutant that finds a way round the control.

If an enduring long-term control on antenna size could be found not only would solar hydrogen production become more efficient – so would the growing of algae in open ponds to make biodiesel (some algae are more than thirty percent oil by weight) or just to make biomass that could be fermented into fuel of some sort. As Benemann, always keen to take the low-tech route, points out, fermenting sun-grown biomass in a reactor that needs no light, and thus does not have to be made of glass, might make up for in cheapness and reliability whatever it lacked in overall efficiency. Small-antenna'd algae could make local, low-tech biomass-energy schemes plausibly cheap. But as yet, the level of genetic control necessary has not been mastered.

Biology is not the only way to tackle the solar hydrogen agenda. In 1972, Akira Fujishima and Kenichi Honda, two university researchers in Japan, showed that hydrogen can be produced in a 'photoelectrochemical cell' in which a photovoltaic electrode is stuck in water along with a normal electrode. When light hits the cell a voltage difference is set up between the two electrodes, and that voltage difference pulls apart the water molecules. The problem was in the efficiency. The voltage generated by a photovoltaic depends on its 'band gap', an intrinsic property of the material involved. Only photons with enough energy to get an electron over that band gap will do any work at all. And in titanium oxide, the material that first photoelectrochemical cell used, the band gap is rather large – three volts. The photons with enough energy to get an electron over a three-volt band gap are those with the shortest of wavelengths, in or close to the ultraviolet. This means that well over ninety percent of the sun's energy does nothing to such an electrode except heat it up.

You don't need three volts to split water. You can do it with a difference of as little as 1.23 volts, and there are lots of photons with enough energy to lift an electron across an electric or redox potential of 1.23 volts. The problem is that the band gap's size is not the only feature that matters. The specific voltages between which the gap is found matter, too. The question is essentially the same as the one which I tried to clarify with talk of ski lifts in Chapter Two. The ski lift does not only have to be long enough. It also has to start at the right place. This is what photosystem II does – it is set up to use the energy of an incoming photon at just the right redox potential to remove hydrogen from water. Titanium dioxide's band gap, in this analogy, represents a much longer ski lift, one which still gets to the right place, but which starts much lower down the mountain. Frustratingly, attempts to re-engineer the band gap to shorten this ski lift mostly lead to the top end of the lift coming down, rather than the bottom end going up. With a lower top end, the system can't split water.

Various semi-plausible candidates have been found with smaller band gaps in roughly the right place, but they have to be able to function when immersed in water, and so far they don't. They either get oxidized themselves, or dissolve, or break down in some other way. An alternative to such novelties is to squirt electrons into the titanium oxide from a dye stuck to the outside of the electrode, rather than using electrons generated within the semiconductor itself. The dye would act, effectively, like the chlorophylls that feed energy into a photosystem, and it would let the electrons on to the ski lift halfway up. This approach hasn't yet proved a good way of making hydrogen-producing photoelectrochemical cells; but its proponents think adapting it to systems which produce straightforward electric currents, as opposed to flows of hydrogen, could provide cheap new solar cells to use as tinting for windows, or as paint.

Another alternative is to take the emulation of nature further and develop systems based on the absorption of light not by a bulk photovoltaics but by individual molecules. This is the sort of work that Tom Moore, his wife Ana and various colleagues have been involved in at Arizona State University for more than twenty years. They have produced clever systems in which light is absorbed by porphyrins and donor molecules oxidized as a result; they have got so far as to stick some of these molecules into biological membranes and to generate ATP with them. (They also, while they were at it, set up what became one of the most influential centres for research into photosynthesis in the world.) But they have not yet managed to couple such a system to a catalyst capable of pulling apart oxygen.

Approaches like Moore's have their place in Europe's Solar-H programme, just as approaches using photobioreactors do. The programme is aimed at exploring any sort of hydrogen production that might be inspired by photosystem II, whether it relies on artifice, like Moore's chemical constructions, or on biology. It brings together scientists from institutions in six countries and as many disciplines. The meeting I attended outside Paris was a four-day get-to-know-you session, in which the senior men (they are mostly men) introduce their younger colleagues to each other, and the collaboration as a whole tries to make sense of what it is, or might be, doing. It is a rich, complex and confusing scientific mish-mash; my impression is that almost all the participants find at least some of the presentations utterly baffling, or of no interest, or both. Does the Dutch expert in growing bacteria in glass bioreactors really need to know about the fine splitting of spectral lines in ruthenium, a metal that the chemists are fond of as an absorber of photons? Do the chemists really care about the various requirements for growing different sorts of bacteria with a range of different hydrogenases, especially since they have recently developed artificial hydrogenase-based systems that seem to put hydrogen molecules together pretty well?

Against these questions, though, one must pose another: will the progress that is needed be made if people aren't forced out of their habitual boxes? To the European Union, which funds it, Solar-H is a high-risk, high-gain sort of project, part of its programme on new and emerging science and technology. It has been put together by Stenbjörn Styring, a Swedish professor with a faint resemblance to Kurt Vonnegut who runs the Paris meeting with the air of a slightly irascible but potentially inspiring headmaster. Styring has been fascinated by photosynthesis throughout his career, and he has allied that fascination to a particular gift for putting together collaborations between disparate groups. Solar-H is not his first EU-funded research network, and he has also gathered a consortium in Sweden that studies artificial analogues to photosynthesis. He's an expert at getting such things funded, and clearly believes in their worth. At the same time, he knows they are, in some ways, unavoidably inefficient.

Science is, in this way, like life – to gets its benefits, you need to pay its maintenance costs. In life, the costs are those of keeping alive the organism you want to exploit – of setting aside pasture for the horses, as it were. In science, the costs are imposed by the getting of money and the training of talent. Only scientists can train new scientists, and the effort they devote to that end is effort that will often be lost to the actual advancement of science. One of Styring's laments is that the best people he has involved in Solar-H all earned their status by being excellent in the laboratory – and all find themselves able to spend little time working in their labs because they have to run them. They spend more time getting a student to do the job than it would have taken them to do it themselves, and end up with a less good result. Listening to Styring, there's little doubt that he thinks that the leaders of the Solar-H effort, with laboratory space, technical back-up, no interruptions and some overall direction from an acknowledged leader, could make more progress more quickly than the collaboration as a whole will. But would the EU be willing to spend more money on fewer researchers? Even if it were, how then would the next generation of scientists be trained? And would the current experts happily abandon their self-motivated research and move into such a dirigiste project?

In Berkeley, sixty-five years ago, the answer was yes. The Rad Lab was turned over to the Manhattan Project, as were large chunks of the University of Chicago. Scientists as diverse as Eugene Rabinowitch and Harold Urey and Bill Arnold and Martin Kamen, working on not just one thing but many interrelated things, were sent to the places where they were needed: to the green hills of Oak Ridge to work on uranium; to the desolate plateau of Hanford, Washington to work on plutonium; to Los Alamos, New Mexico, the eponymously-poplar-flanked mesa thirty miles from Santa Fe, to design the 'gadgets' themselves. They followed not the roads their curiosity would have led them down, but as many roads as seemed feasible to get to the desired result – different bombs fuelled in different ways with different technologies and different designs. Their deliberate journey was grimly and spectacularly successful. They changed the world.

Various scientists have suggested that the challenge of the current carbon/climate crisis deserves a similar response – a vast new programme of energy research directed systematically at all the ways that the technologies needed for harvesting and storing the power of the sun might be developed. We know what the problems are; we need to find ways of mimicking a small number of natural enzymes, of making algae grow as densely as possible, of making photovoltaics that produce current much more cheaply. It's hard to believe that any of these problems wouldn't yield up an answer to a few hundred million dollars of targeted, planned research each, perhaps a few billion for the tough ones. It is inconceivable that none of them could be solved in such a way.

As yet, though, governments are deaf to these calls for Manhattan or, more peacefully, Apollo projects. Remarkably – shamefully – research spending on energy in the world's rich countries has fallen, not risen, over the past three decades. A coherent, long-term programme is nowhere in sight. Yet there are plans afoot for a space telescope that will cost $5 billion and a set of missions to the moon that are budgeted at over $100 billion. Some of the French researchers in Solar-H are from the French nuclear research agency's campus at Cadarache in Provence, where $5 billion at least is to be spent on building ITER, the experimental fusion reactor, and another $5 billion on running it. Yet Bill Rutherford couldn't, that summer, find the money for a new spectrometer, despite the fact that his lab has as good a track record using them as any lab in the world. It is not that fusion research is overfunded; it is an important long-term goal, and deserves significant support even though it is not going to save the world. The point is to raise research into the less glamorous sciences of energy up to a comparable level. But no one is seriously talking about doing so.

And this lack of enthusiasm does not upset the scientists in the field as much as you might expect, or hope. My impression of photosynthesis researchers is that they are happy to think that their work is, in some way, of planetary significance. But most are remarkably uninterested in actually cashing that significance in. They feel about helping the planet rather as researchers in other parts of basic biology feel about healing the sick; they're in favour of it in principle, and in that they think that their work is relevant to that noble end, they feel good about themselves. It provides an inspiring and self-affirming context in which to do what they wanted to do anyway. But it's not why they do what they do; that agenda is set by their own curiosity, and by the judgement of their peers. And while their work's relevance is pleasing, and may even help when selling projects to funding bodies, turning that relevance into practical achievement tends to be seen as someone else's business.

This is not, or not wholly, an abnegation of responsibility. It is a statement of fact. Most scientists really do see themselves as out to understand the world, rather than to change it. They often understand the ways in which the world might be changed far better than their forebears in the Lunar Society did – but the way their profession has developed has robbed them of that magnificent eighteenth-century sense of themselves as actors in the world at question. Instead, they follow their own curiosity, often driven as much by a need for escape from the world as a need for engagement with it. Institutions provide contexts and channels for this enthusiasm, and will not support it arbitrarily. But to a surprising extent scientists get to define their own questions.

This is even true of a political operator like Styring. He would certainly like to help the world develop new technologies. But what really drives him is not the application of the science, but the science itself; the consortium is, in a way, a vehicle to get him there, a magnificent set of compromises designed to feed resources into what he wanted to do anyway. What he wants is the intellectual satisfaction of a molecular device that can mimic photosystem II. What he wants is the first little bubble of hydrogen in a beaker, or the first arcane inscription on some measuring device that proclaims that such bubbling might be under way.

Ordinarily there is much to admire in this way of arranging the world. There is something inspiring about the idea that simple curiosity is the best engine we have for casting light on the darkened rooms of the unknown. But these are not ordinary times.

Finding the various different solutions that undoubtedly lie in the space between photovoltaic stones and living leaves is not impossible. Attracted by a decade of remarkable thirty-percent-per-annum growth, more and more capital is moving into solar energy, especially into new forms of photovoltaics. But to change the world as surely as we would wish there needs to be more, and there need to be better ways of storing solar energy as fuel, too. If it is to produce a terawatt or two by the 2020s, 10TW or more by the 2040s, solar energy must not just maintain its current extraordinary growth but surpass it. The market may manage this; but we cannot assume it will. People invest in duff technologies all the time, or in technologies that require more money than the investors have, or in technologies that might have made it, had a bit more of the basic science underpinning them been understood.

Some maverick seeking glory may crack the problem on his own. Craig Venter, who ran the non-governmental attempt to map the human genome, is fascinated by the possibility of designing radically new types of bacteria that might fix carbon spectacularly well, and possibly produce hydrogen at the same time. (It is not a coincidence that he was an undergraduate at UCSD when Benemann and Kamen published their first biohydrogen paper.) But to rely on mavericks, even capable, well-resourced ones with impressive track records, is not necessarily as productive as it is exhilarating.

If we want to be sure to get the job done, we need to mobilize resources more than commensurate to the task. We need to encourage some – not all – of the world's best scientists to move into a directed programme, or an interconnected set of such programmes, and devote handsome resources to it. There are moves in this direction. At Lawrence Berkeley National Laboratory, the facility into which the original Rad Lab evolved, they have plans for a programme called Helios that will look at all sorts of ways of using sunlight, from nanotechnology to open ponds producing biodiesel (Melis is part of the team). Though scientists follow their curiosity, they will follow the money, too, when there is enough on offer.

Such goals do not have the urgency of the development of a weapon while at war. They are not as dramatic as putting men on the moon, though you might argue they are considerably more dramatic than trying to put men and women back there fifty years later. But they have a human and planetary logic which it would be stupid to ignore. If we meet them as well as history suggests we might – as well as Benson's generation and Hill's opened up the green box of photosynthesis in the first place – relinquishing the use of fossil fuels may end up not even feeling like an effort. The alternatives will be straightforwardly cheaper and better.

Unlike the Lunar men, we no longer believe in planetary providence. But nor do we need to. We have the power to look after ourselves, if only we can find the skill to use it, and the compassion to see that 'ourselves' means everyone. We understand the possibilities for conflict between the needs of the planet and the needs of humanity, but we know that there is enough sunlight to meet all those needs. As Priestley's friend Tom Paine proclaimed, we have it in our power to make the world again. We can provide, like Matthew Boulton, what all men desire.

So we should do it.

Postscript: Incorrigibly plural

Here's what could happen tomorrow. What could really happen. Not tomorrow by the clock on your wall, though; tomorrow by the calendar of the imagination.

Tomorrow the sun will rise and the sun will set; all the time half the earth will be bathed in light and half cast in shadow, a thin ring of dawns and dusks in between. Photons will rain down from the sun on the day side and splash back out into the universe in all directions, their wavelengths diminished but their numbers and entropy increased. Some of the departing photons will be caught in the detectors of the satellites that ceaselessly inspect and monitor the world below, the distributed bioscope that measures water stresses on the prairies and the temperatures of the clouds, that analyses chlorophyll abundances in the oceans and gauges canopy height in the forests with its gentle lasers. The vast majority of the photons, though, will slip past the orbiting watchers and out into the cosmos atlarge, offering signs of an active biosphere to anyone around another star with the instruments to see and the intelligence to understand. In orbit, and on the dark side of the earth, our own instruments will be looking for those same signs in thin traces of light from elsewhere.

If tomorrow is like today, the flow of photons out into space will carry slightly less energy than the flow in from the sun, the difference trapped close to the surface by the infrared-absorbing atmosphere. If so the world will still be warming. But maybe, if tomorrow is far enough away and different enough that we have learned prudence, the energy coming in and going out will be in balance.

Between their arrival and departure the photons will give blue to the sky and shadows to clouds, draw up water for the rain and give breath to the wind, stir the oceans from equator to pole. A small fraction will meet the green machinery of life and fall into its age-old antennae. It will drive electrons through cytochromes and power the Calvin-Benson cycle, a cycle which will be spinning a little more easily than it does today. The atmosphere will be over-endowed with carbon dioxide for centuries to come, even if we drop our emissions down to the level that the biosphere can absorb without complaint, and during those centuries rubisco will run better.

Not all the green machinery that greets the sun, though, will necessarily be as focused on fixing carbon as it was before and is today; nor will it be as natural. Some of it might be making hydrogen – or, if we are clever enough with our genetic engineering, methane. Yesterday we learned that life's substrate is not a spirit or a vital force but the ability to remember and reproduce precisely conceived shapes – shapes that can pull apart oxygen or marry electrons and hydrogen ions into water. Shapes that can make ammonia out of nitrogen with shirt-sleeve ease while we require heat and pressure and catalysts that are both expensive, and, compared to the room temperature cleverness of a cell's enzymes, crass. Today we are learning how the molecules manage this trick, studying the intricate three-dimensional arrangements of atoms in which every one is situated according to its propensities, attempting to grasp quite how a manganese atom placed just so will break the right bond at the right time, how the iron is held precisely close enough for the electron to skip on to where it needs to be.

Tomorrow, having observed and understood, we will ring new variations on the age-old found-sculpture of the proteins. It may be that the best way of making these new catalysts will be to get living things to make them for us. Bacteria are good at making complex little shapes atom by atom, while our fingers are just a bit big for such fine work. Maybe we can make them into systems that will produce our fuels for us. But it could be that the catalysts we design for them to build are best used in systems cells themselves cannot recreate – in fuel cells, or photochemical arrays. Perhaps the new machinery will be grown and then harvested, built by microbes feeding on the sun and then taken from them to run and fuel the engines of our world, just as the life of the pasture is used to raise and feed the draft animals we need for work elsewhere.

Whether it is in living things or in some technology inspired by the workings of life, there will be novelty. For two billion years or more – in some cases ever since biochemistry emerged from geochemistry – the energy that has flowed from the sun into the biosphere and eventually back out to space again has done so through unchanging channels. Creatures and species have come and gone, but the basics of metabolism – of nitrogen fixation, or methane generation, or respiration, or photosynthesis – have remained, atomic architectures endlessly reborn from the genome of the world. Tomorrow those near-eternal verities will be open to change. When our knowledge of life's most fundamental processes allows us to begin redesigning and embellishing them rather than just studying them, a whole range of new chemical technologies will become possible. There will be not one type of solar fuel but dozens. There will be sports of nature and technology: black dragster algae bred for the speed of their cytochromes; slow-growing poet's jasmine with vast fragile antenna complexes that come to life only in the light of the full moon. A glorious plurality of processes will be let loose again.

While we will make use of tomorrow's sunshine by the tens of terawatts, not all of it will flow through these new channels. Much may flow through photovoltaics making nothing more intricate than a voltage difference. Some will flow through hybrid structures, through organic dyes and nanostructures of titanium and conducting organic polymers that assemble themselves into devices not yet dreamt of. Good old-fashioned plants will use yet more. They will use it to feed us, to give us fuels, to draw down carbon from the sky into the soil, to trace lines of beauty and lines of grace across landscapes. And unobserved, unexploited, unthought-of, they will do it just because it is what they do.

The story of photosynthesis is the story of the many ways in which its unconsidered action has shaped our green world, from the choice of the isotopes in our bodies to the range of colours in our forests, from the closed-in wormholes of the chloroplasts' world to our grasslands' predisposition to fire, from the creation of the fuels that power our world to the need for the pollen which makes us sneeze in the summer, from the rate at which the ground waters the sky to the entropy of the world as a whole.

At the same time, though, this story is a celebration of the power of human intelligence, and the scientific culture it has created, a culture which lets us fathom and appreciate the wonders life has hit upon with brilliance but without intent. A culture that connects so many different things, that lets us change the ways in which we see the world – and lets us change the world itself, rewiring those connections. It is because of the creation of this culture that the impact of our species on the world may prove as profound in its effects as the advent of animals – perhaps, in its way, as the invention of photosynthesis itself. It is the addition to the flame not just of memory, but of mind.

Much of what humans are currently doing to the world is terrible, terrible in its lack of foresight if not, as yet, in its outcomes. But we are getting better. We are seeing things differently. We can get to the tomorrow we imagine.

Meanwhile, this afternoon, I put a sprig of mint in an upturned glass of water in a bowl – my own little pneumatic trough, like Priestley's at Bowood. While writing, I checked on it from time to time, sitting in the sun on my kitchen windowsill. And, sure enough, over an hour or so, bubbles like silver beads began to grow on the undersides of the leaves. Simply as a play of the light it is a pretty thing to see. Understood, appreciated, it is beautiful. Entropy was lost; phlogiston removed; a world replenished. An everyday miracle needing nothing but sunlight, air and leaves – and eyes taught to make sense of them.

GLOSSARY

Algae: Photosynthetic organisms of various sorts which live in water. Algae are **eukaryotes** related to plants.

Anoxygenic photosynthesis: photosynthesis which uses electrons taken from some substance other than water, and which as a result produces no oxygen. Practised in a range of forms by various types of **bacteria**.

Archaean: The second eon in the history of the earth (see **geological time**). It lasted from 3.8 billion years ago to 2.5 billion years ago. Life on earth originated near, or possibly before, the beginning of the Archaean. By its end **oxygenic photosynthesis** was firmly established as the primary source of energy for the biosphere.

Archaea: one of the three kingdoms of life. Archaea are single celled organisms superficially similar to **bacteria**, but evolutionarily quite distinct. They do not photosynthesise, as far as is known. Their major known biogeochemical role is the production of methane.

ATP: Adenosine triphosphate, ATP, is a molecule which serves as a store and carrier of energy inside cells. Photosynthesis produces ATP by means of a **chemiosmotic gradient** across a membrane in which **photosystems** are embedded. In plants, algae and cyanobacteria some of this ATP is used to power the **Calvin/Benson cycle**.

Bacteria: one of the three kingdoms of life. **Photosynthesis** first evolved in the bacteria, and is practiced in a range of different forms by various bacteria. The only bacteria involved in **oxygenic photosynthesis** are the **cyanobacteria**.

C3/C4: Two different types of plant metabolism. In C3 plants – the majority – the first product of **photosynthesis** is phosphoglycerate, a sugar with three carbon atoms (hence C3) which is produced in the **Calvin/Benson cycle**. In C4 plants a four-carbon sugar is produced by other means as a precursor to the Calvin/Benson cycle, thus concentrating the supply of carbon to the key enzyme **rubisco**. C4 plants enjoy advantages in situations that encourage **photorespiration**, such as low carbon dioxide and high temperatures; the C4 metabolism is particularly prevalent in tropical grasses.

Calvin/Benson cycle: The process by which, in plants, **algae** and **cyano-bacteria**, photosynthetic energy is used to **reduce** carbon dioxide. The key reaction in the cycle is the reaction of carbon dioxide with ribu-lose biphosphate, a sugar molecule containing five carbon atoms, to produce two molecules of phosphoglycerate; this is the reaction cata-lysed by the key enzyme **rubsico**. Three such reactions produce six molecules of phosphoglycerate. The other enzymes involved in the cycle can, if supplied with energy in the right form, turn five of these three-carbon sugars into three molecules of the original five-carbon sugar ribulose biphosphate. The sixth three-carbon sugar is "profit" that can be channelled off to the rest of the cell's metabolism, or stored as starch.

Carbon cycle: The flow of carbon through the earth system. This book distinguishes two carbon cycles. The biological carbon cycle, which is comparatively fast, is driven by photosynthesis. Carbon dioxide from the atmosphere is **reduced** by photosynthesisers to organic matter; this organic matter is at later times almost entirely **oxidised** back into carbon dioxide through respiration.

In the geological carbon cycle, which is comparatively slow, carbon dioxide is released from the earth's interior through volcanism, and returns to the earth's interior in the form of carbonate minerals and some stored organic carbon buried in the form of sediments. The laying down of fossil fuels feeds carbon from the biological cycle into the geological cycle; burning fossil fuels feeds carbon from the geological cycle to the biological cycle.

Carboniferous: A period of **geological time** lasting from 359 to 299 million years ago, distinguished by great forests and swamps in which the largest part of the world's usable coal reserves were laid down.

Carbon isotopes: Carbon is normally encountered in three isotopes, which differ in the weight of their atomic nuclei. Carbon-12, which is the most common, and carbon-13, much less common, are stable; carbon-14, rarer still, is subject to radioactive decay. It is made naturally in the atmosphere from nitrogen, and also artificially in laboratories. **Rubisco** differentiates slightly between carbon-12 and carbon-13, with the result that plant tissues are enriched in the former. This effect can be used to make various deductions about the state of the **carbon cycle** in the past.

Carbon/climate crisis: The linked changes in the climate and the **carbon cycle** caused by burning large amounts of fossil fuel. A fuller term for the causes and effects of "global warming".

Chemiosmotic gradient: Both **photosynthesis** and **respiration** pump hydrogen ions across membranes, thus building up a chemiosmotic gradient; the flow of hydrogen ions back down this gradient is harnessed to produce **ATP**, which is then used to power the cell's metabolism.

Chlorophyll: A pigment widely used in photosynthesis to absorb light. When chlorophyll molecules are arranged appropriately, light energy absorbed by one can be passed directly to the next: **photosystems** use this property to accumulate energy from many chlorophyll molecules.

Chloroplasts: the specialised structures in which photosynthesis takes place in plants and algae. Chloroplasts contain membranes in which **photosystems** and their **electron transfer chains** are embedded, and across which **chemiosmotic gradients** are set up. Chloroplasts have small genomes of their own, independent of the genome in the nucleus, which are relics of their days as free-living **cyanobacteria**.

Cyanobacteria: the only **bacteria** which practice **oxygenic photosynthesis**. The chloroplasts in algae and plants are derived from **cyanobacteria**.

Cytochrome: A protein involved in electron transport. The key cytochromes in photosynthesis are found in the cytochrome f/cytochrome b6 complex, which sits between **photosystem II** and **photosystem I** in the **electron transfer chain** represented by the **Z-scheme**.

Devonian: A period in **geological time** lasting from 416 to 359 million years ago, distinguished by a great spread of plant life across the continents, and the evolution of the first trees.

Electron transfer chain: Living creatures can turn energy from their environment into a usable form by passing electrons through proteins called **cytochromes**; electron transfer chains are fundamental to setting up **chemiosmotic gradients** which produce energy in both **photosynthesis** and **respiration**.

Endosymbiosis: the process by which one organism takes another into itself, thus gaining a significant biological capacity it would be unlikely to evolve by other means. **Mitochondria** and **chloroplasts** were formed through endosymbiosis.

Entropy: a property defined by thermodynamics which corresponds to the orderliness and likelihood of a given state of affairs. In a closed system, entropy will increase inexorably, and order will decline. In an open system, such as the earth, or a living being, order can be maintained or increased by the export of entropy.

Eukaryotes: One of the three kingdoms of life. The eukaryotes have morphologically complex cells and organelles derived from **endosymbiosis**. The major groups of photosynthetic eukaryotes are the algae and the plants.

Fixation: A chemical process by which an element is made useful to life. **Photosynthesis** fixes carbon by reducing carbon dioxide to organic carbon. Nitrogen fixation turns nitrogen gas into nitrates or ammonia; it is carried out by various forms of bacteria, and increasingly by human industry, by means of the Haber-Bosch process used to make nitrate fertilisers.

Gaia: The ancient Greek word for the world, now used in various senses to express the idea that the earth is, in some ways, akin to a living organism or itself alive. In the work of James Lovelock and others, it is useful to distinguish between the "Gaia hypothesis" of the 1970s, which held that the living parts of the earth regulated the earth system to their own benefit, and "Gaia theory" of the 1980s and onwards, a framework for trying to understand how the earth system manages to stabilise itself in specific states, or oscillates between a small range of comparatively fixed conditions – such as the glacial and interglacial states of the past three million years.

Geological time: The earth is about 4.5 billion years old. Its history is divided into four eons: The **Hadean**, from the beginning to 3,800 million years ago; the **Archaean**, from about 3,800 million years ago to 2,500 million years ago; the **Proterozoic**, from 2,500 million years ago to 543 million years ago; and the Phanerozoic, from 543 million years ago to the present.

The Phanerozoic, marked by an abundance of complex life forms, is further divided into three eras and 11 or 12 periods, depending on convention. Those which play the greatest role in this book are two periods of the upper Palaeozoic era: the **Devonian**, starting 416 million years ago and the **Carboniferous**, starting 359 million years ago and ending beginning 299 million years ago. These are the periods in which land plants started to take a dominant role in the **carbon cycle**.

The current epoch of geological time is increasingly frequently referred to as the Anthropocene, to denote that it is dominated by the activities of human beings.

Hadean: The first eon in **geological time**. It lasted from 4.5 billion years ago to 3.8 billion years ago, and was dominated, at least at its beginning and end, by the massive and destructive impacts of large bodies such as comets and asteroids.

Mitochondria: the specialised structures in eukaryotes in which respiration takes place. Like chloroplasts, they are derived from **endosymbiosis**.

Nucleus: In eukaryotes, the nucleus is where the main genome is stored, though **chloroplasts** and **mitochondria** have small genomes of their own.

Oxidation: A chemical reaction in which electrons are removed from a substance; the necessary counterpart of reduction, in which electrons are added (see **Redox reaction**).

Plants, **algae** and **cyanobacteria** oxidise water in order to access electrons for photosynthesis; since this produces oxygen, the process is called **oxygenic photosynthesis**. **Anoxygenic photosynthesis**, as practised by various bacteria, uses other sources.

Oxygenic photosynthesis: Photosynthesis in which electrons are released from water, and oxygen is thus released. This is the form of photosynthesis found in **cyanobacteria**, **algae** and in plants.

Photon: a packet of light, as described in quantum mechanics. The amount of energy in a photon depends on its wavelength: longer wavelengths – those from the redder end of the spectrum – contain less energy per photon than shorter ones at the bluer end.

Photorespiration: A reaction between organic carbon and oxygen catalysed by **rubisco**. Unlike "normal" **respiration**, it uses up cellular energy, rather than providing more of it, and is thus a source of inefficiency.

When functioning normally rubisco produces two molecules of phosphoglycerate from one molecule of carbon dioxide and one of ribulose biphosphate, a sugar molecule containing five carbon atoms: this is the key reaction of the **Calvin/Benson cycle**. However, rubisco can also catalyse a reaction between oxygen and ribulose biphosphate which produces one molecule of phosphoglycerate and one molecule of phosphoglycolate. This two-carbon-atom molecule is of no use to the Calvin/Benson cycle, and must be converted back into phosphoglycerate through a complex set of reactions.

Photosynthesis: the process by which sunlight is used to generate energy and fix carbon. It relies on **electron transfer chains** within cells, and may or may not produce oxygen. In plants, **algae** and **cyanobacteria** the fixation of carbon dioxide is achieved thanks to the **Calvin/Benson cycle**.

Photosystem: A complex of protein and pigment (notably chlorophyll) in a biological membrane which uses incoming light to remove electrons from a source, thus oxidising it, and passing them on to a receptor.

In plants, algae and cyanobacteria there are two distinct photosystems: Photosystem II, which confusingly comes first, uses light energy to take electrons from water and feed them into an **electron transfer chain** which takes them to photosystem I. Photosystem I uses light energy to pass the electrons on to ferrodoxin, which in turn passes them to the **Calvin/Benson cycle**.

Proterozoic: The third and longest eon in the history of the earth (see **geological time**). It lasted from 2.5 billion years ago to 543 million years ago. Its beginning and ending were marked by severe climatic excursions featuring widespread glaciation. In between it was curiously uneventful, with its middle years sometimes characterised as the "boring billion".

Redox reaction: A reaction in which electrons are transferred from one substance to another – the first being oxidised, the second reduced.

Reduction: A chemical reaction in which electrons are added to a substance (often along with hydrogen ions); the necessary counterpart of oxidation, in which electrons are removed.

Respiration: a biological oxidation of organic matter in which the electrons removed generate **ATP** by way of an **electron transfer chain**. In eukaryotes, respiration is carried out in the **mitochondria**.

Rubisco: the enzyme that plays the central role in the **Calvin/Benson cycle**. When functioning normally rubisco produces two molecules of phosphoglycerate from one molecule of carbon dioxide and one of ribulose biphosphate, a sugar molecule containing five carbon atoms. It can also catalyse a reaction between oxygen and ribulose biphosphate which produces one molecule of phosphoglycerate and one molecule of phosphoglycolate, known as **photorespiration**.

Stomata: the pores in the outer layers of land plants which let carbon dioxide in and water out.

Z-scheme: The realisation that the flow of electrons in **oxygenic photosynthesis** requires an infusion of energy from light at two different stages was given compelling graphical representation in a zig-zagging diagram developed by Robin Hill, of Cambridge, and became known as the Z-scheme. The key point of the scheme is that **photosystem** *II and* **photosystem** I must work in series.

BIBLIOGRAPHY

Adams, J. M. *et al.* (1990) 'Increases in terrestrial carbon storage from the last glacial maximum to the present' *Nature* **348** 711–714

Ainsworth, Elizabeth A. and Long, Stephen P. (2005) 'What have we learned from 15 years of free-air CO_2 enrichment (FACE)? A meta-analytic review of the responses of photosynthesis, canopy properties and plant production to rising CO_2' *New Phytologist* **165** 351–372

Allchin, Douglas (1994) 'James Hutton and phlogiston' *Annals of Science* **51** 615–635

Allen, John F. (2002) 'Photosynthesis of ATP – Electrons, proton pumps, rotors and poise' *Cell* **110** 273–276

Allen, John F. and Martin, William (2007) 'Out of thin air' *Nature* **445** 610–612

Anbar, Ariel and Knoll, Andrew H. (2002) 'Proterozoic ocean chemistry and evolution: A bioinorganic bridge?' *Science* **297** 1137–1142

Arnold, William (1991) 'Experiments' *Photosynthesis Research* **27** 73–82

Arnon, Daniel I. (1955) 'The chloroplast as a complete photosynthetic unit' *Science* **122** 9–16

Arnon, Daniel I. (1984) 'The discovery of photosynthetic phosphorylation' *Trends in Biochemical Sciences* **9** 258–262

Arnon, Daniel I. (1987) 'Photosynthetic CO_2 assimilation by chloroplasts: Assertion, refutation, discovery' *Trends in Biochemical Sciences* **12** 39–42

Arnon, Daniel I. (1991) 'Photosynthetic electron transport: Emergence of a concept, 1949–59' *Photosynthesis Research* **29** 117–131

Atkins, P. W. (1984) *The second law: Energy, chaos and form* (revised edition 1994) Scientific American Library/W. H. Freeman

Barber, James (2004) 'Engine of life and big bang of evolution: a personal perspective' *Photosynthesis Research* 80 137–155

Bassham, J. A. *et al.* (1954) 'The path of carbon in photosynthesis XXI: The cyclic regeneration of carbon dioxide acceptor' *Journal of the American Chemical Society* 76 1760–1770

Baxter, Stephen (2003) *Revolutions in the earth: James Hutton and the true age of the world* Weidenfeld and Nicolson

Beerling, David J. (2007) *The emerald planet: How plants changed Earth's history* Oxford University Press

Beerling, David J. and Berner, Robert A. (2005) 'Feedbacks and the coevolution of plants and atmospheric CO_2' *Proceedings of the National Academy of Sciences* 102 1302–1305

Beerling, David J., Osborne, Colin P. and Chaloner, William G. (2001) 'Evolution of leaf-form in land plants linked to atmospheric CO_2 decline in late Paleozoic era' *Nature* 410 352–354

Bendall, Derek (1994) 'Robert Hill' *Biographical Memoirs of Fellows of the Royal Society* 40 141–171

Benemann, John R. *et al.* (1973) 'Hydrogen evolution by a chloroplast-ferredoxin-hydrogenase system' *Proceedings of the National Academy of Sciences* 70 2317–2320

Benemann, John R. *et al.* (2005) 'A Novel Photobiological Hydrogen Production Process' in van der Est and Bruce (2005)

Benson, Andrew A. (2002a) 'Following the path of carbon in photosynthesis: A personal story' *Photosynthesis Research* 73 29–49, 2002 (also in Govindjee *et al.* 2005)

Benson, Andrew A. (2002b) 'Paving the path' *Annual Review of Plant Biology* 53 1–25

Bernal, J. Desmond (1951) *The physical basis of life* Routledge & Kegan Paul

Berner, Robert A. (1998) 'The carbon cycle and CO_2 over Phanerozoic time: the role of land plants' *Philosophical Transactions of the Royal Society of London B* 353 75–82

Berner, Robert A. *et al.* (2003) 'Phanerozoic Atmospheric Oxygen' *Annual Review of Earth and Planetary Science* 31 105–134

Berner, Robert A. and Kothavala, Zavareth (2001) 'Geocarb III: A revised model of atmospheric CO_2 over Phanerozoic time' *American Journal of Science* 301 182–204

Berner, Robert A., Lasaga, Antonio C. and Garrels, Robert M. (1983) 'The carbonate-silicate geochemical cycle and its effect on atmospheric carbon dioxide over the past 100 million years' *American Journal of Science* 283 641–683

Betts, R. A. *et al.* (2004) 'The role of ecosystem-atmosphere interactions in simulated Amazonian precipitation decrease and forest dieback under global climate warming' *Theoretical and Applied Climatology* doi 10.1007/s00704-004-0050-y

Blankenship, Robert E. (2002) *Molecular mechanisms of photosynthesis* Blackwell Science

Blankenship, Robert E. and Hartman, Hyman (1998) 'The origin and evolution of oxygenic photosynthesis' *Trends in Biological Science* 23 94–97

Bloom, Arnold J. *et al.* (2002) 'Nitrogen assimilation and growth of wheat under elevated carbon dioxide' *Proceedings of the National Academy of Sciences* 99 1730–1735

Bond, W. J., Midgley, G. F. and Woodward, F. I. (2003) 'The importance of low atmospheric CO_2 and fire in promoting the spread of grasslands and savannas' *Global Change Biology* 9 973–982

Bond, W. J., Woodward, F. I. and Midgley, G. F. (2005) 'The global distribution of ecosystems in a world without fire' *New Phytologist* 165 525–538

Brandon, Peter (1999) *The South Downs* Phillimore

Brocks, Jochen J. *et al.* (1999) 'Archaean molecular fossils and the early rise of eukaryotes' *Science* 285 1033–1036

Butterfield, Nicholas J. (2004) 'A vaucheriacean alga from the middle Neoproterozoic of Spitsbergen: implications for the evolution of Proterozoic eukaryotes and the Cambrian explosion' *Paleobiology* 30 231–252

Caldeira, Ken and Kasting, James F. (1992) 'The life span of the biosphere revisited' *Nature* 360 721–723

Calvin, Melvin (1989) 'Forty years of photosynthesis and related activities' *Photosynthesis Research* 21 3–16

Canfield, Donald E. (1998) 'A new model for Proterozoic ocean chemistry' *Nature* 396 450–453

Caroff, Lawrence I. and Des Marais, David J., eds (2000) *Pale blue dot 2 Workshop: habitable and inhabited worlds beyond our own solar system* NASA/CP–2000-209595

Catling, David *et al.* (2005) 'Why O$_2$ is required by complex life on habitable planets and the concept of planetary "oxygenation time"' *Astrobiology* 5 415–438

Catling, David, Zahnle, Kevin J. and McKay, Christopher P. (2001) 'Biogenic methane, hydrogen escape, and the irreversible oxidation of early earth' *Science* 293 839–843

Claire, Mark W., Catling, David C. and Zahnle, Kevin J. (2006) 'Biogeochemical modelling of the rise in atmospheric oxygen' *Geobiology* 4 239–269

Clayton, Roderick (1988) 'Memories of many lives' *Photosynthesis Research* 19 205–224

Conway, Gordon (1997) *The doubly green revolution* Penguin

Conway Morris, Simon (2003) *Life's solution: Inevitable humans in a lonely cosmos* Cambridge University Press

Crowther, J. G. (1962) *Scientists of the industrial revolution* Cresset Press

Davis, Nuel Pharr (1986) *Lawrence and Oppenheimer* Da Capo

Deisenhofer, J. *et al.* (1985) 'Structure of the protein subunits in the photosynthetic reaction centre of *Rhodopseudomonas viridis* at 3Å resolution' *Nature* 318 618–624

Department of Energy (2005) *Basic research needs for solar energy utilization: Report on the basic energy sciences workshop on solar energy utilization* (Chaired by Nathan S. Lewis)

Dick, Steven J. (1996) *The biological universe: The twentieth century extraterrestrial life debate and the limits of science* Cambridge University Press

Dick, Steven J. and Strick, James E. (2004) *The living universe: NASA and the development of astrobiology* Rutgers University Press

Duysens, L. N. M. (1989) 'The discovery of the two photosynthetic systems: a personal account' *Photosynthesis Research* 21 61–79

Dyson, George (2002) *Project Orion: The atomic spaceship 1957–1965* Allen Lane

Edwards, Diane, Duckett, J. G. and Richardson, J. B. (1995) 'Hepatic characters in the earliest land plants' *Nature* 374 635–636

Ehleringer, James R., Cerling, Thure E. and Dearing, M. Denise, eds (2005) *A history of atmospheric CO$_2$ and its effects on plants, animals and ecosystems* Springer

Falkowski, Paul G. and Raven, John A. (1997) *Aquatic photosynthesis* Blackwell

Feher, George (1998) 'Three decades of research in bacterial photosynthesis and the road leading to it: A personal account' *Photosynthesis Research* **55** 1–40

Feher, George (2002) 'My road to biophysics: Picking flowers on the way to photosynthesis' *Annual Review of Biophysics and Biomolecular Structures* **31** 1–44

Feher, George *et al.* (1989) 'Structure and function of bacterial photosynthetic reaction centres' *Nature* **339** 111–116

Ferreira, Kristina *et al.* (2004) 'Architecture of the photosynthetic oxygen-evolving center' *Science* **303** 1831–1838

Friedlingstein, P. *et al.* (2006) 'Climate-carbon cycle feedback analysis: Results from the C⁴MIP Model Intercomparison' *Journal of Climate* **19** 3337–3353

Fujishima, A. and Honda, K. (1972) 'Electrochemical Photolysis of Water at a Semiconductor Electrode' *Nature* **238** 37

Fuller, R. Clinton (1999) 'Forty years of microbial photosynthesis research: Where it came from and what it led to' *Photosynthesis Research* **62** 1–29

Galison, Peter and Hevly, Bruce, eds (1992) *Big science: The growth of large-scale research* Stanford University Press

Garwin, Richard L. and Charpak, George (2002) *Megawatts and megatons: The future of nuclear power and nuclear weapons* University of Chicago Press

Gedney, Nicola *et al.* (2006) 'Detection of a direct carbon dioxide effect in continental river runoff records' *Nature* **439** 835–838

Gest, Howard (1988) 'Sunbeams, cucumbers, and purple bacteria' *Photosynthesis Research* **19** 287–308

Gest, Howard (2000) 'Bicentenary homage to Dr Jan Ingen-Housz, MD (1730–1799), pioneer of photosynthesis research' *Photosynthesis Research* **63** 183–190

Glen, William (1994) *The mass-extinction debates: How science works in a crisis* Stanford University Press

Goldblatt, Colin, Lenton, Timothy and Watson, Andrew J. (2006) 'Bistability of atmospheric oxygen and the Great Oxidation' *Nature* **443** 683–686

Gordon, Line J. *et al.* (2005) 'Human modification of global water vapor flows from the land surface' *PNAS* **102** 7612–7617

Govindjee, Beatty, J. Thomas, Gest, Howard and Allen, John F. (2005)

Discoveries in photosynthesis (*Advances in Photosynthesis and Respiration,* Volume 20) Springer

Grätzel, Michael (2001) 'Photochemical cells' *Nature* **414** 338–344

Grinspoon, David (1997) *Venus revealed: A new look below the clouds of our mysterious twin planet* Basic Books

Harrison, Sandy P. and Prentice, Colin I. (2003) 'Climate and CO_2 controls on global vegetation distribution at the last glacial maximum: analysis based on palaeovegetation data, biome modelling and palaeoclimate simulations' *Global Change Biology* **9** 983–1004

Harvey, Graham (2001) *The forgiveness of nature: The story of grass* Jonathan Cape

Hatch, M. D. (1992) 'I can't believe my luck' *Photosynthesis Research* **33** 1–14

Hay, William W. *et al.* (2003) 'The late Cenozoic uplift-climate change paradox' *International Journal of Earth Science* **91** 746–774

Heaton, Emily, Voigt, Tom and Long, Stephen P. (2004) 'A quantitative review comparing the yields of two candidate C4 perennial biomass crops in relation to nitrogen, temperature and water' *Biomass and Energy* **27** 21–30

Hill, Robert (1939) 'Oxygen produced by isolated chloroplasts' *Proceedings of the Royal Society of London B* **127** 192–210

Hill, Robert (1965) 'The biochemists' green mansions: the photosynthetic electron-transport chain in plants' *Essays in Biochemistry* **1**, 121–151

Hill, Robert (1971) 'Joseph Priestley (1733–1804) and his discovery of photosynthesis in 1771' *Proceedings of the IInd International Congress on Photosynthesis*

Hill, Robert (1975) 'Days of visual spectroscopy' *Annual Review of Plant Physiology* **26** 1–11

Hill, Robert and Bendall, Fay (1960) 'Function of the two cytochrome components in chloroplasts: A working hypothesis' *Nature* **186** 136–137

Hinde, Thomas (1986) *Capability Brown: The story of a master gardener* Hutchinson

Hitchcock, Dian R. and Lovelock, James (1967) 'Life detection by atmospheric analysis' *Icarus* **7** 149–159

Herron, Helen Arnold (1996) 'About Bill Arnold, my father' *Photosynthesis Research* **48** 3–7

Hoffert, Martin I. *et al.* (2002) 'Advanced technology paths to global climate stability: Energy for a greenhouse planet' *Science* **298** 981–987

Imhoff, Marc L. *et al.* (2004) 'Global patterns in human consumption of net primary production' *Nature* **429** 870–873

IPCC (2007) *Intergovernmental Panel on Climate Change Working Group I Fourth Assessment Report*

Jagendorf, André T. (1998) 'Chance, luck and photosynthesis research: An inside story' *Photosynthesis Research* 57 215–229

Johnston, Harold (2003) *A bridge not attacked: Chemical warfare civilian research during World War II* World Scientific

Jones, Jean (1985) 'James Hutton's agricultural research and his life as a farmer' *Annals of Science* 42 573–601

Judson, Horace Freeland (1979) *The eighth day of creation: Makers of the revolution in biology* Jonathan Cape

Juretić, Davor and Županović, Paško (2005) 'The free-energy transduction and entropy production in initial photosynthetic reactions' in Kleidon and Lorenz (2005)

Kamen, Martin D. (1985) *Radiant science, dark politics: A memoir of the nuclear age* University of California Press

Kasting, James F. and Catling, David (2003) 'Evolution of a habitable planet' *Annual Review of Astronomy and Astrophysics* 41 429–463

Keeling, Charles D. (1998) 'Rewards and penalties of monitoring the earth' *Annual Review of Energy and the Environment* 23 25–82

Keilin, David, with Keilin, Joan (1966) *The history of cell respiration and cytochrome* Cambridge University Press

Kellogg, Elizabeth A. (2001) 'Evolutionary history of the grasses' *Plant Physiology* 125 1198–1205

Kenrick, Paul and Davis, Paul (2004) *Fossil plants* The Natural History Museum

Khan, Shahed U. M., Al-Shahry, Mofareh and Ingler, William B. Jr (2002) 'Efficient photochemical water splitting by a chemically modified n-TiO_2' *Science* **297** 2243–2245

Kirchner, J. W. (1991) 'The Gaia hypotheses: Are they testable? Are they useful?' in Schneider and Boston (1991)

Kirschvink, Joseph L. *et al.* (2000) 'Palaeoproterozoic snowball earth: Extreme climatic and geochemical global change and its biological consequences' *Proceedings of the National Academy of Sciences* **97** 1400–1405

Kleidon, Axel (2004) 'Beyond Gaia: Thermodynamics of life and earth system functioning' *Climatic Change* **66** 271–319

Kleidon, Axel and Lorenz, Ralph D. (2005) *Non-equilibrium thermodynamics and the production of entropy* Springer-Verlag

Knoll, Andrew H. (2003a) 'The geological consequences of evolution' *Geobiology* **1** 3–14

Knoll, Andrew H. (2003b) *Life on a young planet: The first three billion years of evolution on earth* Princeton University Press

Kohler, Robert E. (1982) *From medical chemistry to biochemistry: The making of a biomedical discipline* Cambridge University Press

Krebs, Hans, with Schmid, Roswitha (1981) *Otto Warburg: Cell Physiologist, Biochemist and Eccentric* (*trans.* Krebs and Martin) Clarendon Press

Kump, Lee R., Kasting, James F. and Crane, Robert G. (2004) *The Earth System* (second edition) Pearson/Prentice Hall

Kump, Lee R., Pavlov, Alexander and Arthur, Michael A. (2005) 'Massive release of hydrogen sulphide to the surface ocean and atmosphere during intervals of oceanic anoxia' *Geology* **33** 397–[CK]

Lane, Nick (2002) *Oxygen: The molecule that made the world* Oxford University Press

Lane, Nick (2005) *Power, sex, suicide: Mitochondria and the meaning of life* Oxford University Press

Larkin, Philip (1988) *Collected poems* (ed. A. Thwaite) Faber and Faber

Lenton, Timothy M. (1998) 'Gaia and natural selection' *Nature* **394** 439–447

Lenton, Timothy M. and von Bloh, Werner (2001) 'Biotic feedback extends the life span of the biosphere' *Geophysical Research Letters* **28** 1715–1718

Lenton, Timothy M. and Watson, Andrew J. (2004) 'Biological enhancement of weathering, atmospheric oxygen and carbon dioxide in the Neoproterozoic' *Geophysical Research Letters* **31** L05202 [sic]

Lewis, Nathan S. – see Department of Energy (2005)

Lodders, Katharina and Fegly, Bruce Jr (1998) *The planetary scientist's companion* Oxford University Press

Lovelock, James E. (1965) 'A physical basis for life detection experiments' *Nature* **207** 568–570

Lovelock, James E. (1979) *Gaia: A new look at life on earth* Oxford University Press

Lovelock, James E. (1988) *The ages of Gaia: A biography of our living earth* (revised edition 1995) Oxford University Press

Lovelock, James E. (2000) *Homage to Gaia: The life of an independent scientist* Oxford University Press

Lovelock, James E. (2006) *The revenge of Gaia: Why the earth is fighting back – and how we can still save humanity* Allen Lane

Lovelock, James E. and Margulis, Lynn (1974) 'Atmospheric homeostasis by and for the biosphere: the Gaia hypothesis' *Tellus* XXVI 2–9

Lovelock, James E. and Watson, A. J. (1982) 'The regulation of carbon-dioxide and climate: Gaia or geochemistry' *Planetary and Space Science* 30 795–802

Lovelock, James E. and Whitfield, M. (1982) 'Life span of the biosphere' *Nature* 296 561–563

Lunine, Jonathan I. (2004) *Astrobiology: A multidisciplinary approach* Pearson Addison Wesley

MacNeice, Louis (1976) *Collected poems* (ed. E. R. Dodds) Faber

Malone, Thomas F., Goldberg, Edward D. and Munk, Walter H. (1998) 'Roger Randall Dougan Revelle 1909–1991' *Biographical Memoirs of the National Academy of Sciences* 75 3–23

Mann, T. (1964) 'David Keilin' *Biographical Memoirs of Fellows of the Royal Society* 10 183–205

Margulis, Lynn (1998) *The symbiotic planet: A new look at evolution* Weidenfeld and Nicolson

Margulis, Lynn – see also under Sagan, Lynn

Marris, Emma (2006) 'Black is the new green' *Nature* 442 624–626

Martin, William and Russell, Michael J. (2002) 'On the origins of cells; a hypothesis for the evolutionary transitions from abiotic geochemistry to chemoautotrophic prokaryotes, and from prokaryotes to nucleated cells' *Philosophical Transactions of the Royal Society of London B* 358 59–85

Marvell, Andrew (1989) *Selected poems* (ed. Bill Hutchings) Carcanet

Matson, P. A. *et al.* (1997) 'Agricultural intensification and ecosystem properties' *Science* 277 504–509

McCannon, Jock (1987) *The Sick University* Lowlands University Press

McKay, Christopher P. and Hartman, Hyman (1991) 'Hydrogen peroxide and the evolution of oxygenic photosynthesis' *Origins of Life and Evolution of the bIosphere* 21 157–163

Melis, Anastasios and Happe, Thomas (2001) 'Hydrogen production. Green algae as a source of energy' *Plant Physiology* **127** 740–748

Mitchell, Peter (1961) 'Coupling of phosphorylation to electron and hydrogen transfer by a chemiosmotic type of mechanism' *Nature* **191** 144–148

Mitchell, Peter (1979) 'Keilin's respiratory chain concept and its chemiosmotic consequences' *Science* **206** 1148–1159

Morton, Oliver (2002) *Mapping Mars: Science, imagination and the birth of a world* Fourth Estate

Myers, Jack (1994) 'The 1932 experiments' *Photosynthesis research* **40** 303–310

Nash, Leonard K. (1952) *Plants and the atmosphere* Harvard University Press

Niklas, Karl J. (1997) *The evolutionary biology of plants* University of Chicago Press

Nisbet, Euan G. and Fowler, C. Mary R. (2005) 'The early history of life' in *Treatise on geochemistry vol 8* (ed. William H. Schlesinger) Elsevier Science

Oak Ridge National Laboratory (2005) *Biomass as feedstock for a bioenergy and bioproducts industry: The technical feasibility of a billion-ton annual supply*

Oldroyd, David G. (1996) *Thinking about the Earth: A history of ideas in geology* Harvard University Press

Ord, M. G. and Stocken, L. A., eds (1997) *Foundations of Modern Biochemistry Vol. 3* JAI Press

Osborne, Colin P. and Beerling, David J. (2006) 'Nature's green revolution: the remarkable evolutionary rise of C4 plants' *Philosophical Transactions of the Royal Society B* **361** 173–194

Pacala, S. and Socolow, R. (2004) 'Stabilization wedges: Solving the climate problem for the next 50 years with current technologies' *Science* **305** 968–972

Paltridge, Garth (2005) 'Stumbling into the MEP racket: an historical perspective' in Kleidon and Lorenz (2005)

Pavlov, Alexander A. *et al.* (2000) 'Greenhouse warming by CH_4 in the atmosphere of early earth' *Journal of Geophysical Research – Planets* **105** 11981–11990

Pearson, Paul N. and Palmer, Martin R. (2000) 'Atmospheric carbon dioxide concentrations over the past 60 million years' *Nature* **406** 695–699

Perlin, John (2002) *From space to earth: The story of solar electricity* Harvard University Press

Peterson, Kevin J. and Butterfield, Nicholas J. (2005) 'Origin of the Eumetazoa: Testing ecological prediction of molecular clocks against the Proterozoic fossil record' *Proceedings of the National Academy of Sciences* **102** 9547–9552

Petit, J. R. *et al.* (1999) 'Climate and atmospheric history of the past 420,000 years from the Vostok ice core, Antarctica' *Nature* **399** 429–436

Pielke, Roger A. Sr *et al.* (2002) 'The influence of land-use change and landscape dynamics on the climate system: relevance to climate-change policy beyond the radiative effect of greenhouse gases' *Philosophical Transactions of the Royal Society of London A* **360** 1705–1719

Pierpoint, W. S. (1999) 'Norman Wingate Pirie' *Biographical Memoirs of Fellows of the Royal Society* **45** 399–415

Pirie, N. W. (1937) 'The meaninglessness of the terms life and living' in *Perspectives in biochemistry* (ed. J. Needham) Cambridge University Press

Pirie, N. W. (1959) 'Chemical diversity and the origins of life' in Oparin *et al.* (1959)

Pirie, N. W. (1970) 'A nonconformist biologist' *New Scientist* 12 February 15–18

Rabinowitch, Eugene (1961) 'Robert Emerson' *Biographical Memoirs of the National Academy of Sciences* **35** 112–131

Rachmilevitch, Shimon, Cousins, Asaph B. and Bloom, Arnold (2004) 'Nitrate assimilation in plant shoots depends on photorespiration' *Proceedings of the National Academy of Sciences* **101** 11506–11510

Raven, John A. (2002) 'Selection pressures on stomatal evolution' *New Phytologist* **153** 371–386

Raven, John A. – see also Royal Society (2005b)

Raven, John A. and Edwards, Dianne (2001) 'Roots: Evolutionary origins and biochemical significance' *Journal of Experimental Botany* **52** 381–401

Raven, Peter H., Evert, Ray F. and Eichhorn, Susan E. (1999) *Biology of plants* (sixth edition) W. H. Freeman

Raymond, Jason *et al.* (2004) 'The natural history of nitrogen fixation' *Molecular Biology and Evolution* **21** 541–554

Read, Peter and Lermit, Jonathan (2005) 'Bio-energy with carbon storage

(BECS): a sequential decision approach to the threat of abrupt climate change' *Energy* **30** 2654–2671

Revelle, Roger and Suess, Hans E. (1957) 'Carbon dioxide exchange between atmosphere and ocean and the question of an increase of atmospheric CO_2 during the past decades' *Tellus* **9** 18–27

Ridley, Mark (2000) *Mendel's demon: Gene justice and the complexity of life* Weidenfeld and Nicolson

Robinson, Jennifer M. (1991) 'Phanerozoic atmospheric reconstructions: a terrestrial perspective' *Palaeogeography, Palaeoclimatology, Palaeoecology* **97** 51–62

Robinson, Spider (1986) *Night of Power* Berkley

Rojstaczer, Stuart, Sterling, Shannon M. and Moore, Nathan J. (2001) 'Human appropriation of photosynthesis products' *Science* **294** 2549–2552

Royal Society (2005a) 'Food crops in a changing climate: Report of a Royal Society Discussion Meeting held in April 2005' (Policy document 10/05)

Royal Society (2005b) 'Ocean acidification due to increasing atmospheric carbon dioxide' (Policy document 12/05) (Chaired by John A. Raven)

Royer, Dana L., Berner, Robert A. and Beerling, David J. (2001) 'Phanerozoic atmospheric CO_2 change: evaluating geochemical and paleobiological approaches' *Earth-Science Reviews* **54** 349–392

Rutherford, A. W. and Faller, P. (2002) 'Photosystem II: Evolutionary perspectives' *Philosophical Transactions of the Royal Society of London B* **358** 245–253

Ryman, Geoff (1989) *The Child Garden* Tor Books

Sagan, Lynn (1967) 'On the origin of mitosing cells' *Journal of Theoretical Biology* **14** 225–274 (see also Margulis, Lynn)

Sage, Rowan F. (1995) 'Was low atmospheric CO_2 during the Pleistocene a limiting facor for the origin of agriculture' *Global Change Biology* **1** 93–106

Sage, Rowan F. (2004) 'The evolution of C4 photosynthesis' *New Phytologist* **161** 341–370

Sawyer, Kathy (2006) *The Rock From Mars: A detective story on two planets* Random House

Schneider, Stephen H. *et al.* (2004) *Scientists Debate Gaia: The next century* MIT Press

Schneider, Stephen H. and Boston, Penelope J. (1991) *Scientists on Gaia* MIT Press

Schopf, J. William (1999) *Cradle of life: The discovery of earth's earliest fossils* Princeton University Press

Schrödinger, Erwin (1944) *What is life?* (republished with *Mind and matter* and *Autobiographical sketches* 1992) Cambridge University Press

Schwartzmann, David W. and Volk, Tyler (1989) 'Biotic enhancement of weathering and the habitability of Earth' *Nature* 340 457–460

Scott, A. C. (2000) 'The pre-Quaternary history of fire' *Palaeogeography, Palaeoclimatology, Palaeoecology* 164 281–329

Seaborg, Glenn T. and Benson, Andrew A. (1998) 'Melvin Calvin' *Biographical Memoirs of the National Academy of Sciences* 75 96–115

Seidel, Robert (1992) 'The origins of the Lawrence Berkeley Laboratory' in Galison and Hevly (1992)

Skelton, Peter, Spicer, Bob and Rees, Allister (2001) *Evolving life and the earth (S269 Earth and life)* The Open University

Slater, E. C. (1994) 'Peter Dennis Mitchell' *Biographical Memoirs of Fellows of the Royal Society* 40 283–305

Smil, Vaclav (1999) *Energies: An illustrated guide to the biosphere and civilization* MIT Press

Smil, Vaclav (2001) *Enriching the earth: Fritz Haber, Carl Bosch and the transformation of world food production* MIT Press

Smil, Vaclav (2003a) *The earth's biosphere: Evolution, dynamics, and change* MIT Press

Smil, Vaclav (2003b) *Energy at the crossroads: Global perspectives and uncertainty* MIT Press

Spath, Susan B. (1999) 'C. B. van Niel and the culture of microbiology, 1920–1965' PhD dissertation, University of California, Berkeley

Spoehr, H. A. (1926) *Photosynthesis* American Chemical Society

Steinberg-Yfrach, Gali *et al.* (1998) 'Light-driven production of ATP catalysed by F_0F_1-ATP synthase in an artificial photosynthetic membrane' *Nature* 392 479–482

Tilman, David *et al.* (2001) 'Forecasting agriculturally driven global environmental change' *Science* 292 281–284

Tilman, David, Hill, Jason and Lehman, Clarence (2006) 'Carbon negative biofuels from low-input high-diversity grassland biomes' *Science* 314 1598–1600

Tolkien, J. R. R. (1964) *Tree and leaf* Allen & Unwin

Torn, Margaret S. and Harte, John (2006) 'Missing feedbacks, asymmetric uncertainties, and the underestimation of future warming' *Geophysical Research Letters* 33 L10703

Tudge, Colin (2003) *So shall we reap: How everyone who is liable to be born in the next ten thousand years could eat very well indeed; and why, in practice, our immediate descendants are likely to be in serious trouble* Allen Lane

Tudge, Colin (2006) *The tree: A natural history of what trees are, how they live and why they matter* Crown

Turner, Roger (1985) *Capability Brown and the eighteenth-century English landscape* Weidenfeld and Nicolson

Turner, Steven R. (1970) 'Julius Robert Mayer' in *The Dictionary of Scientific Biography* Scribner

Uglow, Jenny (2002) *The lunar men: The friends who made the future 1730–1810* Faber and Faber

Vaitheeswaran, Vijay V. (2005) *Power to the people: How the coming energy revolution will transform an industry, change our lives and maybe even save the planet* Earthscan

van der Est, A. and Bruce, D., eds (2005) *Photosynthesis: Fundamental aspects to global perspectives* (Proceedings of the XIII International Congress, Montreal) Alliance Communications Group

Varley, John (1992) *Steel Beach* Ace

Varley, John (2004) *The John Varley reader: Thirty years of short fiction* Berkeley

Vernadsky, Vladimir I., (1998) *The biosphere* (trans. David Langmuir, ed. Mark McMenamin) Copernicus

Vitousek, Peter M. *et al.* (1986) 'Human appropriation of the products of photosynthesis' *Bioscience* 36 368–373

Volk, Tyler (1998) *Gaia's body: Toward a physiology of earth* Springer-Verlag

Wald, George (1974) 'Fitness in the universe: Choices and necessities' *Origins of Life* 5 7–27

Walker, David Alan (1997) '"Tell me where all the past years are"' *Photosynthesis Research* 51 1–26

Walker, David Alan (2000) *Like clockwork: An unfinished story* Oxygraphics (http://www.oxygraphics.co.uk)

Walker, David Alan (2002) '"And whose bright presence" – an appreciation of Robert Hill and his reaction' *Photosynthesis Research* 73 51–54

Walker, David Alan (2003) 'Chloroplasts in envelopes: CO_2 fixation by fully functional intact chloroplasts' *Photosynthesis Research* 76 319–327

Walker, Gabrielle (2003) *Snowball earth: The story of the great global catastrophe that spawned life as we know it* Bloomsbury

Walker, J. C. G., Hays, P. B. and Kasting, J. F. (1981) 'A negative feedback mechanism for the long-term stabilization of Earth's surface temperature' *Journal of Geophysical Research* 86 9776–9782

Ward, Peter D. and Brownlee, Donald (2000) *Rare earth: Why complex life is uncommon in the universe* Springer-Verlag

Ward, Peter D. and Brownlee, Donald (2002) *The life and death of planet earth: How the new science of astrobiology charts the ultimate fate of our world* Times Books/Henry Holt

Watson, Andrew J. (2004) 'Gaia and observer self-selection' in Schneider *et al.* (2004)

Watson, Andrew J. *et al.* (2000) 'Effect of iron supply on Southern Ocean CO_2 uptake and implications for glacial atmospheric CO_2' *Nature* 407 730–733

Weart, Spencer R. (2003) *The discovery of global warming* Harvard University Press

Weiner, Jonathan (1990) *The next one hundred years: Shaping the fate of our living earth* Rider

Whatley, F. Robert (1995) 'Photosynthesis by isolated chloroplasts: The early work in Berkeley' *Photosynthesis Research* 46 17–26

Wildman, Sam G. (1998) 'Discovery of rubisco' in Kung, S.-D. and Yang, S.-F., eds *Discoveries in plant biology* World Scientific

Wildman, Sam G. (2002) 'Along the trail from Fraction I protein to Rubisco (ribulose bisphosphate carboxylase-oxygenase)' *Photosynthetic Research* 73 243–250

Wilkinson, Bruce H. (2005) 'Humans as geological agents: A deep-time perspective' *Geology* 33 161–164

Wilkinson, Bruce H. and McElroy, Brandon J. (2007) 'The impact of humans on continental erosion and sedimentation' *GSA Bulletin* 119 140–156

Williams, Walter Jon (2003) 'The green leopard plague' *Isaac Asimov's Science Fiction Magazine* (October-November 2003)

Wilson, R. C. L., Drury, S. A. and Chapman, J. L. (2000) *The great ice age: Climate change and life* Routledge/The Open University

Witt, J. L. and Horst, Tobias (2004) 'Steps on the way to building blocks, topologies, crystals and X-ray structural analysis of Photosystems I and II of water oxidizing photosynthesis' *Photosynthesis Research* **80** 85–107

Wolstencroft, R. D. and Raven, John A. (2002) 'Photosynthesis: Likelihood of occurrence and possibility of detection on earth-like planets' *Icarus* **157** 535–548

Woodward, F. Ian (1987) 'Stomatal numbers are sensitive to increases in CO_2 from pre-industrial levels' *Nature* **327** 617–618

Xiong, Jin and Bauer, Carl E. (2002) 'Complex evolution of photosynthesis' *Annual Review of Plant Biology* **53** 503–521

Zachos, James *et al.* (2001) 'Trends, rhythms and aberrations in global climate 65 Ma to present' *Science* **292** 686–693

Zahnle, Kevin J., Claire, Mark W. and Catling, David C. (2006) 'The loss of mass independent fractionation in sulfur due to a Paleoproterozoic collapse of atmospheric methane' *Geobiology* **4** 271–283

Zallen, Doris T. (1993) 'Redrawing the boundaries of molecular biology: the case of photosynthesis' *Journal of the History of Biology* **26** 65–87

Zimmer, Carl (2001) *Parasite rex: Inside the bizarre world of nature's most dangerous creatures* Free Press

FURTHER READING

In general
The two best general texts I have found on photosynthesis are Falkowski and Raven (1997) and Blankenship (2002), but both are written for fairly advanced students. Walker (2000) is fun and approachable. Raven, Evert and Eichhorn (1999) is an excellent undergraduate-level textbook on plant science, and Kump, Kasting and Crane (2004) introduces earth system science at the same level. Weart (2003) is an excellent introduction to the history of global warming as a scientific concern.

Chapter One
Roger Revelle's life is recounted in Malone, Goldberg and Munk (1998) and the pithy summation of his nature is from Feher (2002); his 'great experiment' paper is Revelle and Suess (1957). David Keeling's work is beautifully described in Weiner (1990); his own account is Keeling (1998), and it is put into the context of the overall history of the science of global warming from Arrhenius on in Weart (2003). Andrew Benson has provided two accounts of his career, Benson (2002a) and (2002b); his character is also illuminated in oral history interviews with a number of alumni of Calvin's ORL conducted by Vivian Moses and kindly supplied to the author.

Martin Kamen's story is told in his autobiography, *Radiant Science, Dark Politics* (Kamen 1985), and additional insights into the Rad Lab are to be found in Davis (1986) and Seidel (1992). The account of Sam Ruben's death follows Johnston (2003). Melvin Calvin's contributions are discussed in Calvin (1989) and Seaborg and Benson (1998), and further accounts of the ORL are to be found in Fuller (1999) and Govindjee *et al.* (2005) –

the second of which, indeed, contains a wealth of further reading relevant to all three of the first chapters of this book. The key Calvin-Benson-cycle paper is Bassham *et al.* (1954). The identification of the role of rubisco is dealt with in Wildman (1998 and 2002).

Chapter Two

Mayer's life is sketched in Turner (1970); the Second Law as now understood is very nicely treated in Atkins (1994). For more on Vernadsky's view of the biosphere see Vernadsky (1998), and for the flow of biospheric energy as now understood see Smil (2003a).

Robin Hill's life is evoked in Bendall (1994) and in Walker (1997 and 2002). His 'Hill reaction' paper is Hill (1939) and his Z-scheme is Hill and Bendall (1960); he provides some personal notes in Hill (1965 and 1975). His papers are held in the Cambridge University Library, as are Peter Mitchell's. David Keilin's life is described in Mann (1964) and honoured in Mitchell (1979). Robert Kohler's first-rate history of biochemistry (Kohler, 1982) was my main source on Hopkins' magnificent department, though tea in its tea room helped too. For van Niel's work see Spath (1999). Daniel Arnon's accounts of his own role are to be found in Arnon (1984, 1987, 1991) and his group's key paper is Arnon (1955); Whatley's account is Whatley (1995) and some more of the isolated chloroplast story is to be found in Walker (2003). Mitchell's landmark paper is Mitchell (1961) and his Nobel address is Mitchell (1979); his life story is told in Slater (1994); acceptance of the chemiosmotic approach in the photosynthesis community is discussed in Jagendorf (1998).

Chapter Three

The proceedings of the Montreal congress are van der Est and Bruce (2005). Jim Barber recalls his research trajectory in Barber (2004), and his group's structure for photosystem II is found in Ferreira *et al.* (2004). The early experiments of Emerson and Arnold are brought to life in Myers (1994). The life of Otto Warburg is told with some sympathy in Krebs with Schmid (1981) while Emerson's life story is told with frank affection in Rabinowitch (1961); Arnold tells his own story in Arnold (1991).

Insights into the relationship between photosynthesis and molecular biology, and the role of Delbrück, are drawn from Zallen (1993), which goes some way to explaining why you will find no mention of photosynthesis in the defining popular history of molecular biology, Judson (1979). Biographical accounts of the development of the reaction centre concept can be found in Clayton (1988) and in Feher (1998 and 2002) – for La Jolla in the 1960s see Dyson (2002). The structure paper that won the Nobel prize is Deisenhofer *et al.* (1985) and a summation of the state of play by the late 1980s is Feher (1989). Lynn Margulis's seminal paper is Sagan (1967) and it is discussed in Margulis (1998). An evolutionary understanding of photosystem II is in Rutherford and Faller (2002). The Witt group's work is discussed in Witt and Horst (2004).

Chapter Four

Projects to build telescopes such as those discussed in this chapter include NASA's Terrestrial Planet Finder and the European Space Agency's Darwin, the currently glacial progress of which can be followed on their respective websites. Lovelock's original thinking on life detection is in Lovelock (1965) and Hitchcock and Lovelock (1967); the way in which it led to the 'Gaia hypothesis' of Lovelock and Margulis (1972) and Lovelock (1979) is discussed in Lovelock (2000). Dick (1996) is a magnificent historical source on all thinking about life beyond the earth, and the exobiology/astrobiology transition is a major theme of Dick and Strick (2004). For an inspiring textbook treatment of astrobiology see Lunine (2004).

Dick (1996) is also a fine source on the origin of life. For Bernal's view see Bernal (1951). For Pirie on the meaninglessness of 'life' see Pirie (1937): for *his* life see Pirie (1970) and Pierpoint (1999) and for his hourglass see Pirie (1959). For Russell's geochemical view of life's origins see Martin and Russell (2002).

By far the best guide to the early history of life on earth is Knoll (2003). Early fossils are discussed in Schopf (1999), but see also Sawyer (2006). The evolution of photosynthesis is treated in Blankenship (2002) and more recently in Allen and Martin (2007); see also Xiong and Bauer (2002), Nisbet and Fowler (2005) and McKay and Hartman (1991). Those intrigued by discussions of other planets should consult Grinspoon (1997) on Venus; Morton (2002) deals with some aspects of life on Mars, though it predates the discovery of signs of methane in the atmosphere.

An excellent overview of the issues surrounding the timing and implications of the Great Oxidation is Kasting and Catling (2003). The evidence for cyanobacteria and local oxygen production 300 million years before the Great Oxidation is Brocks *et al.* (1999). The idea of methane-mediated hydrogen escape as a key oxidation mechanism is in Catling *et al.* (2001); the relationship of sulphate to methanogenesis is in Claire, Catling and Zahnle (2006) and Zahnle, Claire and Catling (2006) and the stabilizing role of an ozone layer is in Goldblatt *et al.* (2006). The centrality of oxygen to much of life (and death) on earth is wonderfully dealt with at book length in Lane (2002) and David Catling's specific ideas on the necessity of oxygen for biological complexity are in Catling *et al.* (2005). Ideas about the likelihood or otherwise of oxygen-based biospheres in the universe at large are discussed in Ward and Brownlee (2000) and Caroff and Des Marais (2000). The need for a three-photosystem W-scheme around red dwarfs is mentioned in Wolstencroft and Raven (2002).

Chapter Five

The debate surrounding Raup's impact hypothesis is captured in Glen (1994), though the transatlantic cultural aspects are not explored. The carbonate-weathering thermostat was first described in Walker, Hays and Kasting (1981) and Berner, Lasaga and Garrels (1983). The science of snowball earths, mostly those of the late Proterozoic, is brought to life in Walker (2003); for the snowball attendant on the Great Oxidation see Kirschvink *et al.* (2000). The ideas on endosymbiosis, and particularly the importance of mitochondria for large cells, draw on another terrific book by Nick Lane, Lane (2005); see also Ridley (2000) and Margulis (2002). What a chloroplast needs genes for is analysed in Allen (2003).

For the Canfield ocean see Canfield (1998) and for its possible effects on life see Anbar and Knoll (2002). Nicholas Butterfield's ideas about animals are found in Butterfield (2004) and Peterson and Butterfield (2005). Speculations about fungi and phosphates are in Lenton and Watson (2004). For the development of land plants see first Kenrick and Davis (2004), and for a more theoretical approach Niklas (1997); further details in Edwards, Duckett and Richardson (1995) and Raven and Edwards (2001).

Chapter Six

The feedbacks between land plants and climate in the Devonian and subsequent periods are dealt with fascinatingly in Skelton, Spicer and Rees (2001), with many summarized graphically in Beerling and Berner (2005). David Beerling's recent book, Beerling (2007), tells some of the stories in this chapter and the next, and more besides, from the point of view of a key participant. Life's takeover of the thermostat is mooted in Lovelock and Watson (1982) – see also Schwartzmann and Volk (1989). The story about CO_2 and leaves is from Beerling, Osborne and Chaloner (2001). For Bob Berner's Geocarb model and its relevance to this issue see Berner and Kothavala (2001) and Berner (1998). A fine overview of the debate about the oxygen spike is in Lane (2002), and it is treated in detail in Berner *et al.* (2003). On the return of the Canfield ocean see Kump, Pavlov and Arthur (2005).

The proceedings of the Valencia meeting on Gaia are from Schneider *et al.* (2004), in which all sorts of debates on the nature of Gaia are thrashed out. For backstory of sorts see Schneider and Boston (1991), the proceedings of the American Geophysical Union's previous meeting on the subject. Lenton's views are laid out in Lenton (1998), Watson's in Watson (2004) and Kleidon's in Kleidon (2004b). See also Volk (1998). For much more on maximum entropy, see Kleidon and Lorenz (2005).

Chapter Seven

For more about the landscape, history and charms of the South Downs see Brandon (1999). Harvey (2001) is a wonderful history of and meditation on grass, by no less an authority than the agricultural story editor of *The Archers*; a technical account of grass evolution is Kellogg (2001). The physiology of C4 plants is reviewed in Sage (2004), and some of the story of the pathway's discovery is in Hatch (2002). Carbon-dioxide levels and climate since the Cretaceous are the topic of Pearson and Palmer (2000) and Zachos *et al.* (2001); there is more in Ehleringer, Cerling and Dearing (2005). The 'world without fire' is found in Bond, Woodward and Midgley (2005) – see also Bond, Midgley and Woodward (2003) and, if fire is your big thing, Scott (2000) – and an account of fire and the expansion of C4 grassland is found in Osborne and Beerling (2006) and Beerling (2007). Bill Hay's ideas are to be found in Hay *et al.* (2003).

Wilson, Drury and Chapman (2000) provides a fine overview of the ice ages. The details of the Vostok core are reported in Petit *et al.* (1999) and a field test of the iron-fertilization hypothesis is decribed in Watson *et al.* (2000). The carbon dioxide-starved biosphere of the ice ages is described in Adams *et al.* (1990) and Harrison and Prentice (2003), and a post-glacial increase in CO_2 as a precondition for agriculture is discussed in Sage

(1995). The lifespan of the biosphere is counted out in Lovelock and Whitfield (1982), Caldeira and Kasting (1992) and Lenton and von Bloh (2001), and the whole issue of the earth's long-term future is dealt with in Ward and Brownlee (2002). 'Gotta sing, gotta dance', the story that introduced John Varley's space-dwelling symbionts, can be found in Varley (2004).

Chapter Eight

Capability Brown's philosophy is discussed in Turner (1985) and his work at Bowood in Hinde (1986). On Joseph Priestley and his circle I have followed Jenny Uglow's magnificent account (Uglow 2002), but see also Crowther (1962). The development of Priestley's ideas and those of others on the same subjects are laid out in detail in Nash (1952); Robin Hill's account of Priestley is Hill (1971), and on Ingenhousz see Gest (2000). Hutton's life story is told in Baxter (2002); his agricultural interests are in Jones (1985) and his thoughts on phlogiston in Allchin (1994).

The findings that raised up the hairs on the back of Ian Woodward's neck are in Woodward (1987). In everything written about terrestrial energy flows I am indebted to Vaclav Smil, in particular Smil (2003a), and on nitrogen enrichment Smil (2001). On sedimentation see Wilkinson (2005) and Wilkinson and McElroy (2007). On historical agricultural intensification see Matson et al. (1997), and on human appropriation of net primary productivity see Vitousek et al. (1986), Rojstaczer, Sterling and Moore (2001) and Imhoff et al. (2004). On vapour flows and their impacts, see Gordon et al. (2005) and Pielke et al. (2002).

An overview of FACE experiments is to be found in Ainsworth and Long (2005). On the future of agriculture in general, see Tudge (2003) and Conway (1977), and on its impact on the future environment see Tilman et al. (2001); possibly damaging interactions between climate and agriculture are explored in Royal Society (2005a). Arnold Bloom's ideas about nitrogen availability are laid out in Bloom et al. (2002) and Rachmilevitch, Cousins and Bloom (2004). The current state of play in climate change science is exhaustively detailed in IPCC (2007); the sections on biogeochemistry and palaeoclimate are particularly relevant here. For a much more apocalyptic view, see Lovelock (2006). The comparison between models that shows a likely positive feedback on climate from the biosphere is Friedlingstein et al. (2006). The effects of decreased stomatal conductance

441

on twentieth-century rivers is from Gedney *et al.* (2006) and the die-off of the Amazon is from Betts *et al.* (2004). The dangers of ocean acidification are adumbrated in Royal Society (2005b).

Chapter Nine

Smil (2003b) is probably the surest guide to contemporary energy issues; a more engaged and up-beat reportorial account is in Vaitheeswaran (2005). The wedges made their first appearance in Pacala and Socolow (2004). A very helpful broad-brush approach to long-term energy issues is Hoffert *et al.* (2002), and the best introduction to nuclear energy is Garwin and Charpak (2002). Oak Ridge (2005) covers ambitious plans for biomass, and the charms of miscanthus are spelled out in Heaton, Voigt and Long (2004); an introduction to Peter Read's thinking is in Read and Lermit (2004), the potential of Terra Preta is discussed in Marris (2006) and the prairie approach to carbon-negative biomass can be found in Tilman, Hill and Lehman (2006). An appealing history of photovoltaic cells is Perlin (2002).

The whole range of future technological approaches to the conversion of sunlight into electricity or fuel is discussed in depth in Department of Energy (2005). The original biohydrogen paper is Benemann *et al.* (1972); for Melis see Melis and Happe (2001) and for Benemann's latest see Benemann *et al.* (2005). The first titanium-dioxide photochemical cell is described in Fujishima and Honda (1972), with work brought up to date in Khan, Al-Shahry and Ingler (2002); the use of titanium dioxide and dyes in solar cells is discussed in Grätzel (2001). A significant milestone in Tom Moore's work is Steinberg-Yfrach, Gali *et al.* (1998).

ACKNOWLEDGEMENTS

If this book were a who's who of photosynthetic research, rather than an attempt to tell a number of stories, the absence from its main pages of John Raven and Paul Falkowski would be outrageous, rather than merely surprising. Let me go some way towards rectifying things by naming them first among the many people who were generous with their time, insights and memories in making this book come about. I am also particularly grateful to Andrew Benson, Derek Bendall, Robert and Jean Whatley, George Feher, Jim Barber, John Allen, Jim and Sandy Lovelock, David Catling, Jim Kasting, Euan Nisbet, David Des Marais, Robert Blankenship, Andrew Knoll, Nick Butterfield, Bob Spicer, Paul Kenwick, Robert Berner, David Beerling, Andrew Watson, Tim Lenton, Axel Kleidon, Rowan Sage, Ian Woodward, Stephen Long, Robert Socolow, John Benneman, Tom Moore and Tasios Melis. Govindjee's tireless work getting his colleagues to record in *Photosynthesis Research* their memories of the field provided an unrivalled resource. Many of these people also read this book in part or in whole, for which further thanks. So did Nick Lane, John Morton, Roger Brent and Olivia Judson, and I am more grateful to them than it is easy to express while respecting the parsimony in the matter of adverbs that Olivia insists on.

I am also grateful for help, ideas and hospitality provided by John Kerridge, the late Dave Keeling, Victor Moses, Meredith Belbin and the Marshall family, Peter Rich, Christopher Howe, Jason Raymond, Nobby Gilmore, Howard Gest, Mike Russell, Bill Rutherford, Ally Aukauloo, Kevin Zahnle, Chris McKay, Janet Siefert, Sara Seager, Simon Schaffer, Peter Horton, Colin Prentice, Sandy Harrison, Colin Goldblatt, Ken Nealson, Ralph Lorenz, Des Lambert, Lynn Margulis, Tyler Volk, David

Schwarzmann, Stephan Harding, Larry Joseph, Sherry Stringfield, Stephen Schneider, Penny Boston (not least for the rose between her teeth), Arnold Bloom, Harvey Leifert, Chris Surridge, Peter Cox, Richard Betts, Peter Read, Stenbjörn Styring, Matt Ridley, Henry Gee, Vijay Vaitheeswaran, Mitz Strickland and Kate Fielden. The organisers of the Montreal congress, the AGU Valencia meeting and Jim and Sandy Lovelock's Dartington meeting provided stimulating venues in which to imbibe some understanding. I am grateful to the Mercer's Company and to the staff and inhabitants of Trinity Hospital, Greenwich, for providing me with a place to start writing this book, and a beautiful garden to look at as I did. Terry, who fixed the iBook's voodoo logic boards and other ills, saved me a world of hurt; so did the man who found it left on a train at Victoria station and looked after it. While some artists and writers are acknowledged in the text, others contributed ideas and approaches without such recognition: work by Garry Fabian Miller, Michael Craig-Martin, Olafur Eliasson, Heather Ackroyd and Dan Harvey provided particular inspiration. If my memory serves, I have an essay on Fabian Miller by Mark Haworth-Booth to thank for the title.

This book seemed to be over a long time before it was. Valuable insight into its improvement after the first draft came from my co-conversationalists at Villa d'Este in September 2005, and I am very grateful to Drue Heinz for bringing us together. Another source of improvement was the splendid array of colleagues who deal with manuscripts, write, commission, edit, subedit and even occasionally podcast at *Nature*. I have learned a great deal from them about how best to think about science and how best to express such thoughts; my thanks to Phil Campbell for giving me the chance to work with them.

Sarah Chalfant has been, again, the most supportive of agents, and Tracy Bohan too has been a vital help; at the Wylie agency thanks also to Andrew Woods and Edward Orloff. At 4th Estate, thanks to Mitzi Angel – who among many other things understands about Brussels childhoods – for her insight and patience, and to Robin Harvie for his excellent support. Thanks also to Gail Winston at HarperCollins. Richard Betts provided a very fine piece of copy editing. I am very grateful to Wes Fernandes for his work on the illustrations, and apologise to Roger for the fact that there aren't more.

My thanks to Faber and Faber and the estate of Philip Larkin for kind permission to use the lines from 'The Trees', to Richard Curtis for the lines from *Blackadder* and to Springer Science and Business Media for quotations from 'About Bill Arnold, my father' by Helen Arnold Herron, *Photosynthesis Research* **49** 3–7 (1996); from 'Three decades of research in bacterial photosynthesis and the road leading to it: A personal account' by George Feher, *Photosynthesis Research* **55**, 1–40 (1998); and from 'Forty years of icrobial photosynthesis research: Where it came from and what it led to' by Clinton R. Fuller, *Photosynthesis Research* **62** 1–29 (1999).

Less contractual thanks are due to: Tom, Kirstin, Ella and Miles; Anita, the aforementioned Simon and Eva; Julie, Matt, Kate and Jack; the aforementioned Matt, Anya, Matthew and Iris; among these, Jack, Kate, Eva and Matthew helped me experience a distributed spring through their excellent disposable-camera phenology. Thanks also to John and Kerry; Yo, Stephen and Reuben; and to my various siblings, nieces and nephews – those to whom I am linked by birth and those I have accumulated through marriage, both theirs and mine. Martha, Evan, Katrina and David taught me lessons about myself as a writer, which I fear came at their own expense. Francis Spufford, Jon Turney, Jon Rauch and Roz Kaveney all provided advice and fellowship, as did John Browning, Sean Geer and the formidable team of Simon Ings, Dan Brown, Mateo Willis, Stu Clarke and Peter Tallack. I wish I'd had Mark Legoy to lean on and share with; I feel his loss deeply. I am also much indebted to Michael Knight and Ian Simpson. My thanks to Danielle Carr-Gomm, her sons and the other friendly inhabitants of Wallands House, and to Michael Cockburn, Stephen Bland, Flo Larety and my cousin Simon Loft. Also Tai Chi Eddy. The Heilemann-Rhoten wedding and the Glasgow Worldcon provided fitting punctuations as, in his way, did John Harrison. More abstractly I would like to thank the Cretaceous, the ice ages and untold generations of farmers and sheep for the Downs, and Capability Brown and centuries of gardening staff for the cedar. These things do not just happen.

Nor do marriages, and my greatest debt of thanks is to Nancy Hynes, to whom this long, odd book is dedicated.

INDEX

Page numbers in *italics* denote graphs and tables.

Ages of Gaia: A Biography of Our Living Earth, The (Lovelock) 153, 184, 243–4
agriculture 74, 134, 279, 301, 302–3, 353–6, 387, 436
albedo xiv, 200, 258, 265, 282, 290, 293, 300, 308, 354, 367
algae 207, 234, 404
  antennae complexes 106, 399–400; brown 70, 71; chlorella 101–4, 115; effect of oxygen on carbon fixing ability 251; evolution of 211–12, 272; green 70, 208, 332; low-light growth 234, 234n, 235; red 70, 71, 208; respiration rates 399; role in discovery of photosynthesis 36, 38–9, 104 *see also* phytoplankton
Allen, John 172
Allen, Mary Belle 75, 76
alpha particles 25
Alvarez, Luis 24, 29, 195–6, 383
American Association for the Advancement of Science 44
American Chemical Society 335n
American Geophysical Union 255
American Rubber Company 110

Amsterdam declaration 256
anaerobes 180
Anbar, Ariel 211, 212, 213, 214
angiosperms 273–4, 285, 306
Annual Review of Plant Physiology 128n
Antarctica 6, 147, 283, 294, 298, 299, 392–3
antennae complexes:
  adjustable proteins 208; algae 106, 399–400; arrangements of 234; chlorophyll 116, 119, 120; controlling 399–400, 410; detachable 107; algal productivity and 399–400, 410; geometries 107; integral 106–7
anthropic principle 260–1, 339n
Anthropocene era 348
archaea 71, 169, 182–3, 213
Archaean era 160, 161, 165, 168, 171, 173, 175, 176, 177, 178, 179, 180, 181, 182, 183, 184, 188, 198, 200, 205, 210, 210n, 212, 213, 217, 238, 399
Archaeopteris 225–6, 225n
Arctic 179
Argentina 294

446

Arizona State University 402
Arnold, William 99–100, 101–4, 105, 107–9, 109n, 112, 114, 115, 115n, 116–17, 118, 119, 123n, 127, 135, 381, 404, 433
Arnon, Daniel 72–8, 80, 117, 140, 153, 213n, 397, 432
Arrhenius, Svante 6, 360, 431
asteroid impacts 195–6, 216, 244, 254
astrobiology 154–5, 189–90, 191, 197
AT&T's Bell Labs 123–4
atom 12
  behaviour of 13, 19, 29, 30, 31, 105; bomb 33, 109; splitting 19, 29, 109; structure of 13; radioactivity 16; uranium 29, 30, 31, 109
Atomic Energy Commission 35, 38n, 109
ATP (Adenosine triphosphate) 75–7, 78, 80, 81, 87, 88, 89–90, 91, 92, 116–18, 126, 165, 204, 205, 251, 285, 394, 398
Australian Geological Survey 177

bacteria 71, 166, 410–11
  chloroplast, evolution into 132, 205–10; earth's oxidization and 175–80, 190–1, 204–5, 207–8; evolution of 131–3, 205–10, 212–13; fixing of nitrogen 205, 212–13, 214, 224, 410; green sulphur 71, 171; photosynthetic 71, 72, 106; photosystems and evolution of 71–2, 131–2, 153, 168–94, 205–10; purple sulphur 71, 107, 118, 119, 130, 134, 170, 173, 174 see also cyanobacteria
Baekeland, Leo 16
Barber, Jim 97, 98–9, 127, 130, 131,

133, 134, 135, 136, 137–8, 138n, 432–3
BASF 352–3
Basho 281
Basin and Range (McPhee) 376
Bassham, Al 39, 40, 41, 47
Beck, Charles 225n
beech tree 222, 274–5
Beerling, David 240, 241, 242, 252, 253, 256, 287, 288, 304, 348, 435
Bendall, Derek 57, 79, 80, 83, 84
Bendall, Fay 80
Benemann, John 397, 400, 407
Benson, Andrew 7–9, 11–15, 16, 17, 21, 27, 29, 30, 31, 34, 35, 36, 39, 41, 43, 44–5, 46, 47, 100, 110, 141, 196, 284, 431
Benson, Walter 196
Bernal, John Desmond ('Sage') 128–9, 128n, 132, 158–9, 158n, 159, 164, 265–6, 433
Berner, Robert 238, 239, 240, 242, 245, 248, 249, 252, 253, 256, 282, 435
bibliography 413–30
Big Science 34
Bio Organic Chemistry Group 41, 44
biochemistry 9, 11, 41–2, 164
biodiesel 386, 400, 407
biomass energy 385–93, 399
biophysics 11, 123
biosphere 56, 90, 124, 161, 167, 168, 179, 261, 307, 371
biota 152, 166, 245, 250, 258
Birkbeck College 158n
Black, Joseph 327, 328, 338, 355
Bloom, Arnold 365
Boas, Franz 47
Bohr, Niels 96, 105, 120–1, 122
Boltzman, Ludwig 54, 55, 56, 57, 90, 121, 149, 158, 340
Boring Billion, the 210–18

Bosch, Carl 352–3
Boulton, Matthew 326, 329, 338, 348
Bové, Joseph-Marie 74n
Bowood House 319–24, 325, 330, 331, 334–5, 342, 343, 356, 371, 411
Brasier, Martin 211
Brown, Capability 319–20, 321–2, 323, 324, 335, 342, 347
bryophytes 223
Bulletin of the Atomic Scientists 109
Bulton, Matthew 408
bundle sheath cells 284, 286
burning earth 233, 247–50, 253, 289, 411, 436
Butterfield, Nick 215, 216, 217–18, 310

Caldeira, Ken 307, 308
Caltech (California Institute of Technology) 6, 13–14, 18, 35, 42, 99–100, 101, 102, 103, 108, 110, 118, 120, 135, 201
Calvin, Gen 36
Calvin, Melvin 34–40, 41, 42, 43–4, 45, 47, 60n, 137, 160, 176–7, 332, 431, 432
Calvin-Benson cycle 47, 51–2, 75, 81, 82, 83, 91, 102, 120, 125, 161, 251, 285, 286, 293, 397, 398, 409, 432
Cambrian era 211, 216, 219, 226, 232, 245
Cambridge University 19, 58–66, 73, 75, 79, 84, 113n, 156, 157, 159, 164, 215, 344–6, 432
Canfield, Don 211–12, 213, 238, 245, 248, 252, 435
carbon:
  Archaean cycle 178, 180–1; cycle, general xii, xiii, xiv, xix, 43, 178, 180–1, 199–200, 223, 238–40, 435; fixation of 11; Geocarb

(model of carbon cycle) 238–40, 435; isotopes 5–6, 17, 21–8, 31, 34–6, 37, 38, 43, 47; source of life 9; organic burial of 180–1, 190, 202, 217, 248, 252, 253, 254, 363; photosynthetic path 10, 11, 21–8, 34–40, 106, 169, 233; term 9n *see also* Calvin-Benson cycle *and* rubisco
Carbon-11 17, 21–3, 23n, 24, 26, 31
Carbon-12 17, 23, 24, 43, 161, 253
Carbon-13 17, 23–4, 43, 161, 178, 180
Carbon-14 5–6, 11, 23–8, 30, 31, 47, 109, 284, 383
carbonate weathering thermostat 199–201, 217, 236–54, 282–3, 304–8, 434
carbon dioxide:
  discovery of 10; greenhouse effects of 183–4; photosynthetic process, role in xii, 11, 21–8, 34–40, 106 *see also* Calvin-Benson cycle, photosystem II, rubisco *and* Z-scheme; ocean absorption of xiv, 6, 10, 298–9; sugar, transformation into xii, xiv, 10; weathering feedback loop and 199–200
carbon-dioxide levels:
  annual cycle xii–xiii, xvi, xix, 7; Archaean 200–1, 201–2; Carboniferous 281; Cretaceous 282–3; crops and 364–5; Devonian drop in 235–42, 285n; effect on earth's temperature *see* climate change; fossil fuel use and xv–xvi, xviii, xix, 6, 7, 11, 376–80, 381, 390, 392, 407; future projections 364, 377–80, *379*; human effect on 348–411; ice age 293, 296, 298; Jurassic 282–3; Keeling Curve *7*;

measuring 6–7, 345–8; Permian 281–2; Proterozoic 217; roots and 237; since Carboniferous 293–9, 300; stomata growth and 289, 302, 345–8; Triassic 282–3; volcanoes and 199

Carboniferous period 197, 219n, 227, 238, 239, 245, 246–50, 250n, 252, 253, 271, 281–2, 289

Carbon Mitigation Initiative 378

carboxylic acid 29, 31, 38

catalase 173, 174

Catling, David 182, 184, 186, 187, 191, 434

cedar 320–1, 322–3, 324, 325, 356, 371

cereal crops 364–5

Châtelier, Le 352

chemical weathering 238, 239, 240, 241–2, 244–5, 304–8

chemistry, organic 13–14, 15

chemiosmosis 88–90, 126, 432

chemistry, organic 9–10, 13, 41, 73, 352

Cheniae, George 95, 98

Chernobyl 382

Chicago Tribune 33, 34

Child Garden, The (Ryman) 312

chlorella 101–3, 104, 115

chlorophyll xi, 44, 60, 84, 91
  absorption of photons 90, 102, 104, 105, 107, 117–20; detecting on other worlds 189; fluorescence and 116, 117–20; oxidization of 118–19; pigment colour xi, xv, xvii, 10, 60, 69, 226; pigment pools xi, xii, xvi, xvii, 69, 86; photosynthetic unit and 44, 104–8, 113, 114; molecular structure of 41, *61*, 100, 141; molecules needed to produce one molecule of oxygen 104–5

chloroplast:
  ATP production *see* ATP; carbon dioxide levels and 283; evolution of 131–2, 203–10, 283; genes 133, 207, 208; hydrogen ion transport 87, 88, 91, 265; isolating 68–9, 72, 75, 80; membrane and 85–92, 114, 265; photorespiration and 252, 283, 284; production of oxygen 68–72

chromatogram 36, 37

Claire, Mark 184

Clayton, Roderick 118, 119–22, 125–6, 129, 139, 170, 433

Cleland, John 340

climate change 357–8
  carbon-dioxide levels and 358–71; feedback between land plants and 231–42; fossil fuels and xv–xvi, 376–80, 381, 390, 392, 407; history of 6–7, 231–42, 281–308, 431; ice ages 275, 277, 290–9, 300, 348, 360, 369, 436; interglacials 291, 296, 298, 304, 311; since Carboniferous 281–308; term 357–8

climate modelling 262–6

cloud condensation nuclei 369

coal xv, 273, 227, 246, 249, 250, 253, 259, 341, 342, 343, 360n, 376, 377, 380, 381, 382,390

coccolithophores 272, 273, 278, 369, 370

co-evolution 273–4, 277, 285

Columbia University 23

Colwell, Sidney 375

conifers 273, 274

Cooksey, Donald 33

Copernicus 259

Copley Medal 329, 333

coral 207

Cornell University 119

Crawford, Dorothy 128–9
Cretaceous era 195, 197, 253, 254, 272, 273, 276, 277, 278, 282, 304, 306, 348, 371
Crick, Francis 83n, 97, 120
cyclotrons, first 18–19, 21, 22, 25, 26–7, 28, 30, 33, 35, 37, 109, 160
cyanobacteria 71, 397, 399
  evolution of 172, 173, 175, 176, 177–8, 180, 190, 197, 198, 202, 203, 204–5, 206, 207–8, 209, 210, 212, 213, 217, 218, 234n, 259; lichen and 217; oxidization of earth and 175, 176, 177–8, 180, 190, 197, 198, 202, 203 *see also* bacteria
cytochromes 170, 409
  adjustments to 175, 208, 409, 410; b6 78, 80; definition of 62; electron-transfer pathways 63–5, 67, 86, 87, 88, 89, 91, 126, 140–1; entropy and 92; evolution of 171–2; f 76, 78, 80, 82–3; redox potentials and 63–5, 67; speed of 410; transfer of hydrogen ions 87, 88, 91, 265

Darwin, Charles 195, 198, 338
Darwin, Erasmus 326, 341–2, 391
Dawson, Sir J. W. 225n
de Hevesy, George 105n
de Montfort, Simon 280–1
de Saussure, Nicolas-Théodore 336–7
deep time 339
Deffeyes, Kenneth 376
deforestation 355, 362, 380
Deisenhofer, Johann 134
Delbrück, Max 108, 109n, 118, 120–1, 121n, 122, 433
Department of Agriculture (US) 110, 388
Department of Energy (US) 388

deuterons 17, 25
Devonian era 222, 224–5, 226, 227, 232–42, 244–5, 271, 285n, 289, 290, 435
Dewar, Roderick 264–5
dinoflagellates 272, 278
DNA:
  ATP and 75; chloroplast 207; discovery of 83; photosystems and 128, 165, 168, 169–70; structure of 97
Down House 195, 197
Duysens, Louis 82–3, 84, 114, 115, 118, 126, 134, 140
Dyer, Thomas 321

earth:
  albedo 200, 258, 265, 282, 290, 293, 300, 308, 354, 367; burning 233, 247–50, 253, 289, 411, 436; history of 160–5; hydrogen loss 182; late heavy bombardment 161–2; obliquity 292; orbit 292–3; oxidization of 175–85; photon emission 150–1, 200; self-regulation *see* Gaia; 'snowball' 201–3, 211, 216; temperature and thermostat 199–201, 217, 236–54, 282–3, 304–8, 434
'earth-system science' 255, 256–7
Eastman Kodak 16
Einstein, Albert 101, 101n, 108, 149, 352
electron microscope 85, 135
Electron paramagnetic resonance (EPR) spectrometers 124, 126
electron transfer chain xii, 63–5, 67, 75, 77–8, 77–8n, 82, 88, 90–1, 106, 107, 112, 117, 135, 170, 171, 185, 186, 208, 212, 299, 398
Elements of Agriculture (Hutton) 338
El-Niño events 355–6, 368

Emergency Rubber Project 110
Emerson, Robert 83, 84, 99–104,
    105, 107–8, 109–14, 115, 116–17,
    118, 123n, 127, 140, 358–9, 433
endosymbiosis 132, 133, 153, 205–8,
    209, 435
entropy 54–6, 80, 91–2, 121, 148,
    149, 150, 151, 152, 159, 262–6,
    308, 342, 408, 411
Epica 297
erosion 350 *see also* weathering
ethanol 386–7, 388–9
eukaryotes 70, 71, 85, 132, 166, 169,
    204, 205–6, 209, 210, 211, 212,
    213, 217, 218, 241, 266
Euramerica 239
European Synchrotron Radiation
    Facility (ESRF) 136
European Union 403, 404
excited states 106, 115
excitons 106, 107, 114, 116, 118, 119,
    126, 265
exobiology 149–51, 154, 156, 160,
    167, 186
*Experiments and Observations on
    Different Kinds of Air* (Priestley)
    330, 335n
Experiments Upon Vegetables
    (Ingenhousz) 333

Falkland Islands 294
feedback loops:
    forests and 231–67; negative 199,
    250, 252, 307; positive 184,
    231–67, 258, 288
Feher, George 1, 122–4, 123n, 125,
    126–7, 129, 130, 133, 153, 170,
    315, 431
fire 233, 247–50, 253, 289, 411, 436
fixed air 327–9
Fletcher, Walter 59
fluorescence 115–20

Following the Trail of Light (Calvin)
    44–5
forests 231–67, 280, 362, 368
Förster, Thomas 114
fossils 160–1, 195–228, 253, 310,
    371
fossil fuels:
    climate change and 376–80, 381,
    390, 392, 407; coal xv, 273, 227,
    246, 249, 250, 253, 259, 341, 342,
    343, 360n, 376, 377, 380, 381,
    382, 390; creation of xv; effect of
    human use xv–xvi, xviii, xix,
    376–80, 381, 390, 392, 407; oil
    xv, 177, 375–7, 390, 397; rates of
    human use xv
Franck, James 104–5, 106, 107, 109,
    111, 114, 115n, 139, 360n
Franklin, Benjamin 327, 327n, 329,
    333, 327n
Free-Air Carbon-dioxide Enrichment
    (FACE) 359–60, 363, 364–6
Frisch, Otto 108–9
fuel efficiency 379–80
Fujishima, Akira 400–1
Fuller, Clint 41n, 44–5
further reading 431–6

Gaffron, Hans 106, 108, 114, 121
Gaia 242–5, 433
    and earth's tendency towards
        stability 255–67; creation of
        148–9, 152–4, 155–6, 167, 168;
        first conference on, San Diego
        1988 255; oxygen levels and 249,
        255–67, 296, 308, 309, 314;
        planet cooling photosynthetic
        plankton and 369; positive
        feedbacks and 184, 233, 244, 248,
        255–67; second conference on,
        Valencia, 2002 255–6, 261, 262,
        435; weathering rates and 237

Gaia: A New Look at Life on Earth (Lovelock) 153, 154, 243, 249
'Garden, The' (Marvell) 1, 315
Gee, Henry 312
general circulation models (GCM) 360–2, 368
Geocarb 238–9, 240, 435
geocorona 182, 183
'geophysiology' 255
geothermal energy 384
Glynn House 85
Goldschmidt, Victor 100, 128
Goldsmiths College 195
Gondwana 239
Gran, Haaken Hasberg 293, 294
grasses 276–91, 301, 303, 436
Great Clarification 84, 98, 114
Great Oxidation Event 175–85, 186, 197, 198, 203, 205, 211, 233, 244, 310, 434, 435
'Green Leopard Plague, The' (Williams) 312
greenhouse effect 6, 150, 188, 201–2, 217, 290, 305, 356, 357, 366 *see also under* individual greenhouse gas
Greenland 361
Gulf of Mexico 354
Gulf Stream 291–2
Guns, Germs and Steel (Diamond) 303

Haber, Fritz 352, 353
Hadean era 160, 162, 168, 210
Hadley Centre for Climate Change 367–8
haem 60–1, *61*, 170
haemoglobin 60, 62, 67, 68
Hahn, Otto 109, 109n
Haldane, J. B. S. 61, 159, 160, 164, 259
Hales, Stephen 328, 332

Hamilton, Bill 259
Harlan, Jack 303
Harte, John 300
Hartt, Constance 283–4
Hasenöhr, Friedrich 121
Hassid, Zev 23
Hatch, Hal 284
Hawaii xv–xvi, 6, 346
Hawaiian Sugar Planters Association Experiment Station 284
Hay, Bill 290
heat-death of the universe 304
Helios 407
Henry III, King 280
Herron, Helen Arnold 114n
Hill reaction 72, 76
Hill, Archibald 21
Hill, Fanny 340
Hill, Robin 49n, 57–61, 65–6, 67, 68–70, 71, 73, 75, 76, 77–81, 83, 84, 98, 106, 113, 113n, 128n, 140, 222, 265, 327–8n, 329n, 337, 407
Himalayas 239, 287, 347, 348, 283, 287, 289, 348
Hind, Geoffrey 89
Hoagland. Dennis 74
Holocene era 348
Honda, Kenichi 400
Hooker, Sir John 276
Hopkins, Frederick Gowland 58–9, 60, 61, 67, 70, 157, 159, 432
Horton, Peter 243, 310
House Unamerican Activities Committee 34
Hubble Space Telescope 147
Hubble, Edwin 14n
Huber, Robert 134
Hudson, Esther 20
Hume, David 338
humus theory 336–7
Hutton, James 315, 337–41, 339n, 343, 347, 350, 351, 355, 376n, 377

hydroelectric power 384
hydrogen xi, 182
  cytochrome's transport of ions 87,
    88, 91; earth's daily loss of 182; in
    photosynthesis process xi;
    production 396–404, 405
hydrothermal vents 163, 171
hydroxyl radicals 170, 171, 173, 183
Hynes, Nancy 202n

ice ages 275, 277, 290–9, 300, 348,
    360, 369, 436
ice caps 238, 283, 294–5, 299
ice cores 295–6, 296n, 297–8, 346,
    436
Igwajas 243, 310
Imperial College, London 97, 98,
    127, 135, 136
Ingalls, John James 280
Ingenhousz, Ian 320, 324, 332–5,
    340
Intergovernmental Panel on Climate
    Change (IPCC) 362–3, 371, 377
International Congress of
    Photosynthesis:
  Montpellier 1995 98; Montreal
    2004 95–9, 138, 397, 432
International Geosphere Biosphere
    Program 256
isotopes 17
  carbon 5–6, 17, 21–8, 31, 34–6,
    37, 38, 43, 47, 109, 253, 284, 383;
    distinction between 17;
    radioactive 14, 18, 19–20, 21;
    rates of decay 14, 17; record 165,
    211, 213, 216; sulphur 211; term
    17; tracers 19–20, 21–8
iron 60, 61, 63, 68, 69
  fertilization of oceans 391–3; in
    photosynthesis 293–4; in the
    oceans 202, 213, 294, 295;
    oxidization of 202

ITER 383, 405
Iwata, So 134

Jagendorf, André 89, 90
Jefferson, Thomas 335
John Hopkins University 107
Joliot, Frédéric 19, 115
Journal of the American Chemical
    Society 37
Jupiter 159
Jurassic period 282–3
Juretić, Davor 265

Kamen, Martin 15–22, 23, 24–5,
    26–7, 28, 29, 30, 31, 32–3, 34, 38,
    45, 46, 72n, 96, 103, 105, 109,
    113, 125, 337, 397, 404, 431
Kasting, Jim 176, 182, 183–4, 202–3,
    250, 307, 308, 434
Keeling, Dave 6–8, 11, 124, 346, 431
Keilin, David 61–2, 63–4, 65, 70, 75,
    84, 85, 129, 180n, 432
Keir, James 326
Kendrew, John 129, 136
Kenrick, Paul 220
Kirchner, James 255, 260
Kirschvink, Joe 201, 202
Kleidon, Axel 261–2, 263, 264, 265,
    436
Kluyver, A. J. 70
Knoll, Andy 195, 197, 198, 205, 210,
    211, 212, 213, 214, 215, 217–18,
    241, 434
Kok, Bessel 115, 400
Kornberg, Hans 38n
Kuhn, Thomas 244n
Kump, Lee 250, 253–4

Lansdowne, Marquess of 319
late heavy bombardment 161–2
latent heat 354–5, 356, 360, 366,
    392, 393

Latimer, Wendell 12
Lauber, Alice 36
Lavoisier, Antoine 52, 64, 73–4n,
  331, 334, 335–6, 337, 340
Lawrence Berkeley National
  Laboratory 407
Lawrence, Ernest 18–19, 20, 21, 22,
  24, 25, 28, 30, 33, 34, 35, 135
Lawrence, John 19, 33
'Leaf by Niggle' (Tolkien) 229
leaves:
  effect on earth's climate 237;
  evolution of 226, 227, 240–2;
  growth in relation to carbon
  dioxide levels 240–2; heat stress,
  prone to 241
Lenton, Tim 216–17, 238, 257, 259,
  308
Lepidodendron 247–8
Lewis, Gilbert 13, 14, 20, 26, 28, 34,
  35, 44
lichen 207, 217, 221
Liebig, Justus von 73, 73n, 214
Life's Solution (Conway Morris) 162n
lignins 225, 246–7, 248, 249, 250
Long, Stephen 358, 359, 365, 387
Lord of the Rings (Tolkien) 312
Lorenz, Ralph 263–4
Los Alamos 394, 404
Lovelock, James 148–9, 151–5, 158,
  158n, 161, 166–7, 168, 184, 202n,
  217, 233, 237, 238, 242–5, 249,
  252, 253, 255–6, 257, 258, 260,
  261, 297, 304, 305, 306, 307, 309,
  349, 357, 382, 433, 435
Lovelock, John 202n
Lovelock, Sandy 202n, 309, 349
Lunar Society 326, 329, 335, 337,
  342, 343, 408

manganese 137, 174, 202, 396
malate 284

Manhattan Project 30, 33, 34, 35,
  109, 110, 404, 405
manometry 101, 102
Marat, Jean-Paul 336
Margulis, Lynn 132, 133, 152–3, 205,
  243, 255, 258, 433, 435
Mars 148, 149, 151, 166–7, 168, 182,
  197, 200, 203, 260, 264, 308
Martin, John 294, 295, 298
mass extinctions 195–6, 233, 244,
  253, 254
mass spectrometry 30
Mauna Loa 6
Mayaudon, Jacques 42, 44
Mayer, Robert 52–4, 56, 57, 64, 73n,
  74n, 90, 90n, 340, 432
McKay, Chris 182, 434
McKie, William 220
Medical Research Council (UK) 59,
  153
Meinshausen, Malte 361
Meitner, Lise 109, 109n
Melis, Tasios 397–8, 407
membrane, cell 85–92, 114, 265
methane 399
  detecting on other planets 188;
  greenhouse effects of 183, 357,
  371, 377, 380; origins of life and
  148, 150, 166, 167, 182–3, 184,
  203, 217
Michel, Hartmut 129–30, 134
*Microbial metabolism: evidence for
  life's unity* (Kluyver) 70
microbiology 70
miscanthus 387–8
Mitchell, Peter 84–92, 129, 140, 432
mitochondria 70, 71, 76–7, 85, 86,
  87, 91, 117, 132, 205, 251
molecular biology 11, 42, 96–7, 108,
  120, 135, 154, 251
molybdenum 212, 213, 214
Monbiot, George 389

monochromators 115
monsoon 289, 361–2, 355, 361, 393
Moore, Ana 402
Moore, Tom 402
Mount Palomar Observatory 14, 18
Moyle, Jennifer 85
Mullineaux, Conrad 234n

NADPH 75, 76, 78, 80, 81, 83, 107, 112
NASA 148, 153, 182, 433
Natural History Museum 218–19, 220, 227–8, 247
natural selection 259, 266, 310
Nature 77, 79n, 130, 259, 295, 305
'Nature's Green Revolution' 287
Nernst, Walther 360n
net primary productivity 350–1
neutrons 13, 16, 17–18, 25, 27
New York University 261
Newton, Sir Isaac 93, 259
Niels Bohr Institute, Copenhagen 108
Night of Power (Robinson) 193
Nisbet, Euan 171, 173n, 178–9, 180, 288, 434
nitrates 74, 293, 353–4, 365
nitrogen 73, 74
  artificial fixing of 352–4; bacterial fixing of 205, 212–13, 214, 224; cycle 169, 233, 352–4
nitrogen-13 24
'no till farming' 387
Nobel prize 21, 24, 26, 28, 45, 46, 65, 87, 100, 105, 105n, 107, 125, 127, 134, 157, 334, 432, 433
nuclear:
  fission 29, 30, 31, 63, 109, 109n, 381–2; fusion 381, 382–3; power 242n, 380, 381–3; testing 5–6 *see also* radioactive isotopes
nucleic acids 163

Oak Ridge National Laboratory 115, 118, 404
oceans:
  Canfield 211–14, 238, 435; carbon dioxide levels/absorption xiv, 6, 10, 11, 201–2, 298–9, 370; disappearance of Venus' 188; iron in 202, 293–4, 295; oxygen levels 253–4; planet cooling function 369–7; Proterozoic 211–14; sulphur chemistry 212–14
oil xv, 177, 375–7, 390, 397
Oparin, Alessandr Ivanovich 159, 162
Open University 195
Oppenheimer, Robert 20, 32, 108
Ordovician era 219
organic chemistry 9–10, 13, 41, 73, 352
organic farming 242n
origins of life 155–85
oxygen:
  effect upon plant and algae carbon fixing ability 251; Cambrian levels 245; Carboniferous levels 245, 246; chlorophyll needed to produce one molecule of 104–5; chloroplast production of 68–72; detecting on other worlds 185–92; Devonian levels 245; discovery of 331, 334; Gaia and historical levels of 249, 255–67, 296, 308, 309, 314; Great Oxidation Event 175–85, 186, 197, 198, 203, 205, 211, 233, 244, 310, 434, 435; in haemoglobin 67–8; ocean levels 253–4; Permian levels 245–6 ; Phanerozoic levels 245; spike 242–54, 435; term 9n; water production and 186–7
oxygen-15 24

oxygen-16 24
oxygen-18 24, 72n
ozone layer 170, 254, 366, 434

Pacala, Stephen 379
Paine, Tom 408
Paltridge, Garth 262–3, 264
Panama 291
Pangaea 239, 240, 247, 253, 254, 272, 282
Panthalassia 240
paper chromatography 36
Parasite Rex (Zimmer) 62n
particle accelerators 18
'Path of Carbon in Photosynthesis, Part XXI, The' (1954) 39–40
Paulinella 206n
Pauling, Linus 14–15, 21, 42, 120
Pennsylvania State University 176, 250
Penone, Giuseppe 323–4
Permian era 232, 245–6, 253, 254, 271, 272, 277, 281–2, 310
peroxisome 251
Perutz, Max 129, 136
phage 122
Phanerozoic era 160, 198, 210, 210n, 211, 219, 245
pheophytin 119, 126, 127
Philosophical Transactions of the Royal Society 328
phlogiston 330–1, 334, 335, 336, 337–43, 411
phosgene 31, 32
phosphates 30, 39, 354
phosphoenolpyuvate (PEP) 284–6
phosphoglycerate 38–9, 40, 44, 97, 251, 284
phosphoglycolate 251
phosphorus 20, 30, 33, 73, 74, 214, 250, 253
phosphorylation 86–7, 112, 116

photobioreactor 398–9, 402
photochemical equivalence, law of 101
photometers 115
photomultipliers 75–8, 115, 116, 117, 432
photons 85, 90
chlorophyll and 90, 104, 106, 107; counting 111–14; discovery of 85; earth's emission and reception of 150–1; efficiency of photosynthesis and 111–14; fluorescence and 115–20; number needed to produce oxygen and reduce carbon dioxide 111–14, 115; photosynthetic unit absorbs 102, 104, 105, 111–14; photosynthetic unit emits 115–16; physicists understanding of 115
photophosphorylation 77, 89
photorespiration 251–3, 283, 284–5, 286, 304
photosynthesis xi–xix
chloroplasts role in see chloroplast; detecting on other planets 147–60, 188–92; discovery of 325–43; early visualizations of 10–11; effect of carbon dioxide build up on see carbon dioxide; efficiency of 111–12; energy flows and 51–7, 63–66; evolution of 167–75; feedback loops 231–67; in grasses see grasses; life's origins and 160–85; on planetary scale xiii–xiv, xvii–xviii, 7, 7, 10–11; oxygen production 66–72, 175–85, 242–54; path of carbon in 22–40 see also carbon; products of 10; reaction centre see reaction centre; requirements of 10; rubisco, role of see rubisco;

stomata and *see* stomata; term 10;
unit of *see* photosynthetic unit
Photosynthesis Research 44–5
photosynthetic unit, development of
concept 103–14, 114–15n
photosystem I 82, 107, 112, *131*, 140,
171, 172
photosystem II 82, 83, 85n, 98–114,
127–40, 172, 174–5, 266, 310, 395,
396, 397, 401, 402, 406, 432, 433
photovoltaic cells 384, 394, 399, 401,
402, 404, 406
phytoplankton 96, 272, 272, 277,
278, 293, 294, 307, 362, 369–72
see also algae
Pirie, Bill 156–9, 164–5, 166, 174,
388n, 433, 434
Planck, Max 63, 130
plankton:
absorption of carbon dioxide xiv;
genomes of photosynthetic 96; *see
also* phytoplankton
plants, land:
C4 285–91, 309, 367, 436; current
role in earth's life cycle 271–6;
evolution of land 218–28,
232–42; feedback between climate
and 231–42; 'fix' carbon into
their tissues 10, 11, 251, 325;
stalks 223; stems xii, 223, 225,
227, 233, 241
plate tectonics 163, 196, 201, 342
carbon burial and 181, 190, 240,
253, 283; earth's internal heat and
177, 190; ocean productivity and
213–1; ocean-ridge system 240,
247; power of 349, 350;
Proterozoic 213–14, 216; role in
Devonian era's carbon dioxide
levels 239–40; subduction zone
306
Pleistocene era 348

plutonium 30, 382, 404
Polanyi, Michael 60n
polymers 225
porphyrin 60–1
posthuman future 311–13
Priestley, Joseph 320, 324, 325–36,
337, 343, 344, 408, 411
Princeton University 376, 378
Pringle, Sir John 329, 329n, 333
proteins, reaction centres and action
of 125–6, 128, 129–30, 133–4,
135, 137, 140, 170
Proterozoic era 160, 198, 201, 203,
205, 210–18, 210n, 233, 238,
254
punctuated equilibrium 244, 244n
pyruvate 284, 285

quantum mechanics 13, 14, 63, 85,
97, 105, 106, 108, 116, 121
Queen Mary, University of London
172
quinones 126, 127, 128

Rabinowitch, Eugene 84, 109, 111,
121, 358, 404, 433
Radiant Science, Dark Politics
(Kamen) 28
radioactive tracers 20–8, 30, 31, 33,
34, 35–40, 42, 77
radioisotopes 18, 19–20, 21–8, 34
radiotherapies 19, 20, 33
rainforest 235, 368
Raup, David 195, 196–7, 244, 434
reaction centre:
development of concept 115–27,
129; evolution of 171, 172, 179,
190
Read, Peter 388–9
red fescue 278, 280
redox potential 63, 67, 77, 80, 81, 98,
169, 180, 185, 401

redox reactions 62–3, 69, 70, 71, 72, 137, 148, 166, 167
Revelle, Roger 5, 6, 11, 124, 125, 431
Rhynie Chert 220–2, 224, 225, 227–8, 234, 240, 306
ribulose diphosphate 39, 40, 42, 43, 44, 47
Robinson, Jennifer 246, 247, 249
Rockefeller Foundation 108
Royal Botanic Gardens, Kew 322
Royal Holloway University 171
Royal Society 195, 329, 332
Ruben, Sam 15, 21–3, 24, 25–6, 28, 29, 30, 31–2, 35, 38, 45, 58, 72n, 431
rubisco 41–7, 96, 120, 161, 164, 178, 183, 241, 251–2, 283, 285, 286, 300–1, 365, 367, 398, 409, 432
Russell, Mike 163–4, 179
Rutherford, Bill 95, 96, 97, 132–3, 134, 205, 394, 395–6, 405, 433

Sage, Rowan 301–2, 303
Saturn 263
Schneider, Steve 262
Schrödinger, Erwin 97, 121, 122, 123, 124, 149, 154, 155, 157, 158
Schwartzmann, David 238, 262, 435
Science 99, 112
Scripps Institute of Oceanography 5, 7, 31, 45, 47, 124, 125
seeds 226–7, 273, 276, 301
Segrè, Emilio 19
Senebier, Jean 336
Sepkoski, Jack 196
Shackleton, Nick 345–6
Shanghai Institute of Plant Physiology 89n
sheep's fescue 276, 278
Shelburne, Earl of 330, 334–5
Shen, Yunkang 89n
Silicon Valley 18

Silurian era 219
Slack, Roger 284
Small, William 326, 335
Smil, Vaclav 353, 385
Smith, Adam 338
Smith, John Maynard 259
Snow, C. P. 79
'Snow' (MacNeice) 165
'snowball earth' 201–3, 211, 216, 258, 434–5
Socolow, Robert 378–9
solar power 312–13, 380, 393–6, 406–7, 410
Solar-H 396, 399, 402, 404, 405
South Downs 271–9, 280, 281, 286, 310, 436
SoyFace 366
space exploration 147–9
'spaghetti plots' 377–8
spectrometers 115, 120, 405
Spicer, Bob 195, 197–8, 226, 234, 242, 247, 371, 435
Stanford University 18, 23n, 70, 108, 262
Stapledon, Olaf 312
Stephenson, Marjory 59
steranes, long-chain 177
Stern, Isaac 33
Stoll, Arthur 11, 141
stomata 224, 286
  carbon dioxide levels and growth of 241, 289, 302, 345–8, 358; conductance 366–8; entropy production and 264–5, 267; function of 223–4
Strehler, Bernard 116, 117–18
stroma 87
Styring, Stenbjörn 403–4, 406
sulphur 20
  chemistry of 176, 177; cycle 169, 213–14, 233; oceans and 254, 369–70

sun:
  energy that reaches earth's
    atmosphere xvi; light reflected
    back into space (albedo) xvi, 200,
    258, 265, 282, 290, 293, 300, 308,
    354, 367; solar power 312–13,
    380, 393–6, 406–7, 410; Solar-H
    396, 399, 402, 404, 405;
    temperature of 150
Swift, Jonathan 332
Swiss Federal Institute of Technology
    361
sycamore tree xiii, xiv, xv, 275
symbiotic relationships,
    photosynthetic 207
synchrotrons 35

Technion 122, 123
telescopes and bioscopes 147–8, 155,
    188, 189–90, 191, 405
Teller, Edward 107, 144
Tennyson, Alfred, Lord 269
terrestrial energy flows 349–51, 351n
Theresa, Empress Marie 333
thermodynamics 52, 53–4, 63, 162,
    340, 342, 348
  entropy and 54–6, 80, 91–2, 121,
    148, 149, 150, 151, 152, 159,
    262–6, 308, 342, 408, 411; First
    Law of 52–4; Second Law of 54,
    55, 87, 91, 264, 265, 305, 432;
    Third Law of 360n
thermostat, earth's 199–201, 217,
    236–54, 282–3, 304–8, 434
thylakoid space 86, 87, 89, 91
tidal energy 384
Tilman, David 388, 391
Titan 263
'To his coy mistress' (Marvell) 143
Tolbert, Bert 38n, 41
Tortworth Chestnut 322n
transpiration 224, 241, 282, 366

Treaty of Versailles 353
trees, evolution of 225–7, 225n, 235,
    246–7, 248 see also forests
'Trees, The' (Larkin) 2
Triassic period 282–3
TRIFFID ('Top-down Representation
    of interactive Foliage and Flora
    Including Dynamics) 368–9
tritium (hydrogen-3) 24, 29, 383
Tsiolkovsky, Konstantin 312–13
Tull, Jethro 338

ultraviolet light 173, 176, 182
United Nations Food and
    Agriculture Organization 385
University of Bristol 186
University of California 365
  Berkeley 12, 13, 14, 18, 19, 26, 29,
    31, 34, 35, 45, 60n, 72, 73, 75, 77,
    80, 108, 123, 284, 300, 397; Davis
    365; La Jolla 124, 433; Old
    Radiation Laboratory (ORL)
    35–8, 44, 431, 432; San Diego
    campus 5, 122, 124, 125, 397,
    407; Radiation Laboratory (Rad
    Lab) 15, 18–23, 24, 25–6, 28, 29,
    30, 31, 32, 33, 35, 108, 135, 160,
    383, 404
University of Chicago 15, 18, 105,
    109, 195, 404
University of East Anglia 168, 257
University of Illinois 110, 358–9
University of Manitoba 353, 385
University of Minnesota 388
University of Reading 249
University of Sheffield 240, 287,
    288–9, 344
University of Sydney 177
University of Toronto 285
uranium 29, 30, 31, 63, 109, 381, 404
Urey, Harold 23–4, 25, 124–5, 160,
    199

van Niel, Cornelis 44, 70, 71, 75, 108, 337, 397, 432
Varley, John 311–12, 313
Vatches Farm 51, 57, 58, 66, 67, 68n, 265
Venter, Craig 407
Venus 148, 149, 151, 182, 200, 200n, 258, 260, 308
Vercors 155–6
Vernadsky, Vladimir 56–7, 90, 161, 243, 313, 352, 357, 432
volcanoes 199, 201, 202, 236, 240, 247, 341, 349, 369
Volk, Tyler 238, 250, 261, 262, 435, 436
von Laue, Max 105n
Vostok core 295–6, 296n, 297–8, 436

Wald, George 189
Walker, David 79, 80
Warburg, Emil 101, 104
Warburg, Otto 100–1, 102, 104, 105, 106, 111, 112–13, 114, 121, 251, 332
Washington University, St Louis 33
water xi
cycle 233; oxygen and 186–7; rising carbon dioxide levels and 366–7; term 9n; vapour 188, 201, 235, 258, 290, 293, 307, 356 see also oceans
Watson, Andrew 168, 216–17, 237, 238, 249, 250n, 257, 260–1, 298, 299, 339n

Watson, James D 83n, 97
Watt, James 326, 333, 340, 342, 348
weathering, chemical 238, 239, 240, 241–2, 244–5, 304–8
Weaver, Warren 108
Wedgwood, Josiah 326, 335, 342
What is Life? (Schrödinger) 121, 149, 154, 157
Whatley, Bob 75, 76, 77, 132n, 432
Whitfield, Michael 305, 306, 307
Wildman, Sam 42–3, 120
Willstätter, Richard 11, 100, 128, 141, 352
wind power 384
Witt, Horst 115, 135
Wohl, Kurt 106
wood, evolution of 225, 246, 248, 274
Woodward, Ian 288–9, 304, 344–8
World Climate Research Program 256
Worthington, Priscilla 66
Wurmser, René 69–70

X-ray crystallography 127–30, 134–6

Yale University 238
Yankwich, Peter 32

Zahnle, Kevin 182, 184, 434
Zalessky, Mikhail Dmitrievich 225n
Z-scheme 78–84, 81, 98, 106, 107, 113, 141, 175, 191n, 265, 432
Županović, Paško 265